Proofs, Categories and Computations
Essays in Honor of Grigori Mints

Volume 4
The Way Through Science and Philosophy:
Essays in Honour of Stig Andur Pedersen
H. B. Andersen, F. V. Christiansen, K. F. Jørgensen, and V. F. Hendricks, eds.

Volume 5
Approaching Truth: Essays in Honour of Ilkka Niiniluoto
Sami Pihlström, Panu Raatikainen and Matti Sintonen, eds.

Volume 6
Linguistics, Computer Science and Language Processing.
Festschrift for Franz Guenthner on the Occasion of his 60th Birthday
Gaston Gross and Klaus U. Schulz, eds.

Volume 7
Dialogues, Logics and Other Strange Things.
Essays in Honour of Shahid Rahman.
Cédric Dégremont, Laurent Keiff and Helge Rückert, eds.

Volume 8
Logos and Language.
Essays in Honour of Julius Moravcsik.
Dagfinn Follesdal and John Woods, eds.

Volume 9
Acts of Knowledge: History, Philosophy and Logic.
Essays dedicated to Göran Sundholm
Giuseppe Primiero and Shahid Rahman, eds.

Volume 10
Witnessed Years. Essays in Honor of Petr Hájek
Petr Cintula, Zuzana Haniková and Vítězslav Švejdar, eds.

Volume 11
Heuristics, Probability and Causality. A Tribute to Judea Pearl
Rina Dechter, Hector Geffner and Joseph Y. Halpern, eds.

Volume 12
Dialectics, Dialogue and Argumentation. An Examination of Douglas Walton's Theories of Reasoning and Argument
Chris Reed and Christoher W. Tindale, eds.

Volume 13
Proofs, Categories and Computations. Essays in Honour of Grigori Mints
Solomon Feferman and Wilfried Sieg, eds. With the collaboration of Vladik Kreinovich, Vladimir Lifschitz and Ruy de Queiroz

Tributes Series Editor
Dov Gabbay   dov.gabbay@kcl.ac.uk

Proofs, Categories and Computations
Essays in Honor of Grigori Mints

edited by

Solomon Feferman

and

Wilfried Sieg

With the collaboration of
Vladik Kreinovich, Vladimir Lifschitz and Ruy de Queiroz

© Individual author and College Publications 2010. All rights reserved.

ISBN 978-1-84890-012-7

College Publications
Scientific Director: Dov Gabbay
Managing Director: Jane Spurr
Department of Computer Science
King's College London, Strand, London WC2R 2LS, UK

http://www.collegepublications.co.uk

Original cover design by orchid creative   www.orchidcreative.co.uk
Printed by Lightning Source, Milton Keynes, UK

All rights reserved. No part of this publication may be reproduced, stored in a retrieval system or transmitted in any form, or by any means, electronic, mechanical, photocopying, recording or otherwise without prior permission, in writing, from the publisher.

TABLE OF CONTENTS

Solomon Feferman and Wilfried Sieg   vii
Preface

List of contributors   ix

Toshiyasu Arai   1
Intutionistic fixed point theories over Heyting arithmetic

Jeremy Avigad   15
The computational content of classical arithmetic

Matthias Baaz and Alexander Leitsch   31
Fast cut-elimination by CERES

Lev D. Beklemishev   49
On the Craig interpolation and the fixed point properties for GLP

Guram Bezhanishvili, Leo Esakia and David Gabelaia   61
K4.Grz and hereditarily irresolvable spaces

Andrey Bovykin   71
Unprovable Ramsey-type statements reformulated to talk about primes

Evgeny Dantsin   81
Maximum satisfiability and subexponential time

Alexander Dikovsky   93
On constructive semantics of natural language

Solomon Feferman   109
On the strength of some semi-constructive theories

Ulrich Kohlenbach   131
On the logical analysis of proofs based on nonseparable Hilbert space theory

Nikolai K. Kosovsky   145
Polynomial-time decidability of a bounded universal theory of congruences modulo a prime number

Boris Kushner   153
Forty-five years of friendship

Larisa Maksimova — 159
Weak interpolation in extensions of minimal logic

Diana Ratiu and Helmut Schwichtenberg — 171
Decorating proofs

Wilfried Sieg — 189
Searching for proofs (and uncovering capacities of the mathematical mind)

A.El Khoury, S. Soloviev, L.Mehats and M.Spivakovsky — 217
On varieties of closed categories and dependency of diagrams of canonical maps

W.W.Tait — 231
The substitution method revisited

Michael A. Taitslin — 243
Isomorphisms and strong finite projective classes of commutative semigroups

Henry Towsner — 251
Priority arguments and epsilon substitution

Enn Tyugu — 267
Grigori Mints and computer science

S. Vakulenko and D. Grigoriev — 279
Complexity and stable evolution of circuits

Johan van Benthem — 297
Frame correspondences in modal predicate logic

Preface

SOLOMON FEFERMAN AND WILFRIED SIEG

Grigori ("Grisha") Mints is one the most distinguished proof theorists of our time. He has contributed significantly to the subject in general and to many of its applications. In view of his important work and his worldwide standing, he was recently honored by being elected Member of the Estonian Academy of Sciences and Fellow of the American Academy of Arts and Sciences.

Mints received his PhD in Mathematics from Leningrad University in 1965: his work began with contributions to the Leningrad school of constructive mathematics and has been continuing without letup to this very day. Among his significant results are the conservativeness of one-quantifier induction over primitive recursive arithmetic, the stability of program extraction methods, a continuous cut elimination procedure for infinite derivations, and faithfulness of Gödel's modal translation of intuitionistic arithmetic into S4-arithmetic. His investigation of connections between proof theory and category theory culminated in the proof of a coherence theorem for Cartesian closed categories. In early work he developed methods (together with, among others, N. A. Shanin and S. Yu. Maslov) that were implemented in order to search for natural deduction proofs in sentential logic.

Mints established the fundamental properties of the second-order epsilon calculus and, with S. Tupailo and W. Buchholz, the termination of Hilberts epsilon substitution method for second-order systems. His extension of this method to the theory of inductive definitions eventually led to a termination proof by T. Arai for that system as well as for stronger ones. Mints is currently working on a novel approach to these results with the aim to make them more perspicuous. He originated the study of dynamic topological logic and has led the way by treating several of its important systems.

Mints was born on 7 June 1939 in Leningrad. Seventy years later his birthday was celebrated at Stanford University, where he had joined the faculty of the Philosophy Department in 1991. We had begun planning for that occasion almost a year earlier, in mid-2008. The enthusiastic support by many of Grisha's colleagues worldwide made it possible to present him with a preliminary version of this volume at his 70th birthday.

The full collection of papers in this volume is both a personal tribute to Grisha and a testimony to the breadth and importance of his work. All areas in which he has worked are dealt with: from proof-theoretic reductions through non-classical logics and category theory to automated theorem proving and proof mining, i.e., the extraction of mathematical information from formal proofs. The collection itself is significant for a second reason: it bridges the two logical worlds in which Grisha has worked, the world of the former Soviet

Union and that of the West.

The collection is rich in mathematical results, but it also reflects a highly cooperative endeavor in which we have incurred many debts. Heartfelt thanks go, first of all, to all the contributors and to our co-editors whose general support and specific work was indispensable to getting this volume ready for publication. (Vladik Kreinovich's initiative to reshape all the papers in a uniform style deserves special mention.) Thanks go, secondly, to the anonymous referees of the papers; their work was critical. Finally, we want to express special thanks to the Stanford Department of Philosophy, the Suppes Center for Science and Technology, the Mathematics Research Center, the Center for the Study of Language and Information, the Provost's Office, the Office of the Dean of Humanities and Sciences, and last but not least, Margot and Vaughan Pratt, for support of the Birthday Celebration described above.

June 2010

List of Contributors

Arai, Toshiyasu: Professor; Graduate School of Science, Chiba University, Japan; tosarai@faculty.chiba-u.jp.

Avigad, Jeremy: Professor of Philosophy and Mathematical Sciences; Carnegie Mellon, Pittsburgh, USA; avigad@cmu.edu.

Baaz, Matthias: Head of the Computational Logic Research Unit; Technische Universität Wien, Austria; baaz@logic.at.

Beklemishev, Lev: Principal researcher, V.A. Steklov Institute of Mathematics, Russian Academy of Sciences; Professor of Mathematics, M.V. Lomonossov Moscow State University, Russia; bekl@mi.ras.ru.

Bezhanishvili, Guram: Associate Professor of Mathematics; New Mexico State University, Las Cruces, New Mexico, USA; gbezhani@nmsu.edu.

Bovykin, Andrey: Research Fellow; Department of Mathematics, University of Bristol, England, UK; Andrey.Bovykin@bristol.ac.uk.

Dantsin, Evgeny: Associate Professor of Computer Science; Roosevelt University, Chicago, USA; edantsin@roosevelt.edu.

Dikovsky, Alexander: Professor; LINA, University of Nantes, France; alexandre.dikovsky@univ-nantes.fr.

El Khoury, Antoine: PhD student; Institut de Recherche en Informatique de Toulouse (IRIT), France; elkhoury@irit.fr.

Esakia, Leo: Head, Department of Mathematical Logic; Razmadze Mathematical Institute, Tbilisi, Georgia; esakia@hotmail.com.

Feferman, Solomon: Patrick Suppes Family Professor of Humanities and Sciences, emeritus, Professor of Mathematics and Philosophy, emeritus; Stanford University, Stanford; feferman@stanford.edu.

Gabelaia, David: Researcher, Department of Mathematical Logic; Razmadze Mathematical Institute, Tbilisi, Georgia; gabelaia@gmail.com.

Grigoriev, Dima: Research Director at CNRS; Laboratory Paul Painleve, University of Lille, France; dmitry.grigoryev@math.univ-lille1.fr.

Kohlenbach, Ulrich: Professor of Mathematics; Technische Universität Darmstadt, Darmstadt, Germany; kohlenbach@mathematik.tu-darmstadt.de.

Kosovsky, Nikolai K.: Head, Department of Informatics; St. Petersburg University, Russia; kosov@nkk.usr.pu.ru.

Kushner, Boris: Professor of Mathematics; University of Pittsburgh at Johnstown, Johnstown, Pennsylvania, USA; boris@pitt.edu.

Leitsch, Alexander: Professor of Theoretical Informatics and Logic; Technische Universität Wien, Austria; leitsch@logic.at.

Maksimova, Larisa: Principal Researcher; Sobolev Institute of Mathematics, Siberian Branch of Russian Academy of Sciences, and Novosibirsk State University, Novosibirsk, Russia; lmaksi@math.nsc.ru.

Mehats, Laurent: Postdoc; LaBRI, University of Bordeaux; mehats@labri.fr.

Ratiu, Diana: PhD student in Mathematical Logic at the Ludwig-Maximilians-Universität, Munich, Germany (Supervisor: Helmut Schwichtenberg); ratiu@mathematik.uni-muenchen.de.

Schwichtenberg, Helmut: Professor of Mathematics; Ludwig-Maximilians-Universität, Munich, Germany; schwicht@mathematik.uni-muenchen.de.

Sieg, Wilfried: Patrick Suppes Professor of Philosophy; Carnegie Mellon, Pittsburgh, USA; sieg@cmu.edu.

Soloviev, Sergei: Professor; Institut de Recherche en Informatique de Toulouse (IRIT), France; soloviev@irit.fr.

Spivakovsky, Mark: Directeur de Recherche CNRS; Institute of Mathematics, University of Toulouse; spivakov@math.ups-tlse.fr.

Tait, William W.: Professor of Philosophy, emeritus; University of Chicago, Chicago, USA; williamtait@mac.com.

Taitslin, Michael A.: Professor of Computer Science; Tver State University, Tver, Russia; Michael.Taitslin@tversu.ru.

Towsner, Henry: Hedrick Adjunct Assistant Professor of Mathematics; University of California, Los Angeles, USA; htowsner@gmail.com.

Tyugu, Enn: Principal Research Scientist; Institute of Cybernetics, Tallinn University of Technology, Tallinn, Estonia; tyugu@cs.ioc.ee.

Vakulenko, Sergei: Principal Researcher; Laboratory of Micromechanics of Materials, Institute of Mechanical Engineering Problems, St. Petersburg, Russia; vakulenfr@mail.ru.

Van Benthem, Johan: University Professor of pure and applied logic, University of Amsterdam, and Henry Waldgrave Stuart Professor of philosophy; Stanford, USA; johan@csli.stanford.edu.

Intuitionistic fixed point theories over Heyting arithmetic[1]

TOSHIYASU ARAI[2]

ABSTRACT. In this paper we show that an intuitionistic theory for fixed points is conservative over the Heyting arithmetic with respect to a certain class of formulas. This extends partly the result of mine. The proof is inspired by the quick cut-elimination due to G. Mints

1 Introduction

Fixed points occur frequently in mathematical reasonings. Let us consider in this paper the fixed point predicate $I^\Phi(x)$ for positive formula $\Phi(X,x)$:

(1) $\quad (FP)^\Phi \quad \forall x[I^\Phi(x) \leftrightarrow \Phi(I^\Phi, x)]$

Over the classical logic, the existence of fixed points strengthens theories.

In [2] we have shown that a first-order logic calculus with the axioms (1) has non-elementary speed-ups over the classical first-order predicate logic.

As an extension of the first-order arithmetic PA, the theory $\widehat{ID}$ for fixed points is stronger than PA. In $\widehat{ID}$ one can readily define the truth definition of arithmetic formulas. Moreover $\widehat{ID}$ proves the transfinite induction up to each ordinal less than $\varphi\varepsilon_0 0$ for arithmetic formulas, [4] and [1].

However *intuitionistic* theories for fixed points may be proof-theoretically equivalent to the intuitionistic arithmetic HA.

W. Buchholz [3] showed that an intuitionistic fixed point theory $\widehat{ID}^i(\mathcal{M})$ is conservative over the Heyting arithmetic HA with respect to almost negative formulas (in which $\vee$ does not occur and $\exists$ occurs in front of atomic formulas only). The theory $\widehat{ID}^i(\mathcal{M})$ has the axioms (1) $(FP)^\Phi$ for fixed points for *monotone formula* $\Phi(X,x)$, which is generated from arithmetic atomic formulas and $X(t)$ by means of (first order) monotonic connectives $\vee, \wedge, \exists, \forall$. Namely $\to$ nor $\neg$ does occur in monotone formula. The proof is based on a recursive realizability interpretation.

After seeing the result of Buchholz, we [1] showed that an intuitionistic fixed point (second order) theory is conservative over HA for any arithmetic formulas. In the theory the operator Φ for fixed points is generated from $X(t)$ and any second order formulas by means of first order monotonic connectives and second order existential quantifiers $\exists f(\in \omega \to \omega)$. The proof in [1] is to interpret

---
[1] Dedicated to the occasion of Grisha Mints' 70th birthday
[2] I would like to thank the anonymous referee, who points out the important contribution [8] in the topic.

the fixed points by Σ_1^1-formulas as in [4]. In interpreting the fixed points by Σ_1^1-formulas, we need an axiom of choice AC_{01}. Therefore the proof does not work for strictly positive operators, e.g., $\Phi(X, x) :\Leftrightarrow \neg\exists y \forall z A(x) \to X(x)$ since $\neg\exists y \forall z A(x) \to \exists f R(f, x) \leftrightarrow \exists f[\neg\exists y \forall z A(x) \to R(x)]$ is nothing but the independence of premiss, IP, which is not valid intuitionistically.

The crux is the fact that the axiom of choice adds nothing to HA, i.e., N. Goodman's theorem [5], while the theorem is proved by a combination of a realizability interpretation and a forcing. Also cf. [6] for a proof-theoretic proof of the Goodman's theorem.

I met Grisha first time in Hiroshima, Japan, September, 1995. I explained to him the result in [1]. He soon realized that it be followed from the Goodman's theorem before I told the proof. Then he asked me "Can you prove it by means of proof-transformations, e.g., cut-elimination?". This paper is a partial answer to his question.

Now let $\widehat{ID}^i(\mathcal{HM})$ denote an intuitionistic fixed point theory in which the operator $\Phi(X,x)$ is in a class $\mathcal{HM}$ of formulas, cf. Definition 1 below. The class $\mathcal{HM}$ contains properly the monotone formulas and typically is of the form $H(x) \to M(X, x)$ for a (Rasiowa-)Harrop formula $H($, in which there is no strictly positive occurrence of disjunction nor existential subformulas) and a monotonic formula M.

We show that the theory $\widehat{ID}^i(\mathcal{HM})$ is conservative over HA with respect to the class $\mathcal{HM}$. Thus the result of the paper extends partly one in [1].

On the other side, C. Rüede and T. Strahm [8] extends significantly the results in [3] and [1]. They showed that the intuitionistic fixed point theory for *strictly positive* operators is conservative over HA with respect to negative and Π_2^0-formulas. Moreover they determined the proof-theoretic strengths of intuitionisitic theories for transfinitely iterations of fixed points by strictly positive operators. The class of strictly positive formulas is wider than our class $\mathcal{HM}$. In this respect the result in [8] supersedes ours. A merit here is that the class $\mathcal{HM}$ is wider than the class concerned in [8]. For example any formula in prenex normal form is (equivalent to a formula) in $\mathcal{HM}$, but a Π_3^0-formula is neither negative nor Π_2^0.

Rather, I think that the novelty lies in our *proof technique*, which shows that, cf. Theorem 5, eliminating cut inferences with $\mathcal{HM}$-cut formulas from derivations of $\mathcal{HM}$-end formulas blows up depths of derivations *only by one exponential*, e.g., towers of exponentials are dispensable. This is seen from the fact that there exists an embedding from the resulting tree of cut-free derivation to such a derivation with cut inferences such that the embedding maps the deeper nodes in the tree ordering to larger nodes with respect to *Kleene-Brouwer ordering*. In other words eliminating monotone cut formulas is essentially to linearize the well founded tree as in Kleene-Brouwer ordering. This is an essence of quick cut-elimination in [7].

Let us explain an idea of our proof more closely. First the finitary derivations in $\widehat{ID}^i(\mathcal{HM})$ is embedded to infinitary derivations, and eliminate cuts partially. This results in an infinitary derivation of depth less than ε_0, and in which there occurs cut inferences with cut formulas $I^\Phi(t)$ for fixed points only. Now the constrains on operator Φ and the end formula admits us to invert cut-free derivations of sequents with a Harrop antecedent and a monotonic succedent formula. Therefore the quick cut-elimination (and pruning) technique in Grisha's [7] could work to eliminate cut inferences with cut formulas $I^{\Phi}(t)$. In this way we will get an infinitary derivation of depth less than ε_0, and in which there occurs no fixed point formulas.

By formalizing the arguments we see that the end formula is true in HA.

2 An intuitionistic theory $\widehat{ID}^i(\mathcal{HM})$

L_{HA} denote the language of the Heyting arithmetic. L_{HA} consists of the equality sign $=$, individual constants $0, 1$ for zero and one, function symbols $+, \cdot$ for addition and multiplication, and logical connectives $\vee, \wedge, \rightarrow, \exists, \forall$.

Let X be a fresh predicate symbol, which is assumed to be unary for simplicity. $L_{HA}(X)$ denotes the language $L_{HA} \cup \{X\}$.

DEFINITION 1. Define inductively two classes of formulas $\mathcal{H}$ in L_{HA}, and $\mathcal{HM}$ in $L_{HA}(X)$ as follows.

1. Any atomic formula $s = t$ belongs to both of $\mathcal{H}$ and $\mathcal{HM}$.

2. Any atomic formula $X(t)$ belongs to the class $\mathcal{HM}$.

3. If $H, G \in \mathcal{H}$, then $H \wedge G, \forall x H \in \mathcal{H}$.

4. If $H \in \mathcal{H}$, then $A \rightarrow H \in \mathcal{H}$ for any formula $A \in L_{HA}$.

5. If $R, S \in \mathcal{HM}$, then $R \vee S, R \wedge S, \exists x R, \forall x R \in \mathcal{HM}$.

6. If $L \in \mathcal{H}$ and $R \in \mathcal{HM}$, then $L \rightarrow R \in \mathcal{HM}$.

$\mathcal{H}$ denotes the class of (Rasiowa-)Harrop formulas, in which there occurs no strictly positive existential nor disjunctive subformula.

$\mathcal{HM}$ contains properly the monotone formulas, i.e., the class POS in [3], e.g., $\neg A \rightarrow X(a) \in \mathcal{HM}$ is not intuitionistically equivalent to any monotone formula, but there exists a strongly positive formula with respect to X not in $\mathcal{HM}$, e.g., $(\forall x \exists y A \rightarrow \exists z B) \wedge X(a)$. Any formula in $\mathcal{HM}$ is strictly positive with respect to X.

Let $\widehat{ID}^i(\mathcal{HM})$ denote the following extension of HA. Its language is obtained from L_{HA} by adding a unary set constant I^{Φ} for each $\Phi \equiv \Phi(X, x) \in \mathcal{HM}$, in which only a fixed variable x occurs freely. Its axioms are those of HA in the expanded language(, i.e., the induction axioms are available for any formulas in the expanded language) plus the axiom $(FP)^{\Phi}$, (1) for fixed points.

Now our theorem runs as follows.

THEOREM 2. $\widehat{ID}^i(\mathcal{HM})$ *is conservative over* HA *with respect to formulas in* $\mathcal{HM}$ *(, in which the extra predicate constant X does not occur)*.

3 Infinitary derivations

Given an $\widehat{ID}^i(\mathcal{HM})$-derivation D_0 of an $\mathcal{HM}$-sentence R_0, let us first embed it to an infinitary derivation in an infinitary calculus $\widehat{ID}^{i\infty}(\mathcal{HM})$.

$$\neg A :\Leftrightarrow A \to \bot.$$

Let N denote a number which is big enough so that any formula occurring in D_0 has logical complexity(, which is defined by the number of occurrences of logical connectives) smaller than N. In what follows any formula occurring in infinitary derivations which we are concerned, has logical complexity less than N.

The derived objects in the calculus $\widehat{ID}^{i\infty}(\mathcal{HM})$ are *sequents* $\Gamma \Rightarrow A$, where A is a *sentence* (in the language of $\widehat{ID}^i(\mathcal{HM})$) and Γ denotes a finite set of *sentences*, where each closed term t is identified with its value $\bar{n}$, the nth numeral.

$\bot$ stands ambiguously for false equations $t = s$ with closed terms t, s having different values. $\top$ stands ambiguously for true equations $t = s$ with closed terms t, s having same values.

The *initial sequents* are

$$\Gamma, I(t) \Rightarrow I(t); \quad \Gamma, \bot \Rightarrow A; \Gamma \Rightarrow \top$$

These are regarded as inference rules with empty premiss(upper sequent).

The *inference rules* are $(L\vee)$, $(R\vee)$, $(L\wedge)$, $(R\wedge)$, $(L\to)$, $(R\to)$, $(L\exists)$, $(R\exists)$, $(L\forall)$, $(R\forall)$, (LI), (RI), (cut), and the repetition rule (Rep). These are standard ones.

1.
$$\frac{\Gamma, \Phi(I,t) \Rightarrow C}{\Gamma, I(t) \Rightarrow C} (LI) \ ; \ \frac{\Gamma \Rightarrow \Phi(I,t)}{\Gamma \Rightarrow I(t)} (RI)$$

2.
$$\frac{\Gamma, A_0 \Rightarrow C \quad \Gamma, A_1 \Rightarrow C}{\Gamma, A_0 \vee A_1 \Rightarrow C} (L\vee) \ ; \ \frac{\Gamma \Rightarrow A_i}{\Gamma \Rightarrow A_0 \vee A_1} (R\vee) \ (i = 0, 1)$$

3.
$$\frac{\Gamma, A_0 \wedge A_1, A_i \Rightarrow C}{\Gamma, A_0 \wedge A_1 \Rightarrow C} (L\wedge) \ (i = 0, 1); \ \frac{\Gamma \Rightarrow A_0 \quad \Gamma \Rightarrow A_1}{\Gamma \Rightarrow A_0 \wedge A_1} (R\wedge)$$

4.
$$\frac{\Gamma, A \to B \Rightarrow A \quad \Gamma, B \Rightarrow C}{\Gamma, A \to B \Rightarrow C} (L\to) \ ; \ \frac{\Gamma, A \Rightarrow B}{\Gamma \Rightarrow A \to B} (R\to)$$

5.
$$\frac{\cdots \quad \Gamma, B(\bar{n}) \Rightarrow C \quad \cdots (n \in \omega)}{\Gamma, \exists x B(x) \Rightarrow C} (L\exists) \ ; \ \frac{\Gamma \Rightarrow B(\bar{n})}{\Gamma \Rightarrow \exists x B(x)} (R\exists)$$

6.
$$\frac{\Gamma, \forall x B(x), B(\bar{n}) \Rightarrow C}{\Gamma, \forall x B(x) \Rightarrow C} (L\forall) \ ; \ \frac{\cdots \quad \Gamma \Rightarrow B(\bar{n}) \quad \cdots (n \in \omega)}{\Gamma \Rightarrow \forall x B(x)} (R\forall)$$

7.
$$\frac{\Gamma \Rightarrow A \quad \Delta, A \Rightarrow C}{\Gamma, \Delta \Rightarrow C} \ (cut)$$

8.
$$\frac{\Gamma \Rightarrow C}{\Gamma \Rightarrow C} \ (Rep)$$

The *depth* of an infinitary derivation is defined to be the depth of the well founded tree.

As usual we see the following proposition. Recall that N is an upper bound of logical complexities of formulas occurring in the given finite derivation D_0 of $\mathcal{HM}$-sentence R_0.

PROPOSITION 3.

1. There exists an infinitary derivation D_1 of R_0 such that its depth is less than ω^2 and the logical complexity of any sentence, in particular cut formulas occurring in D_1 is less than N.

2. By a partial cut-elimination, there exist an infinitary derivation D_2 of R_0 and an ordinal $\alpha_0 < \varepsilon_0$ such that the depth of the derivation D_2 is less than α_0 and any cut formula occurring in D_2 is an atomic formula $I(t)($, and the logical complexity of any formula occurring in it is less than N).

Let $\mathcal{HM}(I)$ denote the class of sentences obtained from sentences in $\mathcal{HM}$ by substituting the predicate I^Φ for the predicate X.

DEFINITION 4. The rank $rk(A)$ of a sentence A is defined by

$$rk(A) := \begin{cases} 0 & \text{if } A \in \mathcal{H} \\ 1 & \text{if } A \in \mathcal{HM}(I) \setminus \mathcal{H} \\ 2 & \text{otherwise} \end{cases}$$

For inference rules J, the rank $rk(J)$ of J is defined to be the rank of the cut formula if J is a cut inference. Otherwise $rk(J) := 0$.

For derivations D, the rank $rk(D)$ of D is defined to be the maximum rank of the cut inferences in it.

Let $\vdash_r^\alpha \Gamma \Rightarrow C$ mean that there exists an infinitary derivation of $\Gamma \Rightarrow C$ such that its depth is at most α, and its rank is less than $r($, and and the logical complexity of any formula occurring in it is less than $N)$.

THEOREM 5. Let C_0 denote an $\mathcal{HM}$-sentence, and Γ_0 a finite set of $\mathcal{H}$-sentences. Suppose that $\vdash_2^\alpha \Gamma_0 \Rightarrow C_0$. Then $\vdash_1^{\omega^\alpha+1} \Gamma_0 \Rightarrow C_0$.

Assuming the Theorem 5, we can show the Theorem 2 as follows. Suppose an $\mathcal{HM}$-sentence C_0 is provable in $\widehat{ID}^i(\mathcal{HM})$. By Proposition 3 we have $\vdash_2^{\alpha_0} \Rightarrow C_0$ for a big enough number N and an $\alpha_0 < \varepsilon_0$. Then Theorem 5 yields $\vdash_1^{\beta_0} \Rightarrow C_0$ for $\beta_0 = \omega^{\alpha_0} + 1 < \varepsilon_0$.

Let $Tr_N(x)$ denote a partial truth definition for formulas of logical complexity less than N, cf. [9], 1.5.4. By transfinite induction up to β_0, cf. Lemma 7, we see $\mathrm{Tr}_N(C_0)$. Note that any sentence occurring in the witnessed derivation for $\vdash_1^{\beta_0} \Rightarrow C_0$ has logical complexity less than N, and it is either an $\mathcal{H}$-sentence or an $\mathcal{HM}$-sentence. Specifically there occurs no fixed point formula $I(t)$ in it. Now since everything up to this point is formalizable in HA, we have $\mathrm{Tr}_N(C_0)$, and hence C_0 in HA. This shows the Theorem 2.

A proof of Theorem 5 is given in the next section.

4 Quick cut-elimination for monotone cuts with Harrop side formulas

Our plan of the proof of Theorem 5 is as follows. Pick the leftmost cut J of rank 1:

$$\frac{\begin{array}{c}\vdots\, D_\ell \\ \Gamma \Rightarrow A\end{array} \quad \begin{array}{c}\vdots\, D_r \\ \Delta, A \Rightarrow C\end{array}}{\Gamma, \Delta \Rightarrow C} \ (cut) J$$

with $rk(A) = 1$.

Then $\vdash_1^\alpha \Gamma \Rightarrow A$ and $\vdash_2^\beta \Delta, A \Rightarrow C$ for some α and β. Moreover since for the end sequent $\Gamma_0 \Rightarrow C_0$, $\Gamma_0 \subseteq \mathcal{H}$ and $C_0 \in \mathcal{HM}$, we see that any sentence in $\Gamma \cup \Delta$ is in $\mathcal{H}$, and $C \in \mathcal{HM}$ by $\vdash_2^{\alpha_0} \Gamma_0 \Rightarrow C_0$.

Since Γ consists solely in Harrop formulas, we can invert the derivation D_ℓ, and climb up the derivation D_r with inverted D_ℓ. This results in a derivation of $\Gamma, \Delta \Rightarrow C$ of depth $dp(D_\ell) + dp(D_r)$. Iterating this eliminations, we could get a derivation of rank 0, and of depth at most exponential of the depth of the given derivation.

Though intuitively this would suffice to believe in Theorem 5, we have to prove two facts: first why the iteration eventually terminates? second give a succinct argument to the estimated increase of depth. These are not entirely trivial tasks. It turns out that we need to proceed along the Kleene-Brouwer ordering on well founded trees instead of the depths. Let us explore this.

Let us fix a witnessed derivation D_2 of $\vdash_2^{\alpha_0} \Gamma_0 \Rightarrow C_0$. Let $(T_2, <_{T_2})$ denote the wellordering, where $T_2 \subseteq {}^{<\omega}\omega$ is the naked tree of D_2 and $<_{T_2}$ the Kleene-Brouwer ordering on T_2.

Let us consider infinitary derivations equipped with additional informations as in [6].

DEFINITION 6. An infinitary derivation is a sextuple

$$D = (T, Seq, Rule, rk, ord, kb)$$

which enjoys the following conditions. The naked tree of D is denoted $T = T(D)$.

1. $T \subseteq {}^{<\omega}\omega$ is a tree in the sense that there exists a root $r \in T$ with $\forall a \in T(r \subseteq a)$ and $\forall a, b(r \subseteq a \subseteq b \in T \Rightarrow a \in T)$.

    It is not assumed that the empty node $\emptyset$ is to be the root nor $a * \langle n \rangle \in T \,\&\, m < n \Rightarrow a * \langle m \rangle \in T$.

2. $Seq(a)$ for $a \in T$ denotes the sequent situated at the node a.

If $Seq(a)$ is a sequent $\Gamma \Rightarrow C$, then it is denoted
$$a : \Gamma \Rightarrow C.$$

3. $Rule(a)$ for $a \in T$ denotes the name of the inference rule with its lower sequent $Seq(a)$.

4. $rk(a)$ for $a \in T$ denotes the rank of the inference rule $rk(a)$.

5. $ord(a)$ for $a \in T$ denotes the ordinal$< \varepsilon_0$ attached to a.

6. The quintuple $(T, Seq, Rule, rk, ord)$ has to be locally correct with respect to $\widehat{ID}^{i\infty}(\mathcal{HM})$ and for being well founded tree T.

Besides these conditions an extra information is provided by a labeling function kb.

$kb : T \to T_2$ is a function such that for $a, b \in T$

1. (2) $kb(a) \neq kb(b) \Rightarrow [a <_T b \Leftrightarrow kb(a) <_{T_2} kb(b)]$

    for the Kleene-Brouwer ordering $<_T$ on T.

2. (3) $a \subset b \Rightarrow kb(b) <_{T_2} kb(a)$

    where $a \subset b$ means that a is a proper initial segment of b for $a, b \in {}^{<\omega}\omega$.

3. Let $c \in T$ be a node with $rk(c) = 1$, and $a, b \in T$ nodes such that $c * \langle \ell \rangle \subseteq a$ and $c * \langle r \rangle \subseteq b$ for $\ell < r$. (This means that $Seq(a)$ $[Seq(b)]$ is in the left [right] upper part of the cut inference $Rule(c)$.) Suppose that the right cut formula A in the antecedent of $Seq(c * \langle r \rangle)$ has an ancestor in $Seq(b)$.

    Then

    (4) $kb(a) \neq kb(b)$

The condition (3) to kb ensures us that the depth of T is at most the order type of $<_{T_2}$.

It is easy to see that Kleene-Brouwer ordering $<_{T_2}$ is a well ordering, and its order type is bounded by $\omega^\alpha + 1$ for the depth α of the primitive recursive and wellfounded tree T_2.

LEMMA 7. *The transfinite induction schema (for arithmetical formulas) along the Kleene-Brouwer ordering $<_{T_2}$ is provable in HA.*

Proof. Since the transfinite induction schema along a standard ε_0-ordering is provable in HA up to each $\alpha <_{\varepsilon_0}$, the same holds for the tree ordering $\{(b, a) : a \subset b, a, b \in T_2\}$ (bar induction).

Now let X be a formula, and assume that X is progressive with respect to $<_{T_2}$:
$$\forall a \in T_2 [\forall b <_{T_2} a \, X(b) \to X(a)].$$

Let
$$j[X](a) :\Leftrightarrow \forall y \in T_2 [y \supseteq a \to \forall x <_{T_2} y \, X(x) \to \forall x <_{T_2} a \, X(x)].$$

Then we see that $j[X]$ is progressive with respect to the tree ordering. Therefore $j[X](r)$ for the root $r \in T_2$, and by letting y to be the leftmost leaf in T_2 we have $\forall x <_{T_2} r\, X(x)$. The progressiveness of X with respect to $<_{T_2}$ yields $X(r)$.
∎

The following Lemmas are seen as usual.

LEMMA 8. *Let $D = (T, Seq, Rule, rk, ord, kb)$ be a derivation of rank 1, and of a sequent $\Gamma \Rightarrow A$ such that $\Gamma \subseteq \mathcal{H}$ and $A \in \mathcal{HM}\,\&\,rk(A) = 1$. For any Δ, there exists a derivation $D * \Delta = (T, Seq * \Delta, Rule, rk, ord, kb)$ of the sequent $\Gamma, \Delta \Rightarrow A$.*

LEMMA 9. *(Inversion Lemma)*
Let $D = (T, Seq, Rule, rk, ord, kb)$ be a derivation of rank 1, and of a sequent $\Gamma \Rightarrow A$ such that $\Gamma \subseteq \mathcal{H}$ and $A \in \mathcal{HM}\,\&\,rk(A) = 1$.

1. *If $A \equiv B_0 \vee B_1$, then there exists a derivation*

$$D_i = (T, Seq_i, Rule_i, rk, ord, kb)$$

of rank 1 and of a sequent $\Gamma \Rightarrow B_i$ for an $i = 0, 1$.

2. *If $A \equiv B_0 \wedge B_1$, then there exist derivations*

$$D_i = (T_i, Seq_i, Rule_i, rk, ord, kb)$$

of rank 1 and of sequents $\Gamma \Rightarrow B_i$ for any $i = 0, 1$, where $T_i \subseteq T$ by pruning.

3. *If $A \equiv \exists x B(x)$, then there exists a derivation*

$$D_n = (T, Seq_n, Rule_n, rk, ord, kb)$$

of rank 1 and of a sequent $\Gamma \Rightarrow B(\bar{n})$ for an $n \in \omega$.

4. *If $A \equiv \forall x B(x)$, then there exist derivations*

$$D_n = (T_n, Seq_n, Rule_n, rk, ord, kb)$$

of rank 1 and of a sequent $\Gamma \Rightarrow B(\bar{n})$ for any $n \in \omega$, where $T_n \subseteq T$ by pruning.

5. *If $A \equiv B_0 \to B_1$, then there exist a derivation*

$$D' = (T, Seq', Rule', rk, ord, kb)$$

of rank 1 and of sequents $\Gamma, B_0 \Rightarrow B_1$.

6. *If $A \equiv I(t)$, then there exists a derivation*

$$D' = (T, Seq', Rule', rk, ord, kb)$$

of rank 1 and of sequents $\Gamma \Rightarrow \Phi(I, t)$.

DEFINITION 10. For each cut inference J in a derivation D, $KB(J)$ denotes $kb(a) \in T_2$ with the left upper node a of J:

$$\frac{a : \Gamma \Rightarrow A \quad \Delta, A \Rightarrow C}{\Gamma, \Delta \Rightarrow C} \, J$$

Let us define a cut-eliminating operator $ce_1(D)$ for derivations

$$D = (T, Seq, Rule, rk, ord, kb)$$

of rank 1 and of an end sequent $\Gamma_0 \Rightarrow C_0$ with $\Gamma_0 \subseteq \mathcal{H}$ and $C_0 \in \mathcal{HM}$.

If D is of rank 0, then $ce_1(D) := D$.

Assume that D contains a cut inference of rank 1. Pick the leftmost cut of rank 1:

$$D = \begin{array}{c} \vdots D_\ell \quad \vdots D_r \\ \dfrac{\Gamma \Rightarrow A \quad \Delta, A \Rightarrow C}{\Gamma, \Delta \Rightarrow C} \, (cut) \\ \vdots \\ \Gamma_0 \Rightarrow C_0 \end{array}$$

The leftmostness means that $KB(J)$ is least in the Kleene-Brouwer ordering $<_{T_2}$.

By recursion on the depth[1] of the derivation D_r we define a derivation $ce_2(D_\ell, D_r)$ of $\Gamma, \Delta \Rightarrow C$. Then $ce_1(D)$ is obtained from D by pruning D_ℓ and replacing D_r by $ce_2(D_\ell, D_r)$, i.e., by grafting $ce_2(D_\ell, D_r)$ onto the trunk of D up to $\Gamma, \Delta \Rightarrow C$.

As in Lemma 3.2, [7] the construction of $ce_2(D_\ell, D_r)$ is fairly standard, leaving the resulted cut inferences of rank 0, but has to performed parallely.

Let $\Gamma \cup \Delta \subseteq \mathcal{H}$ and $\boldsymbol{A} = A_1, \ldots, A_k$ be a finite sequence of $\mathcal{HM}$-sentences. Let $\boldsymbol{D_\ell} = D_{\ell,1}, \ldots, D_{\ell,k}$ be rank 0 derivations of $\Gamma \Rightarrow A_i$, and D_r a rank 1 derivation of $\Delta, \boldsymbol{A} \Rightarrow C$. We will eliminate the cuts with the cut formulas A_i in parallel. $ce_2(\boldsymbol{D_\ell}, D_r)$ is defined from the resulting derivation, denoted E by recursion.

$$D_a = \begin{array}{c} \vdots \boldsymbol{D_\ell} \quad \vdots D_r \\ \dfrac{\boldsymbol{b} : \Gamma \Rightarrow \boldsymbol{A} \quad c_1 : \Delta, \boldsymbol{A} \Rightarrow C}{a : \Gamma, \Delta \Rightarrow C} \, (cut) \end{array}$$

denotes the series of cut inferences:

$$\begin{array}{c} \vdots D_{\ell,1} \quad \vdots D_r \\ \dfrac{b_1 : \Gamma \Rightarrow A_1 \quad c_1 : \Delta, \boldsymbol{A} \Rightarrow C}{c_2 : \Gamma, \Delta, A_2, \ldots, A_k \Rightarrow C} \\ \vdots \\ \vdots D_{\ell,k} \\ \dfrac{b_k : \Gamma \Rightarrow A_k \quad c_k : \Gamma, \Delta, A_k \Rightarrow C}{a : \Gamma, \Delta \Rightarrow C} \end{array}$$

---

[1] As in [6] we see that the operators ce_1, ce_2 are primitive recursive. We don't need this fact.

1. If $\Delta, \boldsymbol{A} \Rightarrow C$ is an initial sequent such that one of the cases $C \equiv \top$, $\bot \in \Delta$ or $C \in \Delta$ occurs, then $\Delta \Rightarrow C$, and hence $\Gamma, \Delta \Rightarrow C$ is still the same kind of initial sequent. For example

$$\frac{c_1 : \Gamma, \Delta \Rightarrow \top}{c_2 : \Gamma, \Delta \Rightarrow \top} \; (Rep)$$
$$\vdots$$
$$\frac{c_k : \Gamma, \Delta \Rightarrow \top}{a : \Gamma, \Delta \Rightarrow \top} \; (Rep)$$

   The $T(E)$ is defined by

$$d \in T(E) \Leftrightarrow d \in T(D) \,\&\, \forall i (b_i \not\subseteq d).$$

2. If $\Delta, \boldsymbol{A} \Rightarrow C$ is an initial sequent with the principal formula $\boldsymbol{A} \ni A_i \equiv C \equiv I(t)$, then E is defined to be

$$\vdots \; D_{\ell,i} * \Delta$$
$$\frac{b_i : \Gamma, \Delta \Rightarrow C}{c_{i+1} : \Gamma, \Delta \Rightarrow C} \; (Rep)$$
$$\vdots$$
$$\frac{c_k : \Gamma, \Delta \Rightarrow C}{a : \Gamma, \Delta \Rightarrow C} \; (Rep)$$

   where $D_{\ell,i} * \Delta$ is obtained from $D_{\ell,i}$ by weakening, cf. Lemma 8.

$$d \in T(E) \Leftrightarrow d \in T(D) \,\&\, \forall j \neq i (b_j \not\subseteq d) \,\&\, c_i \not\subseteq d.$$

3. If $A_i \in \boldsymbol{A}$ is of rank 0, then do nothing for the cut inference of A_i.

   In each of the above cases $T(E) \subseteq T(D_a)$. The labeling function kb_E for E is defined to be the restriction of kb_{D_a} to $T(E)$.

   In what follows assume that $\Delta, \boldsymbol{A} \Rightarrow C$ is a lower sequent of an inference rule J.

4. If the principal formula of J is not in $\boldsymbol{A}$, then lift up D_ℓ:

$$\frac{b : \Gamma \Rightarrow \boldsymbol{A} \quad \dfrac{\cdots \; c_{1,i} : \Delta_i, \boldsymbol{A} \Rightarrow C_i \; \cdots}{c_1 : \Delta, \boldsymbol{A} \Rightarrow C}}{a : \Gamma, \Delta \Rightarrow C} \; (J)$$

   where $\boldsymbol{b} = b_k, \ldots, b_1$ with $b_j = c_{j+1} * \langle \ell_j \rangle \; (c_{k+1} := a)$, and $c_j = c_{j+1} * \langle r_j \rangle$ for some $\ell_j < r_j$, and $c_{1,i} = c_1 * \langle n_i \rangle$ for some n_i with $i < j \Rightarrow n_i < n_j$.

   E is defined as follows.

$$\cdots \; \frac{b_i : \Gamma \Rightarrow \boldsymbol{A} \quad c'_{1,i} : \Delta_i, \boldsymbol{A} \Rightarrow C_i}{a_i : \Gamma, \Delta_i \Rightarrow C_i} \; \cdots$$
$$\frac{}{a : \Gamma, \Delta \Rightarrow C} \; (J)$$

where $a_i = a * \langle n_i \rangle$, $\boldsymbol{b}_i = b_{k,i}, \ldots, b_{1,i}$ with $b_{j,i} = c'_{j+1,i} * \langle \ell_j \rangle$ ($c'_{k+1,i} := a_i$), and $c'_{j,i} = c'_{j+1,i} * \langle r_j \rangle$.

The labeling function kb_E is defined by

$$kb_E(a * \langle n_i \rangle) = kb_{D_a}(a * \langle r_k \rangle) = kb_{D_a}(c_k),$$
$$kb_E(a * \langle n_i \rangle * \langle r_k, \ldots, r_{j+1}, \ell_j \rangle * d) = kb_{D_a}(a * \langle r_k, \ldots, r_{j+1}, \ell_j \rangle * d)$$
$$(1 \leq j \leq k),$$
$$kb_E(a * \langle n_i \rangle * \langle r_k, \ldots, r_1 \rangle * d) = kb_{D_a}(a * \langle r_k, \ldots, r_1 \rangle * \langle n_i \rangle * d).$$

5. Finally suppose that the principal formula of J is a cut formula $A_i \in \boldsymbol{A}$ of $rk(A_i) = 1$. Use the Inversion Lemma 9.

   (a) The case when $A_i \equiv \exists x B(x) \in \boldsymbol{A}$. For simplicity suppose $i = 1$.

   $$\frac{\cdots \quad c_{1,n} : \Delta, \boldsymbol{A}_1, B(\bar{n}) \Rightarrow C \quad \cdots}{c_1 : \Delta, \boldsymbol{A} \Rightarrow C} \; (L\exists)$$

   where $A_1 \not\in \boldsymbol{A}_1$.

   By Inversion Lemma 9.3 pick an n such that $\Gamma \Rightarrow B(\bar{n})$ is provable without changing the naked tree.

   E is defined as follows.

   $$\frac{b_1 : \Gamma \Rightarrow \boldsymbol{A}_1 \quad \dfrac{b_1 : \Gamma \Rightarrow B(\bar{n}) \quad \dfrac{c_{1,n} : \Delta, \boldsymbol{A}_1, B(\bar{n}) \Rightarrow C}{c_1 : \Delta, \boldsymbol{A}_1, B(\bar{n}) \Rightarrow C} \; (Rep)}{c_2 : \Gamma, \Delta, \boldsymbol{A}_1 \Rightarrow C}}{a : \Gamma, \Delta \Rightarrow C}$$

   (b) The case when $A_i \equiv H \to A_0 \in \boldsymbol{A}$ with an $H \in \mathcal{H}$ and an $A_0 \in \mathcal{HM}$. For simplicity suppose $i = 1$.

   $$\frac{c_{1,\ell} : \Delta, \boldsymbol{A} \Rightarrow H \quad c_{1,r} : \Delta, \boldsymbol{A}_1, A_0 \Rightarrow C}{c_1 : \Delta, \boldsymbol{A} \Rightarrow C} \; (L \to)$$

   where $A_1 \not\in \boldsymbol{A}_1$, and for $m = \ell, r$, $c_{1,m} = c_1 * \langle j_m \rangle$ with $j_\ell < j_r$.

   E is defined as follows.

   $$\frac{E_1 \quad E_2}{a : \Gamma, \Delta \Rightarrow C}$$

   where E_1 denotes

   $$\frac{b_0 : \Gamma \Rightarrow \boldsymbol{A} \quad \dfrac{c_{1,0,\ell} : \Delta, \boldsymbol{A} \Rightarrow H}{c_{1,0} : \Delta, \boldsymbol{A} \Rightarrow H} \; (Rep)}{c_{k,0} : \Gamma, \Delta \Rightarrow H}$$

E_2 denotes

$$\dfrac{b_1 : \Gamma \Rightarrow \boldsymbol{A}_1 \quad \dfrac{b_{1,1} : \Gamma, H \Rightarrow A_0 \quad \dfrac{c_{1,1,r} : \Delta, \boldsymbol{A}_1 \cup \{A_0\} \Rightarrow C}{c_{1,1} : \Delta, \boldsymbol{A}_1 \cup \{A_0\} \Rightarrow C} (Rep) \quad c_{2,1} : \Gamma, \Delta, H, \boldsymbol{A}_1 \Rightarrow C}{c_{k,1} : \Gamma, \Delta, H \Rightarrow C}}$$

and $\Gamma, H \Rightarrow A_0$ by inversion.

For $m = 0, 1$, $c_{j,m} = c_{j+1,m} * \langle 2r_j + m \rangle$ for $1 \leq j \leq k$ with $c_{k+1,m} = a$, and $\boldsymbol{b}_m = b_{k,m}, \ldots, b_{1,m}$ with $b_{j,m} = a * \langle 2r_k + m, \ldots, 2r_j + m, 2\ell_j + m \rangle$ and $c_{1,0,\ell} = c_{1,0} * \langle j_\ell \rangle$, $c_{1,1,r} = c_{1,1} * \langle j_r \rangle$.

The labeling function is defined by

$$\begin{aligned} kb_E(c_{j,m}) &= kb_{D_a}(c_j) \\ kb_E(b_{j,m} * d) &= kb_{D_a}(b_j * d), \ (m = 0, 1) \end{aligned} \quad (5)$$

(c) The case when $A_i \equiv \forall x B(x) \in \boldsymbol{A}$. For simplicity suppose $i = 1$.

$$\dfrac{\begin{array}{c} \vdots D_{r,n} \\ c_{1,n} : \Delta, \boldsymbol{A}, B(\bar{n}) \Rightarrow C \end{array}}{c_1 : \Delta, \boldsymbol{A} \Rightarrow C} (L\forall)$$

with $c_{1,n} = c_1 * \langle j_n \rangle$.

E is defined as follows:

$$\dfrac{\boldsymbol{b} : \Gamma \Rightarrow \boldsymbol{A} \quad \dfrac{b_{1,1} : \Gamma \Rightarrow B(\bar{n}) \quad \begin{array}{c} \vdots D_{r,n} \\ c'_{1,n} : \Delta, \boldsymbol{A}, B(\bar{n}) \Rightarrow C \end{array}}{c_1 : \Gamma, \Delta, \boldsymbol{A} \Rightarrow C}}{a : \Gamma, \Delta \Rightarrow C}$$

where $b_{1,1} = c_1 * \langle 2j_n \rangle$ and $c'_{1,n} = c_1 * \langle 2j_n + 1 \rangle$.

The labeling function is defined by

(6) $\quad kb_E(b_{1,1} * d) = kb_{D_a}(b_1 * d)$

and

$$kb_E(c'_{1,n}) = kb_{D_a}(c_{1,n}).$$

(d) The case when $A_i \equiv B_0 \vee B_1 \in \boldsymbol{A}$. For simplicity suppose $i = 1$.

$$\dfrac{\begin{array}{cc} \vdots D_{r,0} & \vdots D_{r,1} \\ c_{1,0} : \Delta, \boldsymbol{A}_1, B_0 \Rightarrow C & c_{1,1} : \Delta, \boldsymbol{A}_1, B_1 \Rightarrow C \end{array}}{c_1 : \Delta, \boldsymbol{A} \Rightarrow C} (L\vee)$$

where $A_1 \notin \boldsymbol{A}_1$.

By Inversion Lemma 9.1 pick an $n = 0, 1$ such that $\Gamma \Rightarrow B_n$ is provable without changing the naked tree.

Suppose that $n = 0$. E is defined as follows:

$$\cfrac{b_1 : \Gamma \Rightarrow \boldsymbol{A}_1 \qquad \cfrac{b_1 : \Gamma \Rightarrow B_0 \qquad \cfrac{\cfrac{\vdots\; D_{r,0}}{c_{1,0} : \Delta, \boldsymbol{A}_1, B_0 \Rightarrow C}}{c_1 : \Delta, \boldsymbol{A}_1, B_0 \Rightarrow C}\,(Rep)}{c_2 : \Gamma, \Delta, \boldsymbol{A}_1 \Rightarrow C}}{a : \Gamma, \Delta \Rightarrow C}$$

Note that the new cut inference for B_0 may be of rank 0.

(e) The case when $A_i \equiv B_0 \wedge B_1 \in \boldsymbol{A}$ is treated as in the case 5b for universal quantifier.

Claim 1. The resulting derivation $ce_1(D)$ can be labeled enjoying the conditions (2), (3) and (4).

Proof. Let D_a be the trunk ending with the leftmost cut of rank 1 in D. First observe that the labels $\{kb_E(b) : b \in T(E)\} \subseteq \{kb_{D_a}(b) : b \in T(D_a)\}$. Therefore it suffices to see that E, and hence $ce_2(\boldsymbol{D}_\ell, D_r)$ enjoys the three conditions if D_a does.

Note that (the naked tree of) E is constructed from D_r by appending trees $\boldsymbol{D}_\ell$ only where a right cut formula A_i has an ancestor which is either a formula of rank 0 or a principal formula of an initial sequent $\Phi, I(t) \Rightarrow I(t)$. In the latter case the ancestor has to be the formula $I(t)$ in the antecedent.

E enjoys the first (2) since D_a does the first (2). E enjoys the second (3) since D_a does the third (4). E enjoys the third (4) since D_a does the third (4), and the first (2).

Let us examine cases. Consider the case 4 when $\boldsymbol{D}_\ell$ is lifted up. We have to show
$$kb_E(e) <_{T_2} kb_E(c'_{j,i})$$
for e such that $b_{j,i} \subseteq e$. Then $kb_E(e) = kb_E(a * \langle n_i \rangle * \langle r_k, \ldots, r_{j+1}, \ell_j \rangle * d) = kb_{D_a}(a * \langle r_k, \ldots, r_{j+1}, \ell_j \rangle * d) = kb_{D_a}(b_j * d)$ and $kb_E(c'_{j,i}) = kb_{D_a}(c_j)$.

Since the right cut formula A_i has an ancestor in $c_{1,i} : \Delta, \boldsymbol{A} \Rightarrow C_i$, $kb_E(e) <_{T_2} kb_E(c'_{j,i})$ follows from (4) for D_a.

Next consider the case 5b. Although $b_{j,m}, c_{j,m}$ are duplicated for $m = 0, 1$, and $kb_E(c_{j,0}) = kb_{D_a}(c_j) = kb_E(c_{j,1})$, $kb_E(b_{j,1} * d) = kb_{D_a}(b_j * d) = kb_E(b_{j,0} * d)$ by (5), (6) these are harmless for (4) since the juncture is a cut of rank 0, $rk(H) = 0$.

Finally consider the case 5b.

By (6) we have
$$kb_E(b_{1,1} * d) = kb_{D_a}(b_1 * d) = kb_E(b_1 * d)$$

but the right cut formula A_i has no ancestor in $b_{1,1} : \Gamma \Rightarrow B(\bar{n})$. Thus (4) is enjoyed.

Next for (3) we have
$$kb_{D_a}(c_j) <_{T_2} kb_{D_a}(b_1 * d) = kb_E(b_{1,1} * d)$$

by (3) and (4) in D_a.

Finally for (2) assume $j \neq 1$ and

$$kb_{D_a}(b_1 * d) = kb_E(b_{1,1} * d) \neq kb_E(b_j * e) = kb_{D_a}(b_j * e).$$

Then by (2) in D_a we have $b_j * e <_{T(D_a)} b_1 * d$, and hence $kb_{D_a}(b_j * e) <_{T_2} kb_{D_a}(b_1 * d)$. ∎

This ends the construction of the cut-eliminating operator $ce_1(D)$.

Finally we show Theorem 5. Given a derivation $D_2 = (T_2, Seq, Rule, rk, ord, kb)$ of $\Gamma_0 \Rightarrow C_0$ of rank 1, and assume $\Gamma_0 \subseteq \mathcal{H}$ and $C_0 \in \mathcal{HM}$. $(T_2, <_{T_2})$ denotes the Kleene-Brouwer ordering on the naked tree T_2.

Let $KB(D) := KB(J)$ for the leftmost cut inference J of rank 1 if such a J exists. Otherwise $KB(D)$ denote the largest element in T_2 with respect to $<_{T_2}$, i.e., the root of T_2. Then we see that $KB(D) <_{T_2} KB(ce_1(D))$ if D contains a cut inference of rank 1.

Suppose as the induction hypothesis that any cut inferences J of rank 1 has been eliminated for $KB(J) < a$, and let D denote such a derivation. Also assume that a is a node of a cut inference of rank 1. Then in $ce_1(D)$ the cut inference is eliminated. This proves the Theorem 5 by induction along the Kleene-Brouwer ordering $<_{T_2}$, cf. Lemma 7.

BIBLIOGRAPHY

[1] T. Arai, Some results on cut-elimination, provable well-orderings, induction and reflection, Annals of Pure and Applied Logic vol. 95 (1998), pp. 93–184.
[2] T. Arai, Non-elementary speed-ups in logic calculi, Mathematical Logic Quarterly vol. 6 (2008), pp. 629–640.
[3] W. Buchholz, An intuitionistic fixed point theory, Arch. Math. Logic 37 (1997), pp. 21–27.
[4] S. Feferman, Iterated inductive fixed-point theories:Applications to Hancock's conjecture, in: G. Metakides, ed., Patras Logic Symposion (North-Holland, Amsterdam, 1982), pp. 171–196.
[5] N. Goodman, Relativized realizability in intuitionistic arithmetic of all finite types, J. Symb. Logic 43 (1978), pp. 23–44.
[6] G.E. Mints, Finite investigations of transfinite derivations, in: Selected Papers in Proof Theory (Bibliopolis, Napoli, 1992), pp. 17–72.
[7] G. E. Mints, Quick cut-elimination for monotone cuts, in Games, logic, and constructive sets (Stanford, CA, 2000), CSLI Lecture Notes, 161, CSLI Publ., Stanford, CA, 2003, pp. 75–83.
[8] C. Rüede and T. Strahm, Intuitionistic fixed point theories for strictly positive operators, Math. Log. Quart. 48(2002), pp. 195–202.
[9] A.S. Troelstra, Metamathematical Investigation of Intuitionistic Arithmetic and Analysis. Lecture Notes in Mathematics 344 (Springer, Berlin Heidelberg New York, 1973).

The Computational Content of Classical Arithmetic[1]

JEREMY AVIGAD

ABSTRACT. Almost from the inception of Hilbert's program, foundational and structural efforts in proof theory have been directed towards the goal of clarifying the computational content of modern mathematical methods. This essay surveys various methods of extracting computational information from proofs in classical first-order arithmetic, and reflects on some of the relationships between them. Variants of the Gödel-Gentzen double-negation translation, some not so well known, serve to provide canonical and efficient computational interpretations of that theory.

1 Introduction

Hilbert's program was launched, in 1922, with the specific goal of demonstrating the consistency of modern, set-theoretic methods, using only finitary means. But the program can be viewed more broadly as a response to the radical methodological changes that had been introduced to mathematics in the late nineteenth century. Central to these changes was a shift in mathematical thought whereby the goal of mathematics was no longer viewed as that of developing powerful methods of calculation, but, rather, that of characterizing abstract, possibly infinite, mathematical structures, often in ways that could not easily be reconciled with a computational understanding.

Grisha's work over the years has touched on almost every aspect of proof theory, both of the reductive (foundational) and structural sort, involving a wide range of logical frameworks. But much of his work addresses the core proof-theoretic concern just raised, and has served to provide us with a deep and satisfying understanding of the computational content of nonconstructive, infinitary reasoning. Such work includes his characterization of the provably total computable functions of $I\Sigma_1$ as exactly the primitive recursive functions ([28, 29]); his method of continuous cut elimination, which provides a finitary interpretation of infinitary cut-elimination methods ([32, 11]); and his work on the epsilon substitution method (for example, [35, 37]).

Grisha has also been a friend and mentor to me throughout my career. The characterization of the provably total computable functions of $I\Sigma_1$ just mentioned was, in fact, also discovered by Charles Parsons and Gaisi Takeuti, all independently. I shudder to recall that at a meeting at Oberwolfach in 1998, when I was just two-and-a-half years out of graduate school, I referred to the

---

[1]Dedicated to Grigori Mints in honor of his seventieth birthday.

result as "Parsons' theorem" in a talk before an audience that, unfortunately, included only the other two. Grisha asked the first question after the talk was over, and nothing in his manner or tone even hinted that I had made a *faux pas* (it didn't even occur to me until much later). In fact, I still vividly remember his encouraging and insightful comments, then and in later discussion. (For the record, Gaisi was equally gracious and supportive.)

In this essay, I will discuss methods of interpreting classical first-order arithmetic, often called *Peano arithmetic* (*PA*), in computational terms. Although the study of *PA* was central to Hilbert's program, it may initially seem like a toy theory, or an artificially simple case study. After all, mathematics deals with much more than the natural numbers, and there is a lot more to mathematical argumentation than the principle of induction. But experience has shown that the simplicity of the theory is deceptive: via direct interpretation or more elaborate forms of proof-theoretic reduction, vast portions of mathematical reasoning can be understood in terms of *PA* [4, 13, 43].

Here, I will be concerned with the Π_2, or "computational," consequences of *PA*. Suppose *PA* proves $\forall x\, \exists y\, R(x, y)$, where x and y range over the natural numbers and $R(x, y)$ is a decidable (say, primitive recursive) relation. We would like to understand how and to what extent such a proof provides an *algorithm* for producing such a y from a given x, one that is more informative than blind search. There are four methods that are commonly used to extract such an algorithm:

1. Gödel's *Dialectica* interpretation [18, 6], in conjunction with a double-negation interpretation that interprets *PA* in its intuitionistic counterpart, Heyting arithmetic (*HA*)

2. realizability [22, 23, 27, 45], again in conjunction with a double-negation translation, and either the Friedman *A*-translation [14] (often also attributed to Dragalin and Leivant, independently) or a method due to Coquand and Hofmann ([12, 1]) to "repair" translated Π_2 assertions

3. cut elimination ([15]; see, for example, [41])

4. the epsilon substitution method ([19, 8])

These four approaches really come in two pairs: the Dialectica interpretation and realizability have much in common, and, indeed, Paulo Oliva [39] has recently shown that one can interpolate a range of methods between the two; and, similarly, cut elimination and the epsilon substitution method have a lot in common, as work by Grisha (e.g. [36]) shows. That is not to say that there aren't significant differences between the methods in each pairing, but the differences between the two pairs are much more pronounced.

For one thing, they produce two distinct sorts of "algorithms." The Dialectica interpretation, and Kreisel's "modified" version of Kleene's realizability, extract terms in Gödel's calculus of primitive recursive functionals of finite type, denoted PR^ω in Section 2 below. In contrast, cut elimination and the epsilon substitution method provide iterative procedures, whose termination can be proved by ordinal analysis. Specifically, one assigns (a notation for) an ordinal less than ε_0 to each stage of the computation in such a way that the ordinals decrease as the computation proceeds. Terms in PR^ω and $\prec\varepsilon_0$-recursive algorithms both have computational meaning, and there are various ways to "see" that the computations terminate; but, of course, any means of proving termination formally for all such terms and algorithms has to go beyond the means of *PA*.

Second, as indicated above, the first two methods involve an intermediate translation to *HA*, while the second two do not. It is true that the Dialectica interpretation and realizability can be applied to classical calculi directly (see [42] for the Dialectica interpretation, and, for example, [2, 38] for realizability); but I know of no such interpretation that cannot be understood in terms of a passage through intuitionistic arithmetic [2, 5, 44]. In contrast, cut elimination and the epsilon substitution method apply to classical logic directly. That is not to deny that one can apply cut elimination methods to intuitionistic logic (see, for example, [46]); but the arguments tend to be easier and more natural in the classical setting.

Finally, there is the issue of canonicity. Algorithms extracted from proofs in intuitionistic arithmetic tend to produce canonical witnesses to Π_2 assertions; work by Grisha [33, 34] shows, for example, that algorithms extracted by various methods yield the same results. In contrast, different ways of extracting witnesses from classical proofs yield different results, conveying the impression that there is something "nondeterministic" about classical logic. (There is a very nice discussion of this in [47, 48]. See also the discussion in Section 6 below.) Insofar as one has a natural translation from classical arithmetic to intuitionistic arithmetic, some of the canonicity of the associated computation is transferred to the former theory.

In this essay, I will discuss realizability and the Dialectica interpretation, as they apply to classical arithmetic, via translations to intuitionistic arithmetic. After reviewing some preliminaries in Section 2, I will discuss variations of the double-negation interpretation in Section 3. One, in particular, is very efficient when it comes to introducing negations; in Section 4, I will show that, when combined with realizability or the Dialectica interpretation, this yields computational interpretations of classical arithmetic that are efficient in their use of higher types. In Section 5, I will consider another curious double-negation interpretation, and diagnose an unfortunate aspect of its behavior.

This work has been partially supported by NSF grant DMS-0700174 and a grant from the John Templeton Foundation. I am grateful to Philipp Gerhardy, Thomas Streicher, and an anonymous referee for helpful comments and corrections.

2 Preliminaries

Somewhat imprecisely, one can think of intuitionistic logic as classical logic without the law of the excluded middle; and one can think of minimal logic as intuitionistic logic without the rule *ex falso sequitur quodlibet*, that is, from $\bot$ conclude anything. Computational interpretations of classical logic often pass through minimal logic, which has the nicest computational interpretation. (One can interpret intuitionistic logic in minimal logic by replacing every atomic formula A by $A \vee \bot$, so the difference between these two is small.)

To have a uniform basic to compare the different logics, it is useful to take the first-order logical symbols to be $\forall$, $\exists$, $\wedge$, $\vee$, $\rightarrow$, and $\bot$, with $\neg \varphi$ defined to be $\varphi \rightarrow \bot$. However, when it comes to classical logic, it is often natural to restrict one's attention to formulas in *negation-normal form*, where formulas are built up from atomic and negated atomic formulas using $\wedge$, $\vee$, $\forall$, and $\exists$. A negation operator, $\sim\varphi$, can be defined for such formulas; $\sim\varphi$ is what you get if, in φ, you exchange $\wedge$ with $\vee$, $\forall$ with $\exists$, and atomic formulas with their negations. Note that $\sim\sim\varphi$ is just φ. Classically, every formula φ has a negation-normal form equivalent, φ^{nnf}, obtained by defining $(\theta \rightarrow \eta)^{\mathrm{nnf}}$ to be $\sim\theta^{\mathrm{nnf}} \vee \eta^{\mathrm{nnf}}$, and treating the other connectives in the obvious way. This has the slightly awkward consequence that $(\neg\varphi)^{\mathrm{nnf}}$ translates to $\sim\varphi^{\mathrm{nnf}} \vee \bot$, but simplifying $\theta \vee \bot$ to θ and $\theta \wedge \bot$ to $\bot$ easily remedies this.

There are a number of reasons why negation-normal form is so natural for classical logic. First of all, it is easy to keep track of polarities: if φ is in negation-normal form, then every subformula is a positive subformula, except for, perhaps, atomic formulas; an atomic formula A occurs positively in φ if it occurs un-negated, and negatively if it occurs with a negation sign before it. Second, the representation accords well with practice: any classically-minded mathematician would not hesitate to prove "if φ then ψ" by assuming $\neg\psi$ and deriving $\neg\varphi$, or by assuming φ and $\neg\psi$ and deriving a contradiction; so it is convenient that $\varphi \rightarrow \psi$, $\neg\psi \rightarrow \neg\varphi$, and $\neg(\varphi \wedge \neg\psi)$ have the same negation-normal form representation. Finally, proof systems for formulas in negation-normal form tend to be remarkably simple (see, for example, [41, 46]).

It was Gödel [18] who first showed that the provably total computable functions of arithmetic can be characterized in terms of the primitive recursive functionals of finite type (see [6, 20]). The set of finite types can be defined to be the smallest set containing the symbol N, and closed under an operation which takes types σ and τ to a new type $\sigma \rightarrow \tau$. In the intended ("full") interpretation, N denotes the set of natural numbers, and $\sigma \rightarrow \tau$ denotes the set of all functions from σ to τ. A set of terms, PR^ω, is defined inductively as follows:

1. For each type σ, there is a stock of variables $x, y, z, \ldots$ of type σ.

2. 0 is a term of type N.

3. S (successor) is a term of type $\mathsf{N} \rightarrow \mathsf{N}$.

4. if s is a term of type $\tau \rightarrow \sigma$ and t is a term of type τ, then $s(t)$ is a term of type σ.

5. if s is a term of type σ and x is a variable of type τ, then $\lambda x\, s$ is a term of type $\tau \rightarrow \sigma$.

6. If s is a term of type σ, and t is a term of type $\mathsf{N} \rightarrow (\sigma \rightarrow \sigma)$, then R_{st} is a term of type $\mathsf{N} \rightarrow \sigma$.

Intuitively, $s(t)$ denotes the result of applying s to t, $\lambda x\, s$ denotes the function which takes any value of x to s, and R_{st} denotes the function defined from s

and t by primitive recursion, with $R_{st}(0) = s$ and $R_{st}(S(x)) = t(x, R_{st}(x))$ for every x. In this last equation, I have adopted the convention of writing $t(r, s)$ instead of $(t(r))(s)$.

It will be convenient below to augment the finite types with products $\sigma \times \tau$, associated pairing operations $(\cdot, \cdot)$, and projections $(\cdot)_0$ and $(\cdot)_1$. Product types can be eliminated in the usual way by currying and replacing terms t with sequences of terms t_i. It will also be convenient to have disjoint union types $\sigma + \tau$, an element of which is either an element of σ or an element of τ, tagged to indicate which is the case. That is, for each such type we have insertion operations, inl and inr, which convert elements of type σ and τ respectively to an element of type $\sigma + \tau$; predicates isleft(a) and isright(a), which indicate whether a is tagged to be of type σ or τ; and functions left(a) and right(a), which interpret a as an element of type σ and τ, respectively. References to such sum types can be eliminated by taking $\sigma + \tau$ to be $\mathsf{N} \times \sigma \times \tau$, defining inl($a$) = $(0, a, 0^\tau)$, defining inr(a) = $(1, 0^\sigma, a)$, where 0^σ and 0^τ are constant zero functionals of type σ, τ respectively, and so on.

In the next section, I will describe various double-negation interpretations that serve to reduce classical arithmetic, PA, to intuitionistic arithmetic, HA — in fact, to HA taken over *minimal logic*. These show that if PA proves a Π_2 formula $\forall x \, \exists y \, R(x, y)$, then HA proves $\forall x \, \neg\neg \exists y \, R(x, y)$; in fact, a variant HA' of HA based on minimal logic suffices. This reduces the problem to extracting computational information from the latter proof.

One method of doing so involves using Kreisel's notion of modified realizability, combined with the Friedman A-translation. One can extend HA to a higher-type version, HA^ω, which has variables ranging over arbitrary types, and terms of all the primitive recursive functionals. Fix any primitive recursive relation $A(y)$; then to each formula $\varphi(\bar{x})$ in the language of arithmetic, one inductively assigns a formula "a *realizes* $\varphi(\bar{x})$," as follows.

a *realizes* $\bot$ $\equiv$ $A(a)$

a *realizes* θ $\equiv$ θ, if θ is atomic

a *realizes* $\varphi \wedge \psi$ $\equiv$ $((a)_0$ *realizes* $\varphi) \wedge ((a)_1$ *realizes* $\varphi)$

a *realizes* $\varphi \vee \psi$ $\equiv$ $(\text{isleft}(a) \wedge \text{left}(a)$ *realizes* $\varphi) \vee$
$\qquad\qquad\qquad\qquad\qquad (\text{isright}(a) \wedge \text{right}(a)$ *realizes* $\psi))$

a *realizes* $\varphi \to \psi$ $\equiv$ $\forall b \, (b$ *realizes* $\varphi \to a(b)$ *realizes* $\psi)$

a *realizes* $\forall x \, \varphi(x)$ $\equiv$ $\forall x \, (a(x)$ *realizes* $\varphi(x))$

a *realizes* $\exists x \, \varphi(x)$ $\equiv$ $(a)_1$ *realizes* $\varphi((a)_0)$

Now, suppose classical arithmetic proves $\forall x \, \exists y \, R(x, y)$, for some primitive recursive relation R. Then, using a double-negation translation, HA' proves $\forall x \, \neg\neg \exists y \, R(x, y)$, and hence it proves $\neg\neg \exists y \, R(c, y)$ for a fresh constant c. Fix $A(y)$ in the realizability relation above to be the formula $R(c, y)$. Inductively, one can then extract from the proof of term t of PR^ω such that HA^ω proves that t realizes $\neg\neg \exists y \, R(x, y)$. Now notice that the identity function, id, realizes $\neg \exists y \, R(c, y)$, since a realizer to $\exists y \, R(c, y)$ is simply a value of a satisfying $R(c, a)$. Thus if a realizes $\exists y \, R(c, y)$, then $a(id)$ satisfies $R(c, a(id))$. Viewing a, now, as a function of c, yields the following conclusions:

THEOREM 1. *If classical arithmetic proves $\forall x \, \neg\neg \exists y \, R(x,y)$, there is a term F of PR^ω of type N to N such that HA^ω proves $\forall x \, R(x, F(x))$.*

See [45, 24] for more about realizability, and [14] for the A-translation.

Gödel's Dialectica interpretation provides an alternative route to this result. In fact, one obtains a stronger conclusion, namely that the correctness of the witnessing term can be proved in a quantifier-free fragment PR^ω of HA^ω. To each formula φ in the language of arithmetic, one inductively assigns a formula φ^D of the form $\exists x \, \forall y \, \varphi_D(x,y)$, where x and y are now tuples of variables of appropriate types. Assuming $\psi^D = \exists u \, \forall v \, \psi_D(u,v)$, the assignment is defined as follows:

$$\begin{aligned}
\theta^D &\equiv \theta, \text{ if } \theta \text{ is atomic} \\
(\varphi \wedge \psi)^D &\equiv \exists x, u \, \forall y, v \, (\varphi_D \wedge \psi_D) \\
(\varphi \vee \psi)^D &\equiv \exists z \, \forall y, v \, (\mathrm{isleft}(z) \wedge \varphi_D(\mathrm{left}(z), y) \vee \\
&\qquad (\mathrm{isright}(z) \wedge \psi_D(\mathrm{right}(z), v))) \\
(\varphi \to \psi)^D &\equiv \exists U, Y \, \forall x, v \, (\varphi_D(x, Y(x,v)) \to \psi_D(U(x), v)) \\
(\forall z \, \varphi(z))^D &\equiv \exists X \, \forall z, y \, \varphi_D(X(z), y, z) \\
(\exists z \, \varphi(z))^D &\equiv \exists z, x \, \forall y \, \varphi_D(x, y, z)
\end{aligned}$$

The clause for implication is the most interesting among these, and can be understood as follows: from a witness, x, to the hypothesis, $U(x)$ is supposed to return a witness to the conclusion; and given a purported counterexample, v, to the conclusion, $Y(x,v)$ is supposed to return a counterexample to the hypothesis. Since we have defined $\neg\varphi$ to be $\varphi \to \bot$, notice that $(\neg\varphi)^D$ is $\exists Y \, \forall x \, \neg\varphi_D(x, Y(x))$.

The Dialectica interpretation of $\forall x \, \neg\neg\exists y \, R(x,y)$ is $\exists Y \, \forall x \, \neg\neg R(x, Y(x))$, which is intuitionistically equivalent to $\exists Y \, \forall x \, R(x, Y(x))$, given the decidability of primitive recursive relations. One can show that from a proof of φ in HA, one can extract a term F such that for every x, PR^ω proves $\varphi_D(x, F(x))$, once again yielding Theorem 1.

3 Some double-negation translations

We have seen that one can use modified realizability or the Dialectica interpretation to extract an algorithm from a proof of a Π_2 statement in classical arithmetic, modulo a method of reducing classical arithmetic to intuitionistic arithmetic. Double negation translations provide the latter.

A formula is said to be *negative* if it does not involve $\exists$ or $\vee$ and each atomic formula A occurs in the form $\neg A$; in other words, the formula is built up from negated atomic formula using $\forall$, $\wedge$, $\to$, and $\bot$. Over minimal logic, negative formulas are stable under double negation, which is to say, if φ is any negative formula, then HA proves that $\neg\neg\varphi$ is equivalent to φ (see, for example, [46]).

The Gödel-Gentzen double-negation translation maps an arbitrary first-order formula φ to a negative formula, φ^N:

$$\begin{aligned}
\bot^N &\equiv \bot \\
\theta^N &\equiv \neg\neg\theta, \text{ if } \theta \text{ is atomic} \\
(\varphi \wedge \psi)^N &\equiv \varphi^N \wedge \psi^N \\
(\varphi \vee \psi)^N &\equiv \neg(\neg\varphi^N \wedge \neg\psi^N) \\
(\varphi \to \psi)^N &\equiv \varphi^N \to \psi^N \\
(\forall x\ \varphi)^N &\equiv \forall x\ \varphi^N \\
(\exists x\ \varphi)^N &\equiv \neg\forall x\ \neg\varphi^N
\end{aligned}$$

The translation has the following properties:

THEOREM 2. *For any formula φ and set of sentences Γ:*

1. *Classical logic proves $\varphi \leftrightarrow \varphi^N$*

2. *If φ is provable from Γ in classical logic, then φ^N is provable from Γ^N in minimal logic.*

Since the *HA* proves the $\cdot^N$ translations of its own axioms, we have as a corollary:

COROLLARY 3. *If PA proves φ, then HA proves φ^N.*

In fact, one can strengthen the corollary in three ways:

1. Since *HA* proves $\neg\neg\theta \to \theta$ for atomic formulas θ, one can define θ^N to be θ.

2. Assuming the language of *HA* includes, say, symbols denoting the primitive recursive functions, every negated atomic formula, $\neg\theta$, has an atomic equivalent, $\bar\theta$; so one can define $(\neg\theta)^N$ to be $\bar\theta$.

3. The theorem remains true if one replaces *HA* by a suitable variant, *HA'*, based on minimal logic.

These considerations hold in the theorems that follow as well.

The reason to be concerned about negations is that they are undesirable with respect to the two computational interpretations given in the last section, since they lead to the use of more complicated types in the resulting terms of PR^ω. There is a variant of the double-negation translation known as the *Kuroda translation* that fares slightly better in this regard: for any formula φ, let φ^{Ku} denote the result of doubly-negating atomic formulas, and adding a double negation *after* each universal quantifier, and, finally, adding a double-negation to the front of the formula. Then we have:

THEOREM 4. *For every formula φ, $\varphi^{Ku} \leftrightarrow \varphi^N$ is provable in minimal logic. Hence PA proves φ if and only if HA proves φ^{Ku}.*

Note that intuitionistic logic, rather than minimal logic, is required in the conclusion.

Late in 2005, Grisha asked whether a version of the Dialectica interpretation designed by Shoenfield [42], for classical arithmetic, could be understood as a composition of the usual Dialectica interpretation together with a double-negation translation. I set the question aside and solved it a few months later [5], only to find that Ulrich Kohlenbach and Thomas Streicher had solved it more quickly [44]. In a way that can be made precise, the Shoenfield translation corresponds to the following version of the double-negation interpretation (itself a variant of a translation due to Krivine), expressed for a basis involving the connectives $\neg$, $\wedge$, $\vee$, and $\forall$. We define φ^{Kr} to be $\neg \varphi_{Kr}$, where φ_{Kr} is defined recursively by clauses below. It helps to keep in mind that φ_{Kr} is supposed to represent the *negation* of φ:

$$\begin{aligned}
\theta_{Kr} &\equiv \neg \theta, \text{ if } \theta \text{ is atomic} \\
(\neg \varphi)_{Kr} &\equiv \neg \varphi_{Kr} \\
(\varphi \wedge \psi)_{Kr} &\equiv \varphi_{Kr} \vee \psi_{Kr} \\
(\varphi \vee \psi)_{Kr} &\equiv \varphi_{Kr} \wedge \psi_{Kr} \\
(\forall x\, \varphi)_{Kr} &\equiv \exists x\, \varphi_{Kr}
\end{aligned}$$

Note that we can eliminate either $\vee$ or $\wedge$ and retain a complete set of connectives, but including them both is more efficient. Formulas of the form $\exists x\, \varphi$, however, have to be expressed as $\neg \forall x\, \neg \varphi$ to apply the translation.

THEOREM 5. *For every formula φ, $\varphi^{Kr} \leftrightarrow \varphi^N$ is provable in minimal logic. Hence PA proves φ if and only if HA' proves φ^{Kr}.*

The $\cdot^{Kr}$-translation is particularly good when it comes to formulas in negation-normal form; it only adds two quantifiers for each existential quantifier, as well as one at the beginning. But one can do even better [2]. Taking advantage of the classical negation operator, now φ^M is defined to be $\neg(\sim\varphi)_M$, where the map $\psi \mapsto \psi_M$ is defined recursively as follows:

$$\begin{aligned}
\theta_M &\equiv \theta, \text{ if } \theta \text{ is atomic or negated atomic} \\
(\varphi \vee \psi)_M &\equiv \varphi_M \vee \psi_M \\
(\varphi \wedge \psi)_M &\equiv \varphi_M \wedge \psi_M \\
(\exists x\, \varphi)_M &\equiv \exists x\, \varphi_M \\
(\forall x\, \varphi)_M &\equiv \neg \exists x\, (\sim\varphi)_M.
\end{aligned}$$

Once again, we have

THEOREM 6. *For every formula φ in negation-normal form, $\varphi^M \leftrightarrow \varphi^N$ is provable in minimal logic. Hence PA proves φ if and only if HA' proves φ^M.*

The $\cdot^M$-translation is extremely efficient with respect to negations, introducing, roughly, one at the beginning of the formula, and one for every quantifier alternation after an initial block of universal quantifiers. Alternatively, can define $(\varphi \wedge \psi)_M$ in analogy to $(\forall x\, \varphi)_M$, as $\neg((\sim\varphi)_M \vee (\sim\psi)_M)$. This gives the translation the nice property that given the formulas φ^M and $(\sim\varphi)^M$, one is the negation of the other. But there is a lot to be said for keeping negations to a minimum.

Of course, the $\cdot^M$-translation extends to all classical formulas by identifying them with their canonical negation-normal form equivalents. Since the translation relies on the negation-normal form representation of classical formulas, it shares many nice properties with a more complicated double-negation translation due to Girard [17]. It is this translation that I will use, in the next section, to provide an efficient computational interpretation of classical arithmetic.

Let me close with one more translation, found in [3], which is interesting in its own right. For reasons that will become clear later on, I will call it "the awkward translation": if φ is any formula in negation-normal form, let φ^{awk} denote $\neg(\sim\varphi)$.

THEOREM 7. *For any formula φ, $\varphi^N \to \varphi^{awk}$ is provable in minimal logic. Hence, PA proves φ if and only if HA' proves φ^{awk}.*

Proof. Once can show by induction that if ψ is any formula in negation-normal form, then $\psi \to \psi^N$ is provable in minimal logic. So minimal logic proves that $\sim\varphi$ implies $(\sim\varphi)^N$, and hence that $\neg(\sim\varphi)^N$ implies φ^{awk}. But since $\sim\varphi$ is classically equivalent to $\neg\varphi$, $\neg(\sim\varphi)^N$ is equivalent to $\neg\neg\varphi^N$, which is implied by φ^N. ∎

The $\cdot^{awk}$-translation is almost absurdly efficient with respect to negations: the one classical negation on the inside adds no negations at all (recall that in arithmetic, negated atomic formulas have atomic equivalents), and the translation adds only one negation on the outside. But the attentive reader will have noticed that the first assertion in Theorem 7 is slightly weaker than the corresponding assertions in the the theorems that precede it: only one direction of the equivalence is minimally valid. We will see, in Section 5, that this means that the translation fares very poorly with respect to modus ponens, making it impossible to translate ordinary proofs piece by piece.

For pure first-order logic, an alternative proof of Theorem 7 can be found in [3]. Benno van den Berg (personal communication) later hit upon this same translation, independently. In [3], I claimed that with intuitionistic logic in place of minimal logic, the result is a consequence of a characterization of Glivenko formulas due to Orevkov, described in a very nice survey [30, Section 3.2.5] of Russian proof theory by Grisha.[1] That seems to be incorrect; but van den Berg and Streicher have pointed out to me that in that case the result follows a theorem due to Mints and Orevkov [30, page 404, paragraph 4].

4 Interpreting classical arithmetic

We now obtain direct computational interpretations of classical arithmetic simply by combining the $\cdot^M$-translation of Section 3 with the computational interpretations of HA' given in Section 2. One annoying consequence of the use of the classical negation operator in the $\cdot^M$-translation is that it is impossible to carry out the translation of a formula φ from the inside out: depending on the

---

[1] There are typographical errors on page 401 of that paper, which Grisha has corrected for me. The last class eight lines from the bottom of the page should be $\{\to^-, \sim^-, \vee+, \exists^+\}$; the last class seven lines from the bottom of the page should be $\{\to^-, \sim^-, \vee+, \to^+, \forall^+\}$; and the first class at the bottom line should be $\{\to^+, \sim^+, \vee^-\}$.

context in which a subformula ψ occurs, the computational interpretation of the full formula may depend on either the computational interpretation of ψ or the computational interpretation of $\sim\psi$. In practice, then, it is often more convenient to carry out the interpretation in two steps, applying the $\cdot^M$-translation first, and then one of the two computational interpretations described in Section 2. Nonetheless, it is interesting to see what happens when the steps are composed, which is what I will do here. Both translations apply to formulas in negation-normal form, and we can assume that negated atomic subformulas are replaced by their atomic equivalents.

As in Section 2, the appropriate version of classical realizability is defined relative to a fixed primitive recursive predicate $A(y)$. Most of the clauses look just like ordinary modified realizability:

$$\begin{aligned} a \; \mathit{realizes} \; \theta &\equiv \theta, \text{ if } \theta \text{ is atomic} \\ a \; \mathit{realizes} \; \varphi \wedge \psi &\equiv ((a)_0 \; \mathit{realizes} \; \varphi) \wedge ((a)_1 \; \mathit{realizes} \; \varphi) \\ a \; \mathit{realizes} \; \varphi \vee \psi &\equiv ((\mathrm{isleft}(a) \wedge \mathrm{left}(a) \; \mathit{realizes} \; \varphi) \vee \\ & \quad (\mathrm{isright}(a) \wedge \mathrm{right}(a) \; \mathit{realizes} \; \psi)) \\ a \; \mathit{realizes} \; \exists x \; \varphi(x) &\equiv (a)_1 \; \mathit{realizes} \; \varphi((a)_0) \end{aligned}$$

The only slightly more complicated clause is the one for the universal quantifier. Take $a \; \mathit{refutes} \; \varphi$ to be the formula $\forall b \; (b \; \mathit{realizes} \; \varphi \to A(a(b)))$.

$$a \; \mathit{realizes} \; \forall x \; \varphi(x) \equiv a \; \mathit{refutes} \; \exists x \; \sim\varphi(x)$$

One can then straightforwardly extract, from any proof of a formula φ in classical arithmetic, a term a that refutes $\sim\varphi$. But now notice that the identity function realizes $\forall x \; \bar{A}(x)$, where $\bar{A}$ is the negation of A; so, from a proof of $\exists y \; A(y)$ in classical arithmetic, since the identity function realizes $\forall x \; \bar{A}(x)$, one obtains a term a satisfying $A(a)$. This provides a direct proof of Theorem 1. Details can be found in [2]. A more elaborate realizability relation, based on the A-translation, can be found in [9].

The corresponding variant of the Dialectica translation is similarly straightforward. As with the Shoenfield variant [42, 5], each formula φ is mapped to a formula $\varphi^{D'}$ of the form $\forall x \; \exists y \; \varphi_{D'}(x, y)$, where x and y are sequences of variables. Assuming $\psi^{D'}$ is $\forall u \; \exists v \; \psi_{D'}(u, v)$, the translation is defined recursively, as follows:

$$\begin{aligned} \theta^{D'} &\equiv \theta, \text{ if } \theta \text{ is atomic} \\ (\varphi \wedge \psi)^{D'} &\equiv \forall x, u \; \exists y, v \; (\varphi_{D'}(x, y) \wedge \psi_{D'}(u, v)) \\ (\varphi \vee \psi)^{D'} &\equiv \forall x, u \; \exists y, v \; (\varphi_{D'}(x, y) \vee \psi_{D'}(u, v)) \\ (\forall z \; \varphi)^{D'} &\equiv \forall z, x \; \exists y \; \varphi_{D'}(x, y) \end{aligned}$$

This time, it is the clause for the existential quantifier that is slightly more complicated. If $(\sim\varphi(z))^{D'}$ is $\forall r \; \exists s \; (\sim\varphi)_{D'}(z, r, s)$, define

$$(\exists z \; \varphi)^{D'} \equiv \forall S \; \exists z, r \; \neg(\sim\varphi)_{D'}(z, r, S(z, r)).$$

On can interpret this as saying that for any function $S(z,r)$ that purports to witness $\forall z, r \ \exists s \ (\sim\varphi)_{D'}(z,r,s)$, there are a z and an r denying that claim. Once again, one can straightforwardly extract, from any proof of a formula φ in classical arithmetic, a term a satisfying $\forall x \ \varphi_{D'}(x, a(x))$. This provides another direct proof of Theorem 1.

5 Back to the awkward translation

I would now like to come back to the "awkward translation," discussed at the end of Section 3. I will do this via what at first might seem to be a digression through Kreisel's *no-counterexample interpretation*. Let φ be a formula in prenex form, for example,

$$\exists x \ \forall y \ \exists z \ \forall w \ \theta(x, y, z, w).$$

The *Herbrand normal form* φ^H of φ is obtained by replacing the universally quantified variables of φ by function symbols that depend on the preceding existential variables, to obtain

$$\exists x, z \ \theta(x, f(x), z, g(x, z)).$$

It is not hard to check that, in classical logic, φ implies φ^H. Thus, by a slight variant of Theorem 1 (relativizing it to function symbols), if classical arithmetic proves φ, there will be terms $F_1(f,g)$ and $F_2(f,g)$ of PR^ω such that HA^ω proves

$$\forall f, g \ \theta(F_1(f,g), f(F_1(f,g)), F_2(f,g), g(F_1(f,g), F_2(f,g))).$$

Think of f and g as providing purported counterexamples to the truth of φ, so that F_1 and F_2 effectively foil such counterexamples. The no-counterexample interpretation is simply the generalization of this transformation to arbitrary prenex formulas.

One need not invoke Herbrand normal form to arrive at the previous conclusion. One can check that if φ is a prenex formula, the no-counterexample interpretation of φ is essentially just the Dialectica interpretation of φ^{awk}, so the result follows from Theorem 7 as well.

The no-counterexample interpretation can be viewed as a computational interpretation of arithmetic. But, in a remarkable article [25], Kohlenbach has shown that it is not a very *modular* computational interpretation, in the sense that it does not have nice behavior with respect to modus ponens. To make this claim precise, note that the set of (terms denoting) primitive recursive functionals, PR^ω, can be stratified into increasing subsets PR_n^ω, in such a way that any finite fragment of HA has a Dialectica interpretation (or modified realizability interpretation) using only terms in that set. Kohlenbach [25, Proposition 2.2] shows:

THEOREM 8. *For every n there are sentences φ and ψ of arithmetic such that:*

1. *φ is prenex.*

2. *ψ is a Π_2 sentence, that is, of the form $\forall x \ \exists y \ R(x,y)$ for some primitive recursive relation R.*

3. Primitive recursive arithmetic proves φ.

4. PA proves $\varphi \to \psi$.

5. φ and every prenexation of $\varphi \to \psi$ has a no-counterexample interpretation with functionals in PR_0^ω.

But:

6. There is no term F of PR_n^ω which satisfies the no-counterexample interpretation of ψ; that is, there is no term F such that $\forall x\, R(x, F(x))$ is true in the standard model of arithmetic.

Theorem 8 shows that there is no straightforward way to combine witnesses to the no-counterexample interpretations of φ and $\varphi \to \psi$, respectively, to obtain a witness to the no-counterexample interpretation of ψ.

The problem with the awkward translation is that, similarly, it may behave poorly with respect to modus ponens. Consider pure first-order logic with a single predicate symbol, $A(x)$. Then there are formulas φ and ψ such that ψ^{awk} doesn't follow from φ^{awk} and $(\varphi \to \psi)^{awk}$ in minimal logic: just take φ to be the formula $\forall x\, A(x)$ and ψ to be $\bot$. In that case, $\varphi^{awk} \wedge (\varphi \to \psi)^{awk} \to \psi^{awk}$ is equivalent, over minimal logic, to the double-negation shift, $\forall x\, \neg\neg A(x) \to \neg\neg A(x)$, which is not even provable in intuitionistic logic.

When I showed the awkward translation to Grisha, he remarked right away that its behavior has something to do with Kohlenbach's result. At the time, I had no idea what he meant; but writing this paper finally prodded me to sort it out. Grisha was right: Theorem 8 can, in fact, be used to show that the awkward translation does not provide a modular translation of Peano arithmetic to Heyting arithmetic, in the following sense.

THEOREM 9. *For any fragment T of HA, there are formulas φ and ψ such that the following hold:*

1. *PA proves φ and $\varphi \to \psi$, but*

2. *T together with φ^{awk} and $(\varphi \to \psi)^{awk}$ does not prove ψ^{awk}.*

This shows that modus ponens fails under the $\cdot^{awk}$-translation, in a strong way. Since PA proves φ and $\varphi \to \psi$, it also proves ψ, and so by Theorem 7, HA proves ψ^{awk}. But having the translation φ^{awk} and $(\varphi \to \psi)^{awk}$ may not help much in obtaining such a proof of ψ^{awk}; indeed, obtaining a proof of ψ^{awk} from φ^{awk} and $(\varphi \to \psi)^{awk}$ may be no easier than simply proving ψ^{awk} outright.

Proof. Given T, first let n be large enough so that the Dialectica interpretation of T uses only terms in PR_n^ω, and then let φ and ψ be as in Theorem 8. If there were a proof of ψ^{awk} from φ^{awk} and $(\varphi \to \psi)^{awk}$ in T, applying the Dialectica interpretation, one would obtain terms witnessing the Dialectica interpretation of ψ^{awk} from terms witnessing the Dialectica interpretations of φ^{awk} and $(\varphi \to \psi)^{awk}$. But the Dialectica interpretation of φ^{awk} is the no-counterexample interpretation of φ, and it is not hard to check that the Dialectica interpretation of $(\varphi \to \psi)^{awk}$ is the no-counterexample interpretation of one of the prenexations of $\varphi \to \psi$. Thus there would be a witness to the no-counterexample interpretation of ψ in PR_n^ω, contrary to the choice of φ and ψ. ∎

6 Conclusions

As noted in the introduction, conventional wisdom holds that classical logic is "nondeterministic," in that different ways of extracting algorithms from classical proofs can yield different results. Sometimes, however, nondeterminacy is unavoidable. For example, even minimal logic can prove $\exists x\ A(x) \wedge \exists y\ A(y) \to \exists z\ A(z)$, and any computational interpretation of this formula will have to choose either x or y to witness the conclusion. The difference is that this sentence is typically not taken to be an *axiom* of minimal logic; rather, there are two axioms, $\varphi \wedge \psi \to \varphi$ and $\varphi \wedge \psi \to \psi$, and any proof of the sentence has to choose one or the other. In contrast, standard calculi for classical logic provide cases where there are multiple choices of witnesses, with no principled reason to choose one over the other. For example, starting from canonical proofs of $A(a) \to \exists x\ A(x)$ and $A(b) \to \exists x\ A(x)$, one can weaken the conclusions to obtain proofs of $\varphi \to (A(a) \to \exists x\ A(x))$ and $\neg\varphi \to (A(b) \to \exists x\ A(x))$, for an arbitrary formula, φ. Then, using the law of the excluded middle, $\varphi \vee \neg\varphi$, one can combine these to obtain a proof of $A(a) \wedge A(b) \to \exists x\ A(x)$ where there is little reason to favor a or b as the implicit witness to the existential quantifier. (This example is essentially that given by Lafont [16, p. 150].)

This shows that standard classical calculi are nondeterministic in a way that proofs in intuitionistic and minimal logic are not. Girard [17] has neatly diagnosed the source of the nondeterminacy, and has provided a calculus for classical logic that eliminates it by forcing the prover to make an explicit choice in exactly those situations where an ambiguity would otherwise arise.[2] But note that the realizability interpretation of Section 4 also avoids this nondeterminacy; the translation procedure described in [2] is fully explicit and unambiguous. With respect to the example discussed in the last paragraph, the interpretation chooses a or b based on the logical form of φ.[3] Indeed, the results of [2] show that the witnesses obtained in this way coincide with those obtained using a natural class of cut-elimination procedures.

Yet another response to the example above is that of Urban and Bierman [47, 48], who simply embrace the nondeterminism as an inherent part of the computational interpretation of classical logic. In fact, Urban [47] has provided a nondeterministic programming language to interpret classical logic in a natural way. Passing through a double-negation interpretation, as we have done here, amounts to making specific choices to resolve the nondeterminism. It is

---

[2] One way to understand what is going on is to notice that in minimal logic there are two distinct ways of proving $\neg\neg(\varphi \wedge \psi)$ from $\neg\neg\varphi$ and $\neg\neg\psi$, and this inference is needed to verify the classical axiom $\neg\neg\theta \to \theta$ under the double-negation translation. Another way is to notice that since, in classical logic, φ and $\neg\neg\varphi$ are equivalent, there is little reason to favor φ or $\neg\varphi$ in situations where intuitionistic logic treats them differently. I am grateful to Thomas Streicher for these insights.

[3] More generally, what breaks the symmetry alluded to in the previous footnote is that φ and $\neg\neg\varphi$ have the same negation-normal form, which is distinct from (and dual to) that of $\neg\varphi$.

an open-ended conceptual problem to understand which deterministic instances of the general nondeterministic algorithms can be realized in such a way.

At this point, however, we should be clearer as to the goals of our analysis. Ordinary mathematical proofs are not written in formal languages, and so the process of extracting an algorithm from even a rather constructive mathematical argument can involve nondeterminism of sorts. And, despite some interesting explorations in this direction [10], it is far from clear that classical arithmetic can be used as an effective programming language in its own right. But formal methods are actively being developed in support of software verification [21], and a better understanding of the computational content of classical logic may support the development of better logical frameworks for that purpose [40]. Formal translations like the ones described here have also been effective in "proof mining," the practice of using logical methods to extract mathematically useful information from nonconstructive proofs [7, 24, 26].

Grisha's work has, primarily, addressed the general foundational question as to the computational content of classical methods. In that respect, the general metatheorems described here provide a satisfying answer: for the most part, classical mathematical reasoning *does* have computational content, which is to say, algorithms can be extracted from classical proofs; but by suppressing computational detail, the proofs often leave algorithmic detail underspecified, rendering them amenable to different implementations. Grisha's work has thus contributed to an understanding of the computational content of classical arithmetic that is mathematically and philosophically satisfying, providing a solid basis for further scientific research.

BIBLIOGRAPHY

[1] Jeremy Avigad. Interpreting classical theories in constructive ones. *J. Symbolic Logic*, 65:1785–1812, 2000.

[2] Jeremy Avigad. A realizability interpretation for classical arithmetic. In *Logic Colloquium '98 (Prague)*, pages 57–90. Assoc. Symbol. Logic, Urbana, IL, 2000.

[3] Jeremy Avigad. Algebraic proofs of cut elimination. *J. Log. Algebr. Program.*, 49:15–30, 2001.

[4] Jeremy Avigad. Number theory and elementary arithmetic. *Philosophia Mathematica*, 11:257–284, 2003.

[5] Jeremy Avigad. A variant of the double-negation translation. Carnegie Mellon Technical Report CMU-PHIL 179.

[6] Jeremy Avigad and Solomon Feferman. Gödel's functional ("Dialectica") interpretation. In *Handbook of proof theory*, pages 337–405. North-Holland, Amsterdam, 1998.

[7] Jeremy Avigad, Philipp Gerhardy, and Henry Towsner. Local stability of ergodic averages. To appear in *Transactions of the American Mathematical Society*.

[8] Jeremy Avigad and Richard Zach. The epsilon calculus. Stanford Encyclopedia of Philosophy, 2002.
http://plato.stanford.edu/entries/epsilon-calculus/.

[9] Ulrich Berger, Wilfried Buchholz, and Helmut Schwichtenberg. Refined program extraction from classical proofs. *Ann. Pure Appl. Logic*, 114:3–25, 2002.

[10] Ulrich Berger, Helmut Schwichtenberg, and Monika Seisenberger. The Warshall algorithm and Dickson's lemma: two examples of realistic program extraction. *Journal of Automated Reasoning*, 26:205–221, 2001.

[11] Wilfried Buchholz. Notation systems for infinitary derivations. *Archive for Mathematical Logic*, 30:277–296, 1991.

[12] Thierry Coquand and Martin Hofmann. A new method of establishing conservativity of classical systems over their intuitionistic version. *Mathematical Structures in Computer Science*, 9:323–333, 1999.

[13] Solomon Feferman. Infinity in mathematics: Is Cantor necessary? In G. Toraldo di Francia, editor, *L'infinito nella scienza*, pages 151–209. Istituto della Enciclopedia Italiana, 1987. Reprinted in Solomon Feferman, *In the Light of Logic*, Oxford University Press, New York, 1998, pages 28–73 and 229–248.
[14] Harvey M. Friedman. Classically and intuitionistically provable functions. In H. Müller and D. Scott, editors, *Higher Set Theory*, pages 21–27. Springer, Berlin, 1978.
[15] Gerhard Gentzen. Die Widerspruchsfreiheit der reinen Zahlentheorie. *Mathematische Annalen*, 112:493–465, 1936. Translated as "The consistency of elementary number theory" in Gerhard Gentzen, *Collected Works* (edited by M. E. Szabo), North-Holland, Amsterdam, 1969, pages 132–213.
[16] Jean-Yves Girard. *Proofs and Types*, translated and with appendices by Paul Taylor and Yves Lafont. Cambridge University Press, Cambridge, 1989.
[17] Jean-Yves Girard. A new constructive logic: classical logic. *Math. Structures Comput. Sci.*, 1:255–296, 1991.
[18] Kurt Gödel. Über eine bisher noch nicht benützte Erweiterung des finiten Standpunktes. *Dialectica*, 12:280–287, 1958. Reprinted with English translation in Feferman et al., eds., *Kurt Gödel: Collected Works*, volume 2, Oxford University Press, New York, 1990, pages 241–251.
[19] David Hilbert and Paul Bernays. *Grundlagen der Mathematik*. Springer, Berlin, first volume, 1934, second volume, 1939.
[20] J. Roger Hindley and Jonathan P. Seldin. *Introduction to Combinators and λ-calculus*. Cambridge University Press, Cambridge, 1986.
[21] C. A. R. Hoare. The verifying compiler: A grand challenge for computing research. *J. ACM*, 50:63–69, 2003.
[22] S. C. Kleene. On the interpretation of intuitionistic number theory. *J. Symbolic Logic*, 10:109–124, 1945.
[23] S. S. Kleene. *Introduction to Metamathematics*. North-Holland, Amsterdam, fourth reprint, 1964 edition, 1952.
[24] U. Kohlenbach. *Applied proof theory: proof interpretations and their use in mathematics*. Springer-Verlag, Berlin, 2008.
[25] Ulrich Kohlenbach. On the no-counterexample interpretation. *J. Symbolic Logic*, 64:1491–1511, 1999.
[26] Ulrich Kohlenbach and Paulo Oliva. Proof mining: a systematic way of analyzing proofs in mathematics. *Tr. Mat. Inst. Steklova*, 242(Mat. Logika i Algebra):147–175, 2003.
[27] Georg Kreisel. Interpretation of analysis by means of constructive functionals of finite type. In Arendt Heyting, editor, *Constructivity in Mathematics*, North-Holland, Amsterdam, 1959, pages 101–128.
[28] Grigori Mints (Minc). Quantifier-free and one-quantifier systems. *Zap. Naučn. Sem. Leningrad. Otdel. Mat. Inst. Steklov. (LOMI)*, 20:115–133, 285, 1971.
[29] Grigori Mints (Minc). What can be done in primitive recursive arithmetic. *Zap. Naučn. Sem. Leningrad. Otdel. Mat. Inst. Steklov. (LOMI)*, 60:93–102, 223–224, 1976. Studies in constructive mathematics and mathematical logic, VII.
[30] Grigori Mints. Proof theory in the USSR 1925–1969. *Journal of Symbolic Logic*, 56:385–424, 1991.
[31] Grigori Mints. *Selected Papers in Proof Theory*. Bibliopolis / North-Holland, Naples / Amsterdam, 1992.
[32] Grigori Mints. Normalization of finite terms and derivations via infinite ones. [31], pages 73–76.
[33] Grigori Mints. On E-theorems. [31], pages 105–115.
[34] Grigori Mints. Stability of E-theorems and program verification. [31], pages 117–121.
[35] Grigori Mints. Strong termination for the epsilon substitution method. *Journal of Symbolic Logic*, 61:1193–1205, 1996.
[36] Grigori Mints. Cut elimination for a simple formulation of epsilon calculus. *Ann. Pure Appl. Logic*, 152(1-3):148–160, 2008.
[37] Grigori Mints and Sergei Tupailo. Epsilon substitution method for the ramified language and Δ^1_1-comprehension rule. In Andrea Cantini et al., editor, *Logic and Foundations of Mathematics*, pages 107–130. Kluwer, the Netherlands, 1999.
[38] Chetan R. Murthy. An evaluation semantics for classical proofs. In *Proceedings, Sixth Annual IEEE Symposium on Logic in Computer Science*, pages 96–107, Amsterdam, 1991.
[39] Paulo Oliva. Unifying functional interpretations. *Notre Dame J. Formal Logic*, 47:263–290, 2006.

[40] Frank Pfenning. Logical frameworks. In John Alan Robinson and Andrei Voronkov, editors, *Handbook of Automated Reasoning*, pages 1063–1147. Elsevier and MIT Press, 2001.

[41] Helmut Schwichtenberg. Proof theory: Some aspects of cut-elimination. In Jon Barwise, editor, *Handbook of Mathematical Logic*, North-Holland, Amsterdam, 1977, pages 867–895.

[42] Joseph R. Shoenfield. *Mathematical Logic*. Association for Symbolic Logic, Urbana, IL, 2001. Reprint of the 1973 second printing.

[43] Stephen G. Simpson. *Subsystems of Second-Order Arithmetic*. Springer, Berlin, 1999.

[44] Thomas Streicher and Ulrich Kohlenbach. Shoenfield is Gödel after Krivine. *MLQ Math. Log. Q.*, 53:176–179, 2007.

[45] A. S. Troelstra. Realizability. In *Handbook of proof theory*, volume 137 of *Stud. Logic Found. Math.*, pages 407–473. North-Holland, Amsterdam, 1998.

[46] A. S. Troelstra and Helmut Schwichtenberg. *Basic Proof Theory*. Cambridge University Press, Cambridge, second edition, 2000.

[47] Christian Urban. *Classical Logic and Computation*. PhD thesis, University of Cambridge, 2000.

[48] Christian Urban and Gavin M. Bierman. Strong normalisation of cut-elimination in classical logic. *Fund. Inform.*, 45:123–155, 2001.

Fast Cut-Elimination by CERES

MATTHIAS BAAZ AND ALEXANDER LEITSCH[1]

ABSTRACT. CERES is a method of cut-elimination based on the construction of a resolution refutation γ of a characteristic clause set $\mathrm{CL}(\varphi)$ extracted from an **LK**-proof φ. The complexity of cut-elimination by CERES is directly related to the resolution complexity of $\mathrm{CL}(\varphi)$. Methods of resolution decision procedures for clause classes provide powerful tools to identify classes of proofs admitting elementary cut-elimination, where reductive methods fail. We investigate several classes of proofs for which cut-elimination can be reduced to resolution refutations on well known decidable classes like Herbrand's class and the one-variable class. We also illustrate the limitation of this approach: while the class QMON admits exponential cut-elimination by cut-projection, the characteristic clause sets of QMON, produced by the CERES-method, are not contained in a simple decision class of resolution.

1 Introduction

Proof analysis is a central mathematical activity which proved crucial to the development of mathematics. Many mathematical concepts such as the notion of group or the notion of probability were introduced by analyzing existing arguments. In some sense the analysis and synthesis of proofs form the very core of mathematical progress[12, 13].

Cut-elimination introduced by Gentzen [7] is the most prominent form of proof transformation in logic and plays an important role in automating the analysis of mathematical proofs. The removal of cuts corresponds to the elimination of intermediate statements (lemmas), resulting in a proof which is analytic in the sense, that all statements in the proof are subformulas of the result. Therefore, the proof of a combinatorial statement is converted into a purely combinatorial proof. Cut-elimination is therefore an essential tool for the analysis of proofs, especially to make implicit parameters explicit. In particular, cut free derivations allow

- the extraction of Herbrand disjunctions, which can be used to establish bounds on existential quantifiers (e.g. Luckhardt's analysis of the Theorem of Roth [9]),

- the construction of interpolants, which allow the replacement of implicit definitions by explicit ones according to Beth's Theorem,

- the calculation of generalized variants of the end formula.

[1]supported by the Austrian Science Fund (FWF) proj. nr. P 22028-N13

CERES [4] is a cut-elimination method that is based on resolution. The method roughly works as follows: The structure of the proof containing cuts is mapped to a clause term which evaluates to an unsatisfiable set of clauses $\mathcal{C}$ (the *characteristic clause set*). A resolution refutation of $\mathcal{C}$, which is obtained using a first-order theorem prover, serves as a skeleton for the new proof which contains only atomic cuts. In a final step also these atomic cuts can be eliminated, provided the (atomic) axioms are valid sequents; but this step is of minor mathematical interest only. In the system ceres[1] this method of cut-elimination has been implemented. The system is capable of dealing with formal proofs in **LK**, among them also very large ones. The most substantial proof analyzed by ceres so far is Fürstenberg's proof of the infinity of primes, the 5th proof in the "Book" [1]. But the method CERES proved also useful in proof theoretic investigations: in [5] and [4] we have shown that CERES is, in some sense, more general than reductive Gentzen-type methods and asymptotically more efficient.

By the results of Orevkov [11] and Statman [15] there is no elementary bound on the size of a cut-free proof in terms of the original one. Therefore there exists no fast, i.e. elementary, method of cut-elimination in principle. So also CERES has a nonelementary worst-case complexity. But in [5] we have shown that a nonelementary speed-up of CERES by Gentzen-type methods is impossible; on the other hand there exists a nonelementary speed-up of the Gentzen method by CERES [4]. While the complexity of CERES and Gentzen-type methods were compared via artificial proof sequences (derived from Statman's worst-case sequence) the general behavior of CERES on (interesting) proof classes remained uninvestigated.

In this paper we use CERES as a tool to prove fast cut-elimination for several subclasses of **LK**-proofs. By using the structure of the characteristic clause sets in the CERES-method we characterize fast classes of cut-elimination; these classes are either defined by restrictions on the use of inference rules or on the syntax of formulas occurring in the proofs. In one case we show that CERES is fast on a class of proofs where all Gentzen-type methods of cut-elimination are of nonelementary complexity. Moreover, the simulation of Gentzen type methods by CERES (shown in [5]) indicates that the use of Gentzen type methods in proving the property of fast cut-elimination cannot be successful if CERES fails.

In Section 2 we give some basic definitions and present the key concepts from proof theory and automated deduction needed in this paper. In Section 3 we describe the most important features of the CERES method, mention the central results, give some examples, and define complexity measures for cut-elimination. Section 4 contains new results on fast cut-elimination classes. Cut-elimination on these classes are proven elementary by analyzing the length of resolution refutations of the characteristic clause sets. One of these classes, the class **MC**, cannot be proven elementary by Gentzen-type methods as all of them are of nonelementary complexity on **MC**.

---

[1]available at http://www.logic.at/ceres/

2 Definitions

In this section we give some basic definitions which are used throughout the paper.

For a signature Σ we define $\mathrm{CS}(\Sigma)$ as the set of constant symbols, $\mathrm{FS}(\Sigma)$ as the set of function symbols, and $\mathrm{PS}(\Sigma)$ as the set of predicate symbols in Σ.

DEFINITION 1 (complexity). If F is a formula in PL then the *logical complexity* is the number of occurrences of logical symbols in F. The *symbolic complexity* of F, denoted by $\|F\|$ is defined as number of all symbol occurrences in F. $\|\ \|$ is also used as a measure for **LK**-proofs, sequents, sets of clauses, resolution refutations etc.

DEFINITION 2 (sequent). A sequent is an expression of the form $\Gamma \vdash \Delta$ where Γ and Δ are finite multisets of PL-formulas (i.e. two sequents $\Gamma_1 \vdash \Delta_1$ and $\Gamma_2 \vdash \Delta_2$ are considered equal if the multisets represented by Γ_1 and by Γ_2 are equal and those represented by Δ_1, Δ_2 are also equal). $\vdash$ is called the *empty sequent*.

If $S = \Gamma \vdash \Delta$ and $S' = \Pi \vdash \Lambda$ we define the composition of S and S' by $S \circ S'$, where $S \circ S' = \Gamma, \Pi \vdash \Delta, \Lambda$.

DEFINITION 3 (the calculus **LK**). We choose the version of **LK** used in [5], with just one modification: axioms are of the form $A \vdash A$ where A can be any quantifier-free formula. For the CERES-method we also use a version of **LK** with arbitrary atomic sequents; let $\mathcal{A}$ be a set of such atomic sequents. Then we call an **LK**-derivation φ from $\mathcal{A}$ an *AC-deduction* from $\mathcal{A}$ if all cuts in φ are atomic. The *length* of a proof ω is defined by the number of nodes in ω and is denoted by $l(\omega)$.

Below we give the basic definitions for the resolution calculus.

DEFINITION 4 (clause). A *clause* is an atomic sequent, i.e. a sequent of the form $\Gamma \vdash \Delta$, where Γ and Δ are multisets of atoms.

If $C\colon A_1, \ldots, A_n \vdash B_1, \ldots, B_m$ is a clause we define $C_+ = \vdash B_1, \ldots, B_m$ and $C_- = A_1, \ldots, A_n \vdash$. Moreover we define $|C|$ as $n + m$, the number of atoms in C; for a set of clauses $\mathcal{C}\colon \{C_1, \ldots, C_k\}$ we define $|\mathcal{C}|$ as k, the number of clauses in $\mathcal{C}$.

DEFINITION 5 (unifier). Let $\mathcal{A}$ be a nonempty set of atoms and σ be a substitution with $|\mathcal{A}\sigma| = 1$. Then σ is called a *unifier* of $\mathcal{A}$. σ is called a *most general unifier* (m.g.u.) of $\mathcal{A}$ if σ is a unifier of $\mathcal{A}$ and for every unifier λ of $\mathcal{A}$ there exists a substitution ϑ s.t. $\lambda = \sigma\vartheta$.

Every nonempty unifiable set $\mathcal{A}$ of atoms has a most general unifier σ which can be computed algorithmically. $\|\sigma\|$ and $\|\mathcal{A}\sigma\|$ may be exponential in $\|\mathcal{A}\|$; also the standard unification algorithm is exponential. For details see [8].

DEFINITION 6. Let C be a clause. A *contraction normalization* of C is a clause D obtained from C by omitting multiple occurrences of atoms in C_+ and in C_-.

DEFINITION 7 (factor). Let C be a clause and D be a nonempty subclause of C_+ or of C_- and let σ be an m.g.u. of the atoms of D. Then a contraction normalization of $C\sigma$ is called a *factor* of C.

DEFINITION 8 (resolvent). Let $C = \Gamma \vdash \Delta, A_1, \ldots, A_n$ and $D = B_1, \ldots, B_m, \Pi \vdash \Lambda$ s.t. C and D are variable-disjoint, and σ be a most general unifier of $\{A_1, \ldots, A_n, B_1, \ldots, B_m\}$ Then the clause $\Gamma\sigma, \Pi\sigma \vdash \Delta\sigma, \Lambda\sigma$ is called a *resolvent* of C and D. If $n = 1$ and Γ is empty we speak about a PRF-resolvent (Positive Restricted Factoring). The (single) atom in $\{A_1, \ldots, A_n, B_1, \ldots, B_m\}\sigma$ is called the *resolved atom*.

Resolution is just a combination of most general unification, contraction and cut. Hence, after computation of the m.g.u., resolution becomes a part of **LK**; indeed, a p-resolvent defined below is an *atomic mix* (which consists of contractions and a cut):

DEFINITION 9 (p-resolvent). Let $C = \Gamma \vdash \Delta, A^m$ and $D = A^n, \Pi \vdash \Lambda$ with $n, m \geq 1$. Then the clause $\Gamma, \Pi \vdash \Delta, \Lambda$ is called a *p-resolvent* of C and D.

DEFINITION 10 (resolution deduction). A deduction tree having clauses as leaves and resolution, contraction and weakening as rules is called a *resolution deduction*. If, instead of resolution, we have p-resolution as (the only binary) rule then we call the deduction a *p-resolution deduction*. A (p-)resolution deduction of $\vdash$ is called a (p-)*resolution refutation*.

3 The Method CERES

In [4] we defined a method of cut-elimination which substantially differs from the reductive Gentzen-type methods. It is based on a structural analysis of the whole proof. The key concept of CERES is that of a characteristic clause term of a proof φ, representing an abstract structure of the derivation of the cut formulas in φ. Roughly speaking we compute a clause term from an **LK**-proof φ of S which corresponds to an unsatisfiable set of clauses, compute a resolution refutation of this set, and finally construct an atomic cut normal form of φ.

The CERES-method requires atomic axioms; this is not problematic as, for any sequent $S\colon A \vdash A$, there exists an **LK**-proof $\rho(A)$ of S from atomic axioms of the form $B \vdash B$ s.t. $\|\rho(A)\| \leq k\|A\|$ for a constant k independent of A (see e.g. [3]).

DEFINITION 11. For any **LK**-proof φ, we define $\rho(\varphi)$ as the proof φ after replacement of the axioms $A \vdash A$ by the derivations $\rho(A)$ of $A \vdash A$ from atomic axioms.

DEFINITION 12. The *skolemization* of a closed sequent S is denoted by $sk(S)$. A closed sequent S is called *skolemized* if $sk(S) = S$.

We restrict cut-elimination to derivations with skolemized end-sequents. It is always possible to construct a derivation $sk(\varphi)$ of skolemized end-sequents from the original ones without increase of proof length (see [2] and [3]) – provided the axioms do not contain quantifiers. This condition is fulfilled in our definition of **LK**. After normalization the derivation can be transformed into a derivation of the original (unskolemized) sequent. After skolemization we compute $\rho(sk(\varphi))$ and obtain a skolemized proof from atomic axioms. Clearly

$\rho(sk(\varphi))$ is skolemized and $\|\rho(sk(\varphi))\|$ is polynomial in $\|\varphi\|$.

DEFINITION 13. Let SK be the set of all **LK**-derivations from atomic axioms with skolemized end-sequents; we call these proofs *skolemized*. $SK_\emptyset$ is the set of all cut-free proofs in SK and, for all $i \geq 0$, SK^i is the subset of SK containing all derivations with cut-formulas of logical complexity $\leq i$.

REMARK 14. Proofs in $SK_\emptyset$ do not contain the rules $\forall\colon r$ and $\exists\colon l$. On the other hand, these rules may occur in SK but may not go into the end-sequent.

Our goal is to transform a derivation in SK into a derivation in SK^0. The first step in the corresponding procedure consists in the definition of a clause term corresponding to the sub-derivations of an **LK**-derivation ending in a cut. In particular we focus on derivations of the cut formulas themselves, i.e. on the derivation of formulas having no successors in the end-sequent.

DEFINITION 15. We write Ω for the set of occurrences of all cut-formulas in a proof φ. By $S(\nu, \Omega)$ we define the subsequent of the sequent occurring at the node ν which consists of all ancestors of Ω (for a fully technical definition see [5]). Let A be a formula occurring in a sequent at the node ν. If A occurs in $S(\nu, \Omega)$ then we say that A *goes into a cut*; if A occurs in $S(\nu)$ but not in $S(\nu, \Omega)$ we say that A *goes into the end-sequent*. If A is an auxiliary or principal formula of a rule ρ and A goes into a cut we also say that ρ *goes into a cut*.

DEFINITION 16 (clause term). (Finite) sets of clauses are clause terms.
If X, Y are clause terms then $X \oplus Y$ and $X \otimes Y$ are clause terms.

We define the semantics $|\ |$ of clause terms in the following way:

$$|\mathcal{C}| = \mathcal{C} \text{ for clause sets } \mathcal{C}, \quad |X \oplus Y| = |X| \cup |Y|, \quad |X \otimes Y| = |X| \times |Y|,$$

where $\mathcal{C} \times \mathcal{D} = \{C \circ D \mid C \in \mathcal{C}, D \in \mathcal{D}\}$. We define clause terms to be equivalent if the corresponding sets of clauses are equal, i.e. $X \sim Y$ iff $|X| = |Y|$.

DEFINITION 17 (characteristic term). Let φ be an **LK**-derivation of S in SK and let Ω be the set of all occurrences of cut formulas in φ. We define the *characteristic (clause) term* $\Theta(\varphi)$ inductively:

Let ν be the occurrence of an axiom sequent S' in φ. Let S'' be the subsequent of S' consisting of all atoms which are ancestors of an occurrence in Ω, i.e. $S'' = S(\nu, \Omega)$. Then $\Theta(\varphi)/\nu = \{S''\}$.

Let us assume that the clause terms $\Theta(\varphi)/\nu$ are already constructed for all sequent–occurrences ν in φ with depth$(\nu) \leq k$. Now let ν be an occurrence with depth$(\nu) = k + 1$. We distinguish the following cases:

(a) ν is the conclusion of μ, i.e. a unary rule applied to μ gives ν. Here we simply define $\Theta(\varphi)/\nu = \Theta(\varphi)/\mu$.

(b) ν is the conclusion of μ_1 and μ_2, i.e. a binary rule X applied to μ_1 and μ_2 gives ν.

   (b1) X goes into a cut, i.e. the auxiliary formulas of X occur in $S(\mu_1, \Omega), S(\mu_2, \Omega)$. Then $\Theta(\varphi)/\nu = \Theta(\varphi)/\mu_1 \oplus \Theta(\varphi)/\mu_2$.

(b2) X goes into the end-sequent. In this case we define
$$\Theta(\varphi)/\nu = \Theta(\varphi)/\mu_1 \otimes \Theta(\varphi)/\mu_2.$$

Note that, in a binary inference, either both auxiliary formulas are ancestors of an occurrence in Ω or none of them.

Finally the characteristic term $\Theta(\varphi)$ is defined as $\Theta(\varphi)/\nu$ where ν is the occurrence of the end-sequent.

REMARK 18. If φ is a cut-free proof then there are no occurrences of cut formulas in φ and $|\Theta(\varphi)| = \{\vdash\}$.

DEFINITION 19 (characteristic clause set). Let φ be an **LK**-derivation and $\Theta(\varphi)$ be the characteristic term of φ. Then CL(φ), for CL$(\varphi) = |\Theta(\varphi)|$, is called the *characteristic clause set* of φ.

EXAMPLE 20. Let φ be the **LK**-proof

$$\frac{\varphi_1 \qquad \varphi_2}{(\forall x)(P(x) \to Q(x)) \vdash (\exists y)(P(a) \to Q(y))} \; cut$$

where $\varphi_1 =$

$$\frac{\dfrac{\dfrac{\dfrac{\dfrac{\dfrac{P(u)^\star \vdash P(u) \quad Q(u) \vdash Q(u)^\star}{P(u)^\star, P(u) \to Q(u) \vdash Q(u)^\star} \to: l}{P(u) \to Q(u) \vdash (P(u) \to Q(u))^\star} \to: r}{P(u) \to Q(u) \vdash (\exists y)(P(u) \to Q(y))^\star} \exists: r}{(\forall x)(P(x) \to Q(x)) \vdash (\exists y)(P(u) \to Q(y))^\star} \forall: l}{(\forall x)(P(x) \to Q(x)) \vdash (\forall x)(\exists y)(P(x) \to Q(y))^\star} \forall: r$$

and $\varphi_2 =$

$$\frac{\dfrac{\dfrac{\dfrac{\dfrac{\dfrac{P(a) \vdash P(a)^\star \quad Q(v)^\star \vdash Q(v)}{P(a), (P(a) \to Q(v))^\star \vdash Q(v)} \to: l}{(P(a) \to Q(v))^\star \vdash P(a) \to Q(v)} \to: r}{(P(a) \to Q(v))^\star \vdash (\exists y)(P(a) \to Q(y))} \exists: r}{(\exists y)(P(a) \to Q(y))^\star \vdash (\exists y)(P(a) \to Q(y))} \exists: l}{(\forall x)(\exists y)(P(x) \to Q(y))^\star \vdash (\exists y)(P(a) \to Q(y))} \forall: l$$

The ancestors of the cut are marked by $\star$. The clause terms corresponding to the ancestors of the cut in the axioms are

$$t_1 = \{P(u) \vdash\}, \; t_2 = \{\vdash Q(u)\},$$
$$t_3 = \{\vdash P(a)\}, \; t_4 = \{Q(v) \vdash\}.$$

The uppermost $\to$: l-rule in φ_1 goes into the end sequent and we have to apply $\otimes$. The corresponding clause term is

$$t = t_1 \otimes t_2 = \{P(u)\} \otimes \{Q(u)\}.$$

The $\to$: l-rule in φ_2 goes into the cut and the operation to be applied is $\oplus$. The corresponding clause term is

$$t' = t_3 \oplus t_4 = \{\vdash P(a)\} \oplus \{\; Q(v) \vdash\}.$$

The final binary rule is a cut and its auxiliary formulas are clearly ancestors of the cut-formulas. Therefore we obtain

$$\Theta(\varphi) = t \oplus t' = (\{P(u)\} \otimes \{Q(u)\}) \oplus (\{\vdash P(a)\} \oplus \{\ Q(v) \vdash\}.)$$

The evaluation of $\Theta(\varphi)$ eventually gives

$$\mathrm{CL}(\varphi) = \{P(u) \vdash Q(u); \ \vdash P(a); \ Q(v) \vdash\}.$$

Given a proof φ and a characteristic clause set $\mathrm{CL}(\varphi)$ we can construct an **LK**-derivation φ^* of the empty sequent $\vdash$ from clauses in $\mathrm{CL}(\varphi)$. As φ^* refutes $\mathrm{CL}(\varphi)$ and **LK** is sound we conclude that $\mathrm{CL}(\varphi)$ is unsatisfiable.

PROPOSITION 21. *Let φ be an **LK**-derivation in SK. Then $\mathrm{CL}(\varphi)$ is unsatisfiable.*

Proof. In [4]. ∎

Let φ be a deduction of $S \colon \Gamma \vdash \Delta$ and $\mathrm{CL}(\varphi)$ be the characteristic clause set of φ. Then $\mathrm{CL}(\varphi)$ is unsatisfiable and, by the completeness of resolution (see [14], [8]), there exists a resolution refutation γ of $\mathrm{CL}(\varphi)$. By applying a total m.g.u. to γ we obtain a p-resolution refutation γ' of $\mathrm{CL}(\varphi)$; by our definition of resolution, γ' is also an **LK**-deduction of $\vdash$ from (instances of) $\mathrm{CL}(\varphi)$.

EXAMPLE 22. Let φ be the proof in Example 20. Then

$$\mathrm{CL}(\varphi) = \{C_1 : P(u) \vdash Q(u),\ C_2 : \vdash P(a),\ C_3 : Q(u) \vdash\}.$$

We define a resolution refutation γ of $\mathrm{CL}(\varphi)$:

$$\frac{\dfrac{\vdash P(a) \quad P(u) \vdash Q(u)}{\vdash Q(a)} \quad Q(v) \vdash}{\vdash}$$

From γ we compute γ' by inserting the m.g.u.s into γ. The result is $\gamma' =$

$$\frac{\dfrac{\vdash P(a) \quad P(a) \vdash Q(a)}{\vdash Q(a)} \quad Q(a) \vdash}{\vdash}$$

The substitution applied to γ is $\sigma : \{u \leftarrow a, v \leftarrow a\}$.

This p-resolution refutation γ' may serve as a skeleton of an AC-deduction ψ of $\Gamma \vdash \Delta$ itself. The construction of ψ from γ' is based on *projections* replacing φ by cut-free deductions $\varphi(C)$ of $\bar{P}, \Gamma \vdash \Delta, \bar{Q}$ for clauses $C : \bar{P} \vdash \bar{Q}$ in $\mathrm{CL}(\varphi)$. We merely give an informal description of the projections, for details we refer to [4]. Roughly speaking, the projections of the proof φ are obtained by skipping all the inferences going into a cut. As a "residue" we obtain a characteristic clause in the end sequent. Thus a projection is a cut-free derivation of the end sequent S + some atomic formulas. For the application of projections it is vital to have a skolemized end sequent, otherwise eigenvariable conditions could be violated.

EXAMPLE 23. Let φ be the proof defined in Example 20 and $CL(\varphi) = \{C_1, C_2, C_3\}$ for

$$C_1 = P(u) \vdash Q(u), \quad C_2 = \vdash P(a), \quad C_3 = Q(v) \vdash .$$

By skipping the inferences in φ_1 going to cuts we get

$$\frac{\dfrac{P(u) \vdash P(u) \quad Q(u) \vdash Q(u)}{P(u), P(u) \to Q(u) \vdash Q(u)} \to: l}{P(u), (\forall x)(P(x) \to Q(x)) \vdash Q(u)} \forall: l$$

and from this derivation the projection $\varphi(C_1) =$

$$\frac{\dfrac{\dfrac{P(u) \vdash P(u) \quad Q(u) \vdash Q(u)}{P(u), P(u) \to Q(u) \vdash Q(u)} \to: l}{P(u), (\forall x)(P(x) \to Q(x)) \vdash Q(u)} \forall: l}{P(u), (\forall x)(P(x) \to Q(x)) \vdash (\exists y)(P(a) \to Q(y)), Q(u)} w: r$$

For $C_2 = \vdash P(a)$ we obtain the projection $\varphi(C_2)$:

$$\frac{\dfrac{\dfrac{\dfrac{P(a) \vdash P(a)}{P(a) \vdash P(a), Q(v)} w: r}{\vdash P(a) \to Q(v), P(a)} \to: r}{\vdash (\exists y)(P(a) \to Q(y)), P(a)} \exists: r}{(\forall x)(P(x) \to Q(x)) \vdash (\exists y)(P(a) \to Q(y)), P(a)} w: l$$

Let φ be a proof of S s.t. $\varphi \in$ SK and let γ be a p-resolution refutation of the (unsatisfiable) set of clauses $CL(\varphi)$. Then γ can be transformed into a deduction $\varphi(\gamma)$ of S s.t. $\varphi(\gamma) \in SK^0$. $\varphi(\gamma)$ is a proof with atomic cuts, thus an AC-normal form of φ. $\varphi(\gamma)$ is constructed from γ simply by replacing the clauses at the leaves by the corresponding proof projections. The construction of $\varphi(\gamma)$ is the essential part of the method CERES; the final elimination of atomic cuts is inessential – in the sense that it is of exponential complexity only (note that our goal is the characterization of elementary classes for cut-elimination).

DEFINITION 24 (CERES normal form). Let $\varphi \in$ SK and let γ be a p-resolution refutation of $CL(\varphi)$. Then $\varphi(\gamma)$, defined above, is called a CERES *normal form* of φ. γ is called the *resolution kernel* of $\varphi(\gamma)$.

EXAMPLE 25. Let φ be the proof defined in Example 20 and $\sigma: \{u \leftarrow a, v \leftarrow a\}$ the substitution and γ' be the p-resolution refutation defined in Example 22.

We define $\chi_1 = \varphi(C_1)\sigma$, $\chi_2 = \varphi(C_2)\sigma$ and $\chi_3 = \varphi(C_3)\sigma$.
The end-sequent of φ is $B \vdash C$ for

$$B = (\forall x)(P(x) \to Q(x)), \quad C = (\exists y)(P(a) \to Q(y)).$$

Then $\varphi(\gamma')$ is the following CERES normal form of φ:

$$\cfrac{\cfrac{\cfrac{(\chi_2)}{B \vdash C, P(a)} \quad \cfrac{(\chi_1)}{P(a), B \vdash C, Q(a)}}{B, B \vdash C, C, Q(a)} \, cut \quad \cfrac{(\chi_3)}{Q(a), B \vdash C}}{\cfrac{B, B, B \vdash C, C, C}{B \vdash C} \, \text{contractions}} \, cut$$

The essential source of complexity in the CERES-method is the construction of the resolution refutation γ of $\mathrm{CL}(\varphi)$. In [4] it is shown that $\|\mathrm{CL}(\varphi)\|$ is at most exponential in $\|\varphi\|$. Computing the global m.g.u. σ and a p-resolution refutation γ' from γ is at most exponential in $\|\gamma\|$. Let

$$r(\gamma') = \max\{\|t\| \mid t \text{ is a term occurring in } \gamma'\}.$$

Then obviously $r(\gamma) \leq \|\gamma'\|$ and, for any clause $C \in \mathrm{CL}(\varphi)$ we have $\|C\sigma\| \leq \|C\| * r(\gamma')$. Therefore

$$\|\varphi(C\sigma)\| \leq \|\varphi(C)\| * r(\gamma') \leq \|\varphi\| * r(\gamma').$$

Moreover, as the CERES-normal form is obtained by inserting the projections into the p-resolution refutation γ', we obtain

$$\|\varphi(\gamma')\| \leq \|\gamma'\| * \|\varphi\| * r(\gamma').$$

So if $\|\gamma'\|$ is elementarily bounded in $\|\varphi\|$, the CERES normal form $\varphi(\gamma')$ is too. Note that also the computing time spent on the search for a resolution refutation of $\mathrm{CL}(\varphi)$ is elementary in the length of its result $\|\gamma\|$.

DEFINITION 26. Let $\mathcal{C}$ be an unsatisfiable set of clauses. Then the *resolution complexity* of $\mathcal{C}$ is defined as

$$rc(\mathcal{C}) = \min\{\|\gamma\| \mid \gamma \text{ is a resolution refutation of } \mathcal{C}\}.$$

Clearly, by the undecidability of clause logic, there is no recursive bound on $rc(\mathcal{C})$ in terms of $\|\mathcal{C}\|$. However, the resolution complexity of characteristic clause sets is always bounded by a primitive recursive function; this because CERES cannot be outperformed by Tait's method (see [16] and [5]) for which there exists a primitive recursive, though nonelementary, bound.

DEFINITION 27. Let $\mathcal{K}$ be a class of skolemized proofs. We say that CERES is *fast on* $\mathcal{K}$ if there exists an elementary function f s.t. for all φ in $\mathcal{K}$:

$$rc(\mathrm{CL}(\varphi)) \leq f(\|\varphi\|).$$

By the discussion above, the run time of the whole algorithm CERES which constructs $\mathrm{CL}(\varphi)$, computes the refutations γ and γ', the projections and eventually $\varphi(\gamma)$, is bounded by an elementary function – provided CERES is fast as defined above.

The main goal of this paper is to identify classes $\mathcal{K}$ where CERES is fast, thus giving proofs of elementary cut-elimination on $\mathcal{K}$.

4 Proof Classes with Fast Cut-Elimination

In this section we demonstrate, that the characteristic clause set of a proof yields crucial information on the nature of cuts and, in particular, about their elimination.

It is well known that the complexity of cut-elimination does not only depend on the complexity of cut formulas but also on the syntactic form of the end sequent. The first class we are presenting here is known to have an elementary cut-elimination, but the proof of this property is trivial via the CERES-method, thus giving a flavor of our approach.

DEFINITION 28. **UIE** is the class of all skolemized **LK**-proofs with atomic axioms where all inferences going into the end-sequent are unary.

PROPOSITION 29. *Cut-elimination is at most exponential on* **UIE**.

Proof. Let φ be a proof of $\Gamma \vdash \Delta$ in **UIE**. As there are no binary inferences going into the end-sequent there are no products in the characteristic term $\Theta(\varphi)$. Therefore the characteristic clause set $\mathrm{CL}(\varphi)$ contains just the union of all cut-ancestors in the axioms of φ; hence every $C \in \mathrm{CL}(\varphi)$ is of the form (1) $\vdash A$, (2) $A \vdash$, or (3) $A \vdash A$ for atoms A. Clauses of the form (3) are tautologies and can be deleted. We are left with a finite set of unit clauses which are contained in the Herbrand class. As $\mathrm{CL}(\varphi)$ is unsatisfiable $\mathrm{CL}(\varphi)$ contains two clauses of the form $C_1 \colon \vdash A$ and $C_2 \colon B \vdash$ s.t. $\{A, B\}$ is unifiable (by some m.g.u. ϑ). Let $\varphi[C_1\vartheta]$ and $\varphi[C_2\vartheta]$ be the projections to $C_1\vartheta$ and $C_2\vartheta$. Then the proof ψ:

$$\frac{\dfrac{\varphi[C_1\vartheta]}{\Gamma \vdash \Delta, A\vartheta} \quad \dfrac{\varphi[C_2\vartheta]}{A\vartheta, \Gamma \vdash \Delta}}{\dfrac{\Gamma, \Gamma \vdash \Delta, \Delta}{\Gamma \vdash \Delta} c^*} \text{cut}$$

is a CERES-normal form of φ. Note that the unification ϑ can lead to an exponential increase (in fact $\|A\vartheta\|$ can be exponential in $\|A\|$). Therefore there exists a number k s.t.

$$\|\varphi[C_i\vartheta]\| \leq 2^{k*\|\varphi\|} \text{ for } i = 1, 2.$$

Finally we obtain

$$\|\psi\| \leq 2^{r*\|\varphi\|} \text{ for some } r \in \mathbb{N}.$$

As there are no binary rule applications in the proofs $\varphi[C_1\vartheta]$ and $\varphi[C_2\vartheta]$, transforming ψ into a cut-free proof is only linear. Thus cut-elimination on **UIE** can be done in exponential time. ∎

In [3] we have shown that cut-elimination on proofs with a single montone cut is nonelementary. In a first step the cut can be transformed into negation normal norm. In a second one the negations in the cut formula (by the NNF-transformation they are immediately above atoms) can be eliminated by generalized disjunctions added at the left-hand side of the end-sequent; for details see [3]. We show now that a further restriction on the arity of inferences in the proofs leads to an elementary cut-elimination class.

DEFINITION 30. **UILM** is the class of all skolemized **LK**-proofs φ with atomic axioms, s.t. φ contains only one cut which is monotone, and all inferences in the left cut-derivation which go into the end-sequent are unary.

We need some notation for replacing subproofs within proofs: let ν be a node in the proof φ and $\varphi.\nu$ the subproof with end-sequent $S(\nu)$ (the sequent occurring at node ν). Let ψ be a(nother) proof of $S(\nu)$. Then $\varphi[\psi]_\nu$ denotes φ after replacement of $\varphi.\nu$ by ψ.

For cut-elimination via CERES on **UILM** we use a refinement of resolution, the so-called hyperresolution method. Hyperresolution is not only one of the most efficient refinements in theorem proving but is also a powerful tool for the decision problem of first-order clause classes [8].

DEFINITION 31 (condensation). Let C be a clause and D be a factor of C s.t. D is a proper subclause of C; then we say that D is obtained from C by *condensation*. A clause which does not admit condensations is called *condensed*. A condensation of a clause C is a clause D which is condensed and can be obtained by (iterated) condensation from C.

DEFINITION 32. Let C be a clause and D be a condensation of C. $N_c(C)$, the *condensation normal form*, is the clause which is obtained from D by renaming the variables to $\{x_1, \ldots, x_n, \ldots\}$ and by ordering the atoms in C_+ and in C_- by a total ordering.

DEFINITION 33 (hyperresolution). Let $\mathcal{C}$ be a set of clauses, $C_1, \ldots, C_n$ positive clauses in $\mathcal{C}$, and D be a nonpositive clause. Then the sequence $\lambda \colon (D; C_1, \ldots, C_n)$ is called a *clash sequence*. We define

$$E_0 = D, \quad E_{i+1} \text{ is a PRF-resolvent of } E_i \text{ and } C_{i+1} \text{ for } i < n.$$

If E_n is positive then the normalization $N_c(E_n)$ is called a *hyperresolvent* of λ over $\mathcal{C}$. Let $\rho_H(\mathcal{C})$ be the set of all hyperresolvents over $\mathcal{C}$. We define

$$RH(\mathcal{C}) = \mathcal{C} \cup \rho_H(\mathcal{C}),$$
$$RH^{i+1}(\mathcal{C}) = RH(RH^i(\mathcal{C})) \text{ for } i \in \mathbb{N}, \quad RH^*(\mathcal{C}) = \bigcup_{i=0}^{\infty} RH^i(\mathcal{C}).$$

By $RH^*_+(\mathcal{C})$ we denote the set of positive clauses in $RH^*(\mathcal{C})$.

THEOREM 34. *Cut-elimination is elementary on* **UILM**.

Proof. Let φ be a proof in **UILM** then φ is of the form $\varphi[\psi]_\nu$ where ψ is the only cut-derivation in φ (occurring at the node ν in the proof). Assume that $\psi =$

$$\frac{(\psi_1) \qquad (\psi_2)}{\Gamma \vdash \Delta, A \quad A, \Pi \vdash \Lambda} \; cut$$
$$\overline{\Gamma, \Pi \vdash \Delta, \Lambda}$$

As there is only this single cut in φ the end-sequent $S \colon \Gamma, \Pi \vdash \Delta, \Lambda$ of ψ is skolemized. Therefore we may apply CERES to ψ itself and replace ψ in φ by the obtained CERES-normal form ψ'.

We compute $\mathrm{CL}(\psi)$: first consider the ancestors of the cut in the axioms. Note that the cut is monotone. In ψ_1 the ancestors of the cut in the axioms are of the form $\vdash A_i$ $(i = 1, \ldots n)$ for some atoms A_i, in ψ_2 they are of the form $B_j \vdash$ $(j = 1, \ldots m)$ for some atoms B_j. As the left derivation ψ_1 does not contain binary inferences going into S, the characteristic term $\Theta(\psi)$ is of the form $t_1 \oplus t_2$ where t_1 does not contain products. Therefore the clause set $\mathrm{CL}(\psi)$ is of the form $\mathcal{C}_1 \cup \mathcal{C}_2$, where

$$\mathcal{C}_1 = \{\vdash A_1;\ \ldots;\ \vdash A_n\}, \text{ and}$$
$$\mathcal{C}_2 \subseteq \bigcup \{B_{j_1}, \ldots, B_{j_k} \vdash \mid \{j_1, \ldots, j_k\} \subseteq \{1, \ldots, m\},\ k \leq m\}.$$

$\|\mathcal{C}_2\|$ is at most exponential in $\|\Theta(\psi)\|$, for which we have $\|\Theta(\psi)\| \leq \|\psi\|$.

Hence $\mathrm{CL}(\psi)$ is a set of Horn clauses consisting of positive unit clauses and negative clauses only. As $\mathrm{CL}(\psi)$ is unsatisfiable there exists a refutation ρ by hyperresolution of $\mathrm{CL}(\psi)$. Moreover there are no mixed clauses in $\mathrm{CL}(\psi)$, so ρ must consist of a single hyperresolvent, based on the clash sequence

$$\gamma\colon (B_{j_1}, \ldots, B_{j_k} \vdash;\ \vdash A_{i_1}, \ldots, \vdash A_{i_k}),$$

of the form

$$\frac{\vdash A_{i_k} \quad \dfrac{\vdash A_{i_1} \quad B'_{j_1}, \ldots, B'_{j_k} \vdash}{\vdots \quad \vdots \atop B'_{j_k} \vdash}}{\vdash}$$

Where the B'_{j_k} are instances of the B_{j_k}. A p-resolution refutation ρ' can be obtained from ρ by applying a global m.g.u. ϑ to ρ. ρ' can be used as a resolution kernel of a CERES-normal form ψ' of ψ. As the application of ϑ can lead to an exponential blow up w.r.t. the size of the clash we obtain

$$\|\rho'\| \leq 2 * m * 2^{k*a}$$

where a is the maximal complexity of an atom in γ. Clearly $a \leq \|\psi\|$ and so

$$\|\rho'\| \leq 2^{r*\|\psi\|} \leq 2^{r*\|\varphi\|}$$

for some constant r; we see that the CERES-normal form ψ' of ψ is at most exponential in ψ.

Now $\varphi' = \varphi[\psi']_\nu$ is a proof of the same end-sequent with only atomic cuts and

$$\|\varphi'\| \leq \|\varphi\| + \|\psi'\| \leq \|\varphi\| + 2^{r*\|\varphi\|}.$$

Therefore CERES is fast on **UILM** and thus cut-elimination is elementary on **UILM**. ∎

REMARK 35. Specific strategies of Gentzen's method can be shown to behave nonelementarily on **UILM**. The fully nondeterministic Gentzen method, however, is elementary on **UILM**. However it is very intransparent to use Gentzen's method as a method to *prove* fast cut-elimination for this class.

DEFINITION 36. **UIRM** is the class of all skolemized **LK**-proofs φ with atomic axioms, s.t. φ contains only one cut which is monotone, and all inferences in the right cut-derivation which go into the end-sequent are unary.

THEOREM 37. *Cut-elimination is elementary on* **UIRM**.

Proof. Like for **UILM**. ∎

COROLLARY 38. *Cut-elimination is elementary on* **UIRM** ∪ **UILM**

Proof. obvious. ∎

DEFINITION 39. A proof φ is in the class **AXDC** if φ is skolemized with atomic axioms and different axioms in φ are variable-disjoint.

THEOREM 40. CERES *is fast on* **AXDC**. *Therefore cut-elimination is elementary on* **AXDC**.

Proof. Let φ be in **AXDC** and let

$$\{\vdash A_1, \ldots, \vdash A_n,\ B_1 \vdash, \ldots, B_m \vdash\}$$

be the set of subsequents of axioms which are cut-ancestors (tautologies are omitted). Then no two different subsequences of axioms in the set share variables. Let $\mathcal{C} = \mathrm{CL}(\varphi)$. Then any clause $C \in \mathcal{C}$ is disconnected (two different atoms occurring in C do not share variables). We prove that, by using hyperresolution with condensing, we can find a refutation of $\mathcal{C}$ within exponential time relative to $\|\mathcal{C}\|$.

To this aim it is sufficient to show that the total size of derivable positive clauses $RH^*_+(\mathcal{C})$ is exponential in $\|\mathcal{C}\|$ (note that all clauses in $RH^*(\mathcal{C}) \setminus \mathcal{C}$ are positive, and thus are contained in $RH^*_+(\mathcal{C})$). We will see below that there is no exponential increase in the size of atoms.

First we observe that all clauses in $RH^*_+(\mathcal{C})$ (i.e. the clauses which are actually derivable) are also variable disjoint. This is easy to see as all clauses are disconnected and have to be renamed prior to resolution; in fact, if a clause is used twice in a clash sequence, a renamed variant has to be constructed.

Now let, for $i = 1, \ldots, k$, be $\mathcal{A}_i$ the set of atoms occurring in the clause head of the i-th clause in $\mathcal{C}$.

By definition of hyperresolution the clauses in $RH^*_+(\mathcal{C})$ are "accumulations" of renamed instances of subsets of the $\mathcal{A}_i$. As all clauses are disconnected no unification substitution is ever stored in the resolvents and in the hyperresolvent. Moreover the hyperresolvents are condensed and normalized, i.e. they do not contain different variants of atoms. Therefore every hyperresolvent is of the form

$$\vdash A_1, \ldots, A_r$$

where the A_i are variants of atoms in $\mathcal{A}_1 \cup \ldots \cup \mathcal{A}_k$ and $r \leq h$ for $h = |\mathcal{A}_1| + \ldots + |\mathcal{A}_k|$. Therefore the number of possible hyperresolvents is $\leq 2^h$. But $2^h < 2^{\|\mathcal{C}\|}$.

So we have shown that computing the contradiction by RH is at most exponential in $\|\mathrm{CL}(\varphi)\|$.

Note that deciding unification is of linear complexity only, and computing the most general unifiers explicitly is not necessary for the computation of $RH^*_+(\mathcal{C})$.

As $\|\mathrm{CL}(\varphi)\|$ may be exponential in $\|\varphi\|$, computing the CERES normal form is at most double exponential. ∎

We used the refinement of hyperresolution to construct fast cut-elimination procedures by CERES for the classes **UILM**, **UIRM** and **AXDC**. For the analysis of next class **MC** to be defined below we need another refinement, namely ordered resolution.

Let A be an atom; then $\tau(A)$ denotes the maximal term depth in A. For clauses C we define $\tau(C) = \max\{\tau(A) \mid A \text{ in } C\}$. $\tau_{\max}(x, A)$ denotes the maximal depth of the variable x in A. $V(A)$ defines the set of variables in A.

DEFINITION 41 (depth ordering). Let A and B be atoms; we define $A <_d B$ if (1) $V(A) \subseteq V(B)$, (2) $\tau(A) < \tau(B)$ and (3) for all $x \in V(A)$: $\tau_{\max}(x, A) < \tau_{\max}(x, B)$.

In [8] it is shown that $<_d$ is an atom ordering.

DEFINITION 42 (ordered resolution). Let C and D be clauses in a clause set $\mathcal{C}$ and E be a resolvent of C, D with resolved atom A. We define $N_c(E) \in \rho_{<_d}(\mathcal{C})$ iff there is no atom B in E s.t. $A <_d B$. The corresponding resolution operator is defined by:

$$R_{<_d}(\mathcal{C}) = \mathcal{C} \cup \rho_{<_d}(\mathcal{C}),\ R^*_{<_d}(\mathcal{C}) = \bigcup_{i=0}^{\infty} R^i_{<_d}(\mathcal{C}).$$

$R_{<_d}$ is complete (see [8]), i.e. $\vdash \in R^*_{<_d}(\mathcal{C})$ if $\mathcal{C}$ is unsatisfiable.

DEFINITION 43. Let **MC** be the set of all skolemized **LK**-proofs containing only unary function symbols and all predicate symbols occurring in cut-formulas are monadic.

Note that the end-sequents of **MC** define an undecidable class and so the class is nontrivial for cut-elimination. As the satisfiability problem of the prefix class $\forall \exists \forall$ is undecidable (see [6]), the provability of sequents of the form

$$(\forall x)(\forall z) A(x, f(x), z) \vdash$$

(where $A(x, f(x), z)$ is a quantifier free matrix over the terms $x, f(x), z$) is undecidable too. Therefore there is no way to find cut-free proofs of elementary size within **MC** by exhaustive search.

In order to show that cut-elimination in **MC** is elementary we need some preparatory steps.

DEFINITION 44. The class K is the set of all finite condensed sets of clauses $\mathcal{C}$ s.t. for all $C \in \mathcal{C}$: $|V(A)| \leq 1$ for all atoms A occurring in C.

LEMMA 45. $R^*_{<_d}(\mathcal{C})$ is finite for each $\mathcal{C} \in K$ and $\tau(R^*_{<_d}(\mathcal{C})) \leq 2 * \tau(\mathcal{C})$.

Proof. In [8], Theorem 5.2.1. ∎

DEFINITION 46. K_{mon} is the subclass of K containing only monadic predicate symbols and monadic function symbols.

Clearly Lemma 45 holds also for K_{mon}, but we may obtain sharper complexity bounds on the deductive closure.

LEMMA 47. Let $\mathcal{C} \in K_{mon}$. Then

(1) $|R^*_{<_d}(\mathcal{C})| \leq 2^{3r^2}$, and

(2) $\max\{\|C\| \mid C \in R^*_{<_d}(\mathcal{C})\} \leq 2r(\tau(\mathcal{C}) + 2)$

for $r = 2|\mathrm{PS}(\Sigma)|\|\mathrm{FS}(\Sigma)|^{2\tau(\mathcal{C})}(|\mathrm{CS}(\Sigma)| + 1)$ where Σ is the signature of $\mathcal{C}$.

Proof. Let $t = \tau(\mathcal{C})$ and Σ be the signature of $\mathcal{C}$. By Lemma 45 $R^*_{<_d}(\mathcal{C})$ is finite and $\tau(R^*_{<_d}(\mathcal{C})) \leq 2t$.

Now let A be an atom occurring in a clause in $R^*_{<_d}(\mathcal{C})$; then A is of the form $P(f_1 \ldots f_n s)$ where $s \in V \cup \mathrm{CS}(\Sigma)$, $P \in \mathrm{PS}(\Sigma)$, $f_i \in \mathrm{FS}(\Sigma)$, and $n \leq 2t$.

The number $g(2t, \Sigma)$, the number of ground atoms over Σ (or the number of atoms containing a fixed variable v in case $\mathrm{CS}(\Sigma) = \emptyset$) of depth $\leq 2t$ can be estimated by

$$g(2t, \Sigma) \leq |\mathrm{PS}(\Sigma)|\|\mathrm{FS}(\Sigma)|^{2t}(|\mathrm{CS}(\Sigma)| + 1).$$

As, by definition, all clauses in K_{mon} are are condensed, the atoms $P(f_1 \ldots f_n c)$ and $P(f_1 \ldots f_n v)$ (for $c \in \mathrm{CS}(\Sigma)$, $v \in V$) cannot appear in the same clause at the same side of the sequent sign: in fact, condensing would eliminate the atom $P(f_1 \ldots f_n v)$. The same situation holds for the atoms $P(f_1 \ldots f_n v_1)$ and $P(f_1 \ldots f_n v_2)$ for different variables v_1, v_2. For this reason we have for every $C \in R^*_{<_d}(\mathcal{C})$

$$\max\{|C_+|, |C_-|\} \leq r$$

for $r = |\mathrm{PS}(\Sigma)|\|\mathrm{FS}(\Sigma)|^{2t}(|\mathrm{CS}(\Sigma)| + 1)$.

and therefore

$$\max\{|C| \mid C \in R^*_{<_d}(\mathcal{C})\} \leq 2r.$$

As predicate symbols and function symbols are monadic we also have

$$\max\{|V(C)| \mid C \in R^*_{<_d}(\mathcal{C})\} \leq r.$$

Therefore, by standard renaming (enforced by the normalization operator N_c) the only variables which can occur in a clause in $R^*_{<_d}(\mathcal{C})$ are $x_1, \ldots, x_r$. Therefore the number of possible atoms $a(R^*_{<_d}(\mathcal{C}))$ occurring in a clause in $R^*_{<_d}(\mathcal{C})$ is bounded by the number

$$a(R^*_{<_d}(\mathcal{C})) \leq |\mathrm{PS}(\Sigma)|\|\mathrm{FS}(\Sigma)|^{2t}(|\mathrm{CS}(\Sigma)| + r) \leq r(r + 1).$$

As the clause length is at most r we obtain

$$|R^*_{<_d}(\mathcal{C})| \leq (r(r + 1))^r \leq 2^{3r^2}.$$

This proves (1).

As the maximal number of atoms occurring in a clause is $2r$ and $\|A\| \leq 2 + \tau(\mathcal{C})$ for every atom occurring in $R^*_{<_d}(\mathcal{C})$ we have

$$\max\{\|C\| \mid C \in R^*_{<_d}(\mathcal{C})\} \leq 2r(2 + \tau(\mathcal{C})).$$

This proves (2). ∎

THEOREM 48. *CERES is fast on* **MC**. *As a consequence, cut-elimination is elementary on* **MC**.

Proof. Let $\psi \in$ **MC** and $\varphi = \rho(\psi)$ for the transformation ρ in Definition 11. As the cut formulas contain only monadic function symbols and predicate symbols, the ancestors of the cuts in the axioms are of the form $\vdash A$ or $A \vdash$ where A is of the form $P(f_1 \ldots f_n s)$ for $s \in \mathrm{CS} \cup V$. Note that, as always, we may omit tautologies in the construction of $\mathrm{CL}(\varphi)$. Therefore the clause set $\mathrm{CL}(\varphi)$ (defined by union and product) only consists of clauses built from atoms of this type. $\mathrm{CL}(\varphi)$ itself need not be in $\mathbf{K}_{mon}$, but its condensation $\mathcal{C} \colon N_c(\mathrm{CL}(\varphi))$ is in $\mathbf{K}_{mon}$. By Lemma 47 we get

$$|R^*_{<_d}(\mathcal{C})| \leq 2^{3r^2}$$

for $r = 2|\mathrm{PS}(\Sigma)||\mathrm{FS}(\Sigma)|^{2\tau(\mathcal{C})}(|\mathrm{CS}(\Sigma)|+1)$ and $\Sigma = \Sigma(\mathcal{C})$.

So, as $\mathcal{C}$ is unsatisfiable, there exists a resolution refutation containing at most $\leq 2^{3r^2}$ different clauses. If the refutation is represented as a proof tree γ we obtain

$$l(\gamma) \leq r * 2^{2^{3r^2}}$$

Note that we also counted the applications of the condensations which are nothing else than repeated factors. The global unifier of γ which yields a propositional resolution refutation γ^* does not insert terms deeper than $2\tau(\mathcal{C})$ (this follows from the property that terms of depth greater than $\tau(\mathcal{C})$ are ground — see [8] Theorem 5.2.1). So also after global unification the clauses in γ^* are still of depth ≤ 2 and so

$$\|\gamma^*\| \leq \max\{\|C\| \mid C \in R^*_{<_d}(\mathcal{C})\} * r * 2^{2^{3r^2}}.$$

By Lemma 47 we get

$$\|\gamma^*\| \leq 2r(\tau(\mathcal{C})+2) * r * 2^{2^{3r^2}}.$$

Obviously $\tau(\mathcal{C}), |\mathrm{PS}(\Sigma)|, |\mathrm{FS}(\Sigma)|, |\mathrm{CS}(\Sigma)|$ are all bound by $\|\varphi\|$ and so $\|\gamma^*\|$ is elementary in $\|\varphi\|$. Eventually we obtain for the CERES normal form φ^* corresponding to γ^*

$$\|\varphi^*\| \leq k\|\gamma^*\|\|\varphi\|$$

for a constant k measuring additional contractions in the CERES normal form. So $\|\varphi^*\|$ is elementary in $\|\varphi\|$. ∎

THEOREM 49. *Gentzen's cut-elimination method is nonelementary on* **MC**.

Proof. We consider Gentzen's cut-elimination method in the version defined in [17], with the exception that we admit fully nondeterministic applications of the rules (like in [5]). We take the worst-case proof sequence $(\rho_n)_{n \in \mathbb{N}}$ of V.P. Orevkov [11]. The skolemization of $(\rho_n)_{n \in \mathbb{N}}$ yields a new proof sequence $(\xi_n)_{n \in \mathbb{N}}$ with only one unary function symbol f at the term level and only one ternary predicate symbol P at the level of atomic formulas. The end sequent of ξ_n is

$(\forall w)P(w, c, f(w)),$
$(\forall u, v, w)((\exists y)(P(y, c, u) \wedge (\exists z)(P(v, y, z) \wedge P(z, y, w))) \to P(v, u, w))$
$\vdash$
$(\exists v_n)(P(c, c, v_n) \wedge (\exists v_{n-1})(P(c, v_n, v_{n-1}) \wedge \ldots \wedge (\exists v_0)P(c, v_1, v_0)\ldots)),$

and the (only) cut formula $A_n(c)$ in ξ_n is defined inductively as

$A_0(\alpha) \equiv (\forall w_0)(\exists v_0)P(w_0, \alpha, v_0), \ \bar{A}_0(\alpha, \delta) \equiv (\exists v_0)P(\alpha, \delta, v_0),$
$\bar{A}_{i+1}(\alpha, \delta) \equiv (\exists v_{i+1})(A_i(v_{i+1}) \wedge P(\alpha, \delta, v_{i+1})),$
$A_{i+1}(\alpha) \equiv (\forall w_{i+1})(A_i(w_{i+1}) \to \bar{A}_{i+1}(w_{i+1}, \alpha)).$

Also the new skolemized sequence $(\xi_n)_{n \in \mathbb{N}}$ is of nonelementary complexity for cut-elimination. Now we replace the predicate $\lambda x, y, z.P(x, y, z)$ by the following conjunction of new unary predicates:

$$\lambda x, y, z((Q_1(x) \wedge Q_2(y)) \wedge Q_3(z)).$$

everywhere in the proof sequence and obtain a new sequence $(\varphi_n)_{n \in \mathbb{N}}$ of proofs belonging to the class **MC**. Therefore CERES is fast on $\{\rho(\varphi_n) \mid n \in \mathbb{N}\}$. But every step of Gentzen's procedure on φ_n is completely isomorphic to the corresponding step performed on ξ_n. Therefore the sequence of cut-elimination steps on $(\varphi_n)_{n \in \mathbb{N}}$ is as long as that on the original sequence $(\xi_n)_{n \in \mathbb{N}}$. As a consequence the Gentzen procedure is of nonelementary complexity on $(\rho_n)_{n \in \mathbb{N}}$. ∎

Finally we want to illustrate the limitations of the CERES-method as a tool to prove fast cut-elimination.

DEFINITION 50. A formula A is called *monotone* if the only logical symbols occurring in A are $\wedge, \vee, \forall, \exists$.

DEFINITION 51. The set of *quasi-monotone* formulas is defined inductively in the following way:

1. Atomic formulas and $\bot$ (representing falsum) are quasi-monotone.

2. If A and B are quasi-monotone then $(\forall x)A', (\exists x)A'$ and $A \wedge B$ are quasi-monotone (where x is a bound variable and A' a variant of A containing x in place of a free variable)

3. If A is quasi-monotone and B is monotone then $B \to A$ is quasi-monotone.

A sequent $\Gamma \vdash \Delta$ is called a *QM-sequent* if Γ is quasi-monotone and Δ is monotone.

DEFINITION 52. **LK**$\bot$ is **LK** extended by the additional axiom $\bot \vdash$.

DEFINITION 53. $\mathcal{QMON}$ is the class of all **LK**$\bot$-proofs ω s.t. (1) the end sequent of ω is a QM-sequent, and (2) all cut formulas are monotone.

THEOREM 54. *Cut-elimination is at most exponential on QMON.*

Proof. In [3] and in [10]. ∎

QMON is an essentially intuitionistic proof class, a feature which was used in the proof projection method defined in [3]. The CERES-method which is a method for classical logic does not distinguish between $\vee: l$ and $\rightarrow: l$ in the construction of the characteristic clause set, thus "eliminating" the intuitionistic character of the proof. In fact CERES does not yield characteristic clause sets which belong to well-known decidable clause classes. That does not mean that CERES is not fast on QMON; instead we do not know how to prove it by using only refutations of the characteristic clause sets in QMON.

BIBLIOGRAPHY

[1] M. Aigner, G. M. Ziegler. *Proofs from THE BOOK*. Springer 1998.
[2] M. Baaz, A. Leitsch: On skolemization and proof complexity, *Fundamenta Informaticae*, 20(4), pp. 353–379, 1994.
[3] M. Baaz, A. Leitsch: Cut normal forms and proof complexity, *Annals of Pure and Applied Logic*, 97, pp. 127-177, 1999.
[4] M. Baaz, A. Leitsch: Cut-elimination and Redundancy-elimination by Resolution, *Journal of Symbolic Computation*, 29(2), pp. 149-176, 2000.
[5] M. Baaz, A. Leitsch: Towards a Clausal Analysis of Cut-Elimination, *Journal of Symbolic Computation*, 41, pp. 381–410, 2006.
[6] E. Börger, E. Grädel, and Y. Gurevich. The Classical Decision Problem. *Perspectives in Mathematical Logic*. Springer, 1997.
[7] G. Gentzen: Untersuchungen über das logische Schließen, *Mathematische Zeitschrift* 39, pp. 405–431, 1934–1935.
[8] A. Leitsch: The Resolution Calculus, *Texts in Theoretical Computer Science*, Springer, 1997.
[9] H. Luckhardt (1989). Herbrand-Analysen zweier Beweise des Satzes von Roth: polynomiale Anzahlschranken. *The Journal of Symbolic Logic* **54**, 234–263.
[10] G. Mints: Quick Cut-Elimination for Monotone Cuts, in *Games, Logic, and Constructive Sets*, The University of Chigaco Press, 2003.
[11] V.P. Orevkov. Lower bounds for increasing complexity of derivations after cut elimination. *Journal of Soviet Mathematics*, 20:2337–2350, 1982.
[12] G. Polya. *Mathematics and plausible reasoning, Volume I: Induction and Analogy in Mathematics*. Princeton University Press, Princeton, New Jersey, 1954.
[13] G. Polya. *Mathematics and plausible reasoning, Volume II: Patterns of Plausible Inference*. Princeton University Press, Princeton, New Jersey, 1954.
[14] J.A. Robinson: A Machine Oriented Logic Based on the Resolution Principle, Journal of the ACM 12(1), pp. 23–41, 1965.
[15] R. Statman: Lower bounds on Herbrand's theorem, in *Proc. of the Amer. Math. Soc.* 75, pp. 104–107, 1979.
[16] W.W. Tait: Normal derivability in classical logic, in *The Syntax and Semantics of Infinitary Languages*, ed. by J. Barwise, Springer, Berlin, pp. 204–236, 1968.
[17] G. Takeuti: Proof Theory, North-Holland, Amsterdam, 2nd edition, 1987.

On the Craig interpolation and the fixed point properties for GLP

Lev D. Beklemishev[1]

ABSTRACT. We study the polymodal provability logic **GLP** introduced by Japaridze. We prove the Craig interpolation and the fixed point properties for **GLP** by finitary methods formalizable in primitive recursive arithmetic — thereby answering an open question posed by Ignatiev in 1992.

An important provability logic **GLP** was introduced by Giorgi Japaridze and shown to be arithmetically complete as early as in 1986 [10]. The language of **GLP** extends that of classical propositional logic by unary modalities $[n]$, for each $n \in \omega$. A natural provability interpretation of **GLP** — somewhat different from the original one — is to translate $[n]\varphi$ into the language of Peano arithmetic PA as the statement "*φ is provable from the axioms of* PA *together with all true arithmetical Π_n-sentences.*" Dual modalities $\langle n \rangle \varphi := \neg[n]\neg\varphi$ then correspond to the standard uniform Σ_n-reflection schemata for the theory $\mathsf{PA} + \varphi$.

Modal logical properties of **GLP** have been extensively studied by Konstantin Ignatiev [8, 9] who, in addition, considerably simplified the proof of Japaridze's arithmetical completeness theorem. Main results on **GLP** have been incorporated by George Boolos into his 1993 book [6] where one can find a very readable exposition of this provability logic.

Ignatiev, among other things, established the Craig interpolation and the fixed point properties for **GLP**. However, it remained open if these results could be established by finitary methods formalizable in PA. (The question about the Craig interpolation property was stated e.g. in ref. [8].)

Possible concerns about formalizability in PA of various results on **GLP** appear more justified nowadays than they have been in 1993. It was established that the closed fragment of **GLP** yields a natural ordinal notation system for the proof-theoretic ordinal ϵ_0. Moreover, **GLP** was used as a basis for a simple proof-theoretic analysis of Peano arithmetic [1]. Based on **GLP**, some interesting combinatorial statements independent from PA were also suggested [3]. See ref. [2] for a survey of these developments.

In view of the above, it is not surprising that there are certain facts on **GLP** itself that cannot be established within PA. One such result has recently been stated in ref. [7]: any extension of **GLP** by closed formulas of the form $[n]\varphi$,

---

[1] Research supported by the Russian Foundation for Basic Research (RFBR) and Russian Presidential Council for Support of Leading Scientific Schools.

for any n and φ, is finitely axiomatizable. This fact, however, cannot be proved in ACA_0 — a second order conservative extension of PA.

On the other hand, certain other aspects of **GLP** are completely finitary. A finitary treatment of the closed fragment of **GLP** was given in ref. [5]. In particular, a normal form theorem for the closed fragment and various facts concerning the associated ordinal notation system are formalizable in elementary arithmetic, which is useful for applications of **GLP** in proof theory. Recently, Ilya Shapirovsky [13] has shown that **GLP** is decidable in polynomial space, and hence that **GLP** is PSPACE-complete, just like many other standard modal logics.

In this note we provide finitary proofs of the Craig interpolation and of the fixed point properties for **GLP**. These proofs are based on our previous paper [4] where a complete Kripke semantics for **GLP** is given. In that paper, using only finitary methods, the system **GLP** is reduced to a certain natural subsystem, denoted **J**.[1] **J** is sound and complete w.r.t. a natural class of finite Kripke frames [4]. (It is well-known that **GLP** is not complete w.r.t. *any* class of Kripke frames.) We establish the Craig interpolation and the fixed point properties for **J**, which also enables us to extend them to **GLP**. Apart from the reduction of **GLP** to **J** established in ref. [4] our methods are quite standard and somewhat simpler than those used by Ignatiev.

1 Preliminaries

The system **J** is given by the following axiom schemata and inference rules.

Axioms: (i) Boolean tautologies;
 (ii) $[n](\varphi \to \psi) \to ([n]\varphi \to [n]\psi)$;
 (iii) $[n]([n]\varphi \to \varphi) \to [n]\varphi$;
 (iv) $[m]\varphi \to [n][m]\varphi$, for $m \leq n$.
 (v) $\langle m \rangle \varphi \to [n]\langle m \rangle \varphi$, for $m < n$.
 (vi) $[m]\varphi \to [m][n]\varphi$, for $m \leq n$.

Rules: modus ponens, $\vdash \varphi \Rightarrow \vdash [n]\varphi$.

GLP is obtained from **J** by adding the *monotonicity schema*

$$[m]\varphi \to [n]\varphi, \quad \text{for } m \leq n.$$

Ignatiev's logic **I** is obtained from **J** by deleting axiom schema (vi).

The system **J** is sound and complete w.r.t. the class of standard Kripke frames $(\mathcal{W}; R_0, R_1, \dots)$ satisfying the following conditions:

- R_k is a upwards well-founded, transitive ordering relation on $\mathcal{W}$, for each $k \geq 0$;

- $\forall x, y \, (xR_ny \Rightarrow \forall z \, (xR_mz \Leftrightarrow yR_mz))$ if $m < n$; (I)

---
[1] Similarly, Ignatiev [9] used a reduction of **GLP** to a weaker subsystem **I**, however the reducibility has not been established by finitary methods.

- $\forall x, y \, (xR_m y \,\&\, yR_n z \Rightarrow xR_m z)$ if $m \leq n$. \hfill (J)

Such frames will be called J-frames.

One of the main results of ref. [4] states that **GLP** is reducible to **J** as follows. Let
$$M(\varphi) := \bigwedge_{i<s}([m_i]\varphi_i \to [m_i + 1]\varphi_i),$$
where $[m_i]\varphi_i$ for $i < s$ are all subformulas of φ of the form $[k]\psi$. Let
$$\Box^+\varphi := \varphi \wedge \bigwedge_{i \leq n}[i]\varphi,$$
where $n := \max_{i<s} m_i$, and let $M^+(\varphi) := \Box^+ M(\varphi)$. Then,
$$\mathbf{GLP} \vdash \varphi \iff \mathbf{J} \vdash M^+(\varphi) \to \varphi. \hfill (Red)$$

This result is proved by finitary methods based on Kripke semantics for **GLP** developed in [4].

2 Craig interpolation theorem for J

In the proof of the Craig interpolation theorem we shall use notation similar to Tait-style sequent calculus, that is:

- *Formulas* are built-up from constants $\bot$, $\top$, propositional variables p_i, $i \geq 0$, and their negations $\bar{p}_i$ using $\wedge$, $\vee$, and modalities $\langle n \rangle$, $[n]$, for each $n \geq 0$;

- *Sequents* are finite sets of formulas (denoted Γ, Δ, etc.) understood as disjunctions of their elements. We write $\vdash \Gamma$ if $\mathbf{J} \vdash \bigvee \Gamma$.

- *Negation* $\neg \varphi$ of a formula φ is defined by de Morgan's rules and the following identities: $\neg[n]\varphi := \langle n \rangle \neg \varphi$, $\neg \langle n \rangle \varphi := [n]\neg \varphi$. Implication $\varphi \to \psi$ is defined by $\neg \varphi \vee \psi$.

As usual we write Γ, Δ for $\Gamma \cup \Delta$ and Γ, φ for $\Gamma \cup \{\varphi\}$. We use the following abbreviations: $\langle n \rangle \Gamma := \{\langle n \rangle \varphi : \varphi \in \Gamma\}$, $[n]\Gamma := \{[n]\varphi : \varphi \in \Gamma\}$.

Let $\Diamond_{\geq n}\Gamma$ denote any set of formulas of the form $\langle m \rangle \varphi$ such that $\varphi \in \Gamma$ and $m \geq n$. $\Box_{\geq n}\Gamma$ is similarly defined.

LEMMA 1. *Suppose Δ is a set of formulas of the form $\langle m \rangle \psi$ and $[m]\psi$ with $m < n$. If*
$$\vdash \Delta, \Diamond_{\geq n}\Gamma, \Gamma, \langle n \rangle \neg \varphi, \varphi$$
then $\vdash \Delta, \langle n \rangle \Gamma, [n]\varphi$.

Proof. Assume $\mathbf{J} \vdash \bigvee(\Delta, \Gamma, \Diamond_{\geq n}\Gamma, \langle n \rangle \neg \varphi, \varphi)$, then by propositional logic
$$\mathbf{J} \vdash (\bigwedge \neg \Delta \wedge \bigwedge \neg \Gamma \wedge \bigwedge \Box_{\geq n}\neg \Gamma) \to ([n]\varphi \to \varphi).$$
Denoting $\varphi_1 := \bigwedge \neg \Delta$, $\varphi_2 := \bigwedge \neg \Gamma$, and $\varphi_3 := \bigwedge \Box_{\geq n}\neg \Gamma$ we obtain:
$$\mathbf{J} \vdash [n](\varphi_1 \wedge \varphi_2 \wedge \varphi_3) \to [n]([n]\varphi \to \varphi) \hfill (1)$$
$$\to [n]\varphi, \quad \text{by (iii)}. \hfill (2)$$

However, if $[m]\psi \in \neg\Delta$ then $\mathbf{J} \vdash [m]\psi \to [n][m]\psi$ by (iv), and if $\langle m\rangle\psi \in \neg\Delta$, then $\mathbf{J} \vdash \langle m\rangle\psi \to [n]\langle m\rangle\psi$ by (v). Hence,

$$\mathbf{J} \vdash \varphi_1 \to [n]\varphi_1.$$

Similarly, if $[k]\psi \in \square_{\geq n}\neg\Gamma$ then $\mathbf{J} \vdash [n]\psi \to [n](\psi \wedge [k]\psi)$, by (vi). Hence,

$$\mathbf{J} \vdash [n]\varphi_2 \to [n](\varphi_2 \wedge \varphi_3).$$

We conclude

$$\begin{aligned}\mathbf{J} \vdash \varphi_1 \wedge [n]\varphi_2 &\to [n](\varphi_1 \wedge \varphi_2 \wedge \varphi_3) \\ &\to [n]\varphi, \quad \text{by (2)}.\end{aligned}$$

It follows that $\mathbf{J}$ derives $\bigvee(\Delta, \langle n\rangle\Gamma, [n]\varphi)$, as required. ∎

Let $\mathrm{Var}(\varphi)$ denote the set of variables occurring in φ and let

$$\mathrm{Var}(\Gamma) := \bigcup_{\varphi \in \Gamma} \mathrm{Var}(\varphi).$$

Similarly, let $\mathrm{Op}(\varphi)$ denote the set of all $n \in \omega$ such that $[n]$ or $\langle n\rangle$ occurs in φ, and let $\mathrm{Op}(\Gamma) := \bigcup_{\varphi \in \Gamma} \mathrm{Op}(\varphi)$.

We say that θ *interpolates* a pair of sequents $(\Gamma; \Delta)$ if $\mathrm{Var}(\theta) \subseteq \mathrm{Var}(\Gamma) \cap \mathrm{Var}(\Delta)$ and

$$\vdash \Gamma, \theta \text{ and } \vdash \neg\theta, \Delta.$$

$(\Gamma; \Delta)$ is *inseparable* if it does not have an interpolant.

The following theorem subsumes both the completeness theorem for $\mathbf{J}$ and the Craig interpolation theorem.

THEOREM 2. *The following statements are equivalent:*

(i) $(\Gamma; \Delta)$ *has an interpolant;*

(ii) $\vdash \Gamma, \Delta$;

(iii) For all (finite) $\mathbf{J}$-models $\mathcal{W}$, $\mathcal{W} \vDash \bigvee(\Gamma \cup \Delta)$.

Proof. The implications (i)⇒(ii) and (ii)⇒(iii) are easy. We prove (iii)⇒(i).

Call a finite set Φ of formulas *adequate* if it is closed under subformulas, negation, the following operation:

$$[n]\varphi, [m]\psi \in \Phi \Rightarrow [m]\varphi \in \Phi,$$

and for each variable $p \in \Phi$ contains $p \wedge \bar{p}$. Clearly, every finite set of formulas Ψ can be extended to a finite adequate set $\Phi \supseteq \Psi$ such that $\mathrm{Op}(\Phi) = \mathrm{Op}(\Psi)$ and $\mathrm{Var}(\Phi) = \mathrm{Var}(\Psi)$.

Let us fix some finite adequate Φ. Below we shall only consider sequents Γ over Φ, that is, $\Gamma \subseteq \Phi$. An inseparable pair $(\Gamma_1; \Gamma_2)$ is *maximal* if for any other inseparable pair $(\Delta_1; \Delta_2)$ such that $\Gamma_1 \subseteq \Delta_1$ and $\Gamma_2 \subseteq \Delta_2$ one has $\Gamma_1 = \Delta_1$ and $\Gamma_2 = \Delta_2$.

LEMMA 3. *Suppose $(\Gamma_1; \Gamma_2)$ is maximal inseparable. Then, for all $\varphi, \psi \in \Phi$, and $i = 1, 2$:*

(i) $(\varphi \wedge \psi) \in \Gamma_i \Rightarrow \varphi \in \Gamma_i$ or $\psi \in \Gamma_i$;

(ii) $(\varphi \vee \psi) \in \Gamma_i \Rightarrow \varphi \in \Gamma_i$ and $\psi \in \Gamma_i$;

(iii) If $\mathrm{Var}(\varphi) \subseteq \mathrm{Var}(\Gamma_i)$ then either $\varphi \in \Gamma_i$ or $\neg \varphi \in \Gamma_i$;

(iv) For no φ both $\varphi, \neg\varphi \in \Gamma_i$.

Proof. By the obvious symmetry it is sufficient to prove both claims for $i = 1$.

(i) Assume $\varphi, \psi \notin \Gamma_1$. We claim that at least one of the following two pairs is inseparable: $(\Gamma_1, \varphi; \Gamma_2)$ and $(\Gamma_1, \psi; \Gamma_2)$. Indeed, if θ_1 interpolates the first pair and θ_2 interpolates the second pair, then

$$\vdash \Gamma_1, \varphi, \theta_1 \qquad \vdash \neg\theta_1, \Gamma_2$$
$$\vdash \Gamma_1, \psi, \theta_2 \qquad \vdash \neg\theta_2, \Gamma_2,$$

whence

$$\vdash \Gamma_1, \varphi \wedge \psi, \theta_1 \vee \theta_2 \qquad \vdash \neg\theta_1 \wedge \neg\theta_2, \Gamma_2.$$

Hence, $\theta_1 \vee \theta_2$ interpolates $(\Gamma_1; \Gamma_2)$, a contradiction. It follows that $(\Gamma_1; \Gamma_2)$ is not maximal.

(ii) Assume $\varphi \notin \Gamma_1$ then $(\Gamma_1, \varphi; \Gamma_2)$ is inseparable. Otherwise, if θ interpolates this pair, then

$$\vdash \Gamma_1, \varphi, \theta \text{ and } \vdash \neg\theta, \Gamma_2$$

hence

$$\vdash \Gamma_1, \varphi \vee \psi, \theta \text{ and } \vdash \neg\theta, \Gamma_2,$$

that is, θ interpolates $(\Gamma_1; \Gamma_2)$, a contradiction. It follows that $(\Gamma_1; \Gamma_2)$ is not maximal. The case $\psi \notin \Gamma_1$ is similar.

(iii) Assume $\mathrm{Var}(\varphi) \subseteq \mathrm{Var}(\Gamma_1)$ and $\varphi, \neg\varphi \notin \Gamma_1$. Then one of the pairs $(\Gamma_1, \varphi; \Gamma_2)$ and $(\Gamma_1, \neg\varphi; \Gamma_2)$ is inseparable. Otherwise, if

$$\vdash \Gamma_1, \varphi, \theta_1 \qquad \vdash \neg\theta_1, \Gamma_2,$$
$$\vdash \Gamma_1, \neg\varphi, \theta_2 \qquad \vdash \neg\theta_2, \Gamma_2,$$

then

$$\vdash \Gamma_1, \theta_1 \vee \theta_2 \qquad \vdash \neg\theta_1 \wedge \neg\theta_2, \Gamma_2.$$

Hence, $\theta_1 \vee \theta_2$ interpolates $(\Gamma_1; \Gamma_2)$, a contradiction. It follows that $(\Gamma_1; \Gamma_2)$ is not maximal.

(iv) If $\varphi, \neg\varphi \in \Gamma_1$ then $\vdash \Gamma_1, \bot$ and $\vdash \top, \Gamma_2$, which is impossible. ∎

Consider the following Kripke frame. Let

$$\mathcal{W} := \{(\Gamma_1; \Gamma_2) : (\Gamma_1; \Gamma_2) \text{ is maximal inseparable over } \Phi\}.$$

For any $x = (\Gamma_1; \Gamma_2)$ and $y = (\Delta_1; \Delta_2)$ in $\mathcal{W}$ let $xR_n y$ if the following conditions hold, for $i = 1, 2$:

1. $\mathrm{Var}(\Gamma_i) = \mathrm{Var}(\Delta_i)$;

2. For any $\langle n\rangle\varphi \in \Gamma_i$, $\varphi \in \Delta_i$ and
$$\forall k \geq n\ (k \in \mathrm{Op}(\Phi) \Rightarrow \langle k\rangle\varphi \in \Delta_i);$$

3. For any $m < n$,
$$\langle m\rangle\varphi \in \Gamma_i \iff \langle m\rangle\varphi \in \Delta_i;$$

4. For some $j \in \{1,2\}$, there is a $\langle n\rangle\varphi \in \Delta_j$ such that $\langle n\rangle\varphi \notin \Gamma_j$.

LEMMA 4. *$\mathcal{W}$ is a **J**-frame.*

Proof. Condition 4 guarantees the irreflexivity of the relations R_n.

Assume $(\Gamma_1;\Gamma_2)R_n(\Delta_1;\Delta_2)$, $(\Gamma_1;\Gamma_2)R_m(\Sigma_1;\Sigma_2)$ and $m < n$; we prove
$$(\Delta_1;\Delta_2)R_m(\Sigma_1;\Sigma_2).$$

Indeed, $\mathrm{Var}(\Delta_i) = \mathrm{Var}(\Gamma_i) = \mathrm{Var}(\Sigma_i)$, for $i=1,2$. If $\langle m\rangle\varphi \in \Delta_i$ then $\langle m\rangle\varphi \in \Gamma_i$, since $m < n$. Hence $\varphi, \langle k\rangle\varphi \in \Sigma_i$, for $k \geq m$, $k \in \mathrm{Op}(\Phi)$. If $k < m$ then $\langle k\rangle\varphi \in \Delta_i \Leftrightarrow \langle k\rangle\varphi \in \Gamma_i \Leftrightarrow \langle k\rangle\varphi \in \Sigma_i$. Finally, we have $\langle m\rangle\psi \in \Sigma_j$, $\langle m\rangle\psi \notin \Gamma_j$, for some ψ, j. Hence, $\langle m\rangle\psi \notin \Delta_j$ because $m < n$.

Assume $(\Gamma_1;\Gamma_2)R_n(\Delta_1;\Delta_2)$, $(\Delta_1;\Delta_2)R_m(\Sigma_1;\Sigma_2)$ and $m \leq n$; we prove $(\Gamma_1;\Gamma_2)R_m(\Sigma_1;\Sigma_2)$. Indeed, if $\langle m\rangle\varphi \in \Gamma_i$ then $\langle m\rangle\varphi \in \Delta_i$, since $m \leq n$. Hence $\varphi, \langle k\rangle\varphi \in \Sigma_i$, for $k \geq m$, $k \in \mathrm{Op}(\Phi)$. If $k < m$ then $\langle k\rangle\varphi \in \Gamma_i \Leftrightarrow \langle k\rangle\varphi \in \Delta_i \Leftrightarrow \langle k\rangle\varphi \in \Sigma_i$. Finally, we have $\langle m\rangle\psi \in \Sigma_j$, $\langle m\rangle\psi \notin \Delta_j$, for some ψ, j. Hence, $\langle m\rangle\psi \notin \Gamma_j$ because $m \leq n$.

Assume $(\Gamma_1;\Gamma_2)R_m(\Delta_1;\Delta_2)$, $(\Delta_1;\Delta_2)R_n(\Sigma_1;\Sigma_2)$ and $m \leq n$; we prove $(\Gamma_1;\Gamma_2)R_m(\Sigma_1;\Sigma_2)$. Let $k \in \mathrm{Op}(\Phi)$. If $m \leq k \leq n$, then $\langle m\rangle\varphi \in \Gamma_i$ implies $\langle k\rangle\varphi \in \Delta_i$ and $\langle k\rangle\varphi \in \Sigma_i$. If $k \geq n$, then $\langle m\rangle\varphi \in \Gamma_i$ implies $\langle n\rangle\varphi \in \Delta_i$ and $\varphi, \langle k\rangle\varphi \in \Sigma_i$. Finally, there is a ψ such that $\langle m\rangle\psi \in \Delta_j$, $\langle m\rangle\psi \notin \Gamma_j$. Since $m \leq n$ we also have $\langle m\rangle\psi \in \Sigma_j$, and we are done. ∎

We define the evaluation of propositional variables on $\mathcal{W}$ by letting
$$(\Gamma_1;\Gamma_2) \Vdash p \iff p \notin \Gamma_1 \cup \Gamma_2. \qquad (*)$$

LEMMA 5. *For any $\varphi \in \Gamma_1 \cup \Gamma_2$ one has $(\Gamma_1;\Gamma_2) \nVdash \varphi$.*

Proof. Induction on the length of φ. We consider the following cases.

CASE 1: $\varphi = \top$. If $\top \in \Gamma_1$ then $\vdash \Gamma_1, \bot$ and $\vdash \top, \Gamma_2$, hence $(\Gamma_1;\Gamma_2)$ is not inseparable. Thus, $\top \notin \Gamma_1$ and similarly $\top \notin \Gamma_2$.

CASE 2: $\varphi = \bot$. We always have $(\Gamma_1;\Gamma_2) \nVdash \bot$.

CASE 3: $\varphi = p$. By $(*)$.

CASE 4: $\varphi = \bar{p}$. Suppose $\bar{p} \in \Gamma_1$. If $p \in \Gamma_1$, then $\vdash \Gamma_1, \bot$ and $\vdash \top, \Gamma_2$, a contradiction. If $p \in \Gamma_2$, then $\vdash \Gamma_1, p$ and $\vdash \bar{p}, \Gamma_2$, also contradicting the inseparability of $(\Gamma_1;\Gamma_2)$. Hence, $p \notin \Gamma_1 \cup \Gamma_2$ which entails $(\Gamma_1;\Gamma_2) \Vdash p$ and $(\Gamma_1;\Gamma_2) \nVdash \bar{p}$.

CASE 5: $\varphi = \varphi_1 \wedge \varphi_2$. If $\varphi \in \Gamma_i$ then by Lemma 3 either $\varphi_1 \in \Gamma_i$ or $\varphi_2 \in \Gamma_i$. Hence, $(\Gamma_1; \Gamma_2) \not\vdash \varphi_1$ or $(\Gamma_1; \Gamma_2) \not\vdash \varphi_2$. Therefore, $(\Gamma_1; \Gamma_2) \not\vdash \varphi_1 \wedge \varphi_2$.

CASE 6: $\varphi = \varphi_1 \vee \varphi_2$. This is established dually by the same lemma.

CASE 7: $\varphi = \langle n \rangle \varphi_0$. Assume $\varphi \in \Gamma_1$. If $(\Gamma_1; \Gamma_2) R_n(\Delta_1; \Delta_2)$ then $\varphi_0 \in \Delta_1$ and by the induction hypothesis $(\Delta_1; \Delta_2) \not\vdash \varphi_0$. Since this holds for all such $(\Delta_1; \Delta_2)$, we have $(\Gamma_1; \Gamma_2) \not\vdash \langle n \rangle \varphi_0$.

CASE 8: $\varphi = [n]\varphi_0$. This is the central case.

Assume $[n]\varphi_0 \in \Gamma_1$. Let Δ_i for $i = 1, 2$ denote the union of the following sets of formulas:

1. $\Phi_1^i := \{\langle m \rangle \psi : \langle m \rangle \psi \in \Gamma_i, \ m < n\}$;

2. $\Phi_2^i := \{[m]\psi : [m]\psi \in \Gamma_i, \ m < n\}$;

3. $\Phi_3^i := \{\langle k \rangle \psi, \psi : \langle n \rangle \psi \in \Gamma_i, \ k \geq n, k \in \mathrm{Op}(\Phi)\}$;

4. $\Phi_4^i := \{p \wedge \bar{p} : p \in \mathrm{Var}(\Gamma_i)\}$.

We show that the pair
$$(\Delta_1, \langle n \rangle \neg \varphi_0, \varphi_0 \,;\, \Delta_2)$$
is inseparable. Assume otherwise, then for some θ,
$$\vdash \Delta_1, \langle n \rangle \neg \varphi_0, \varphi_0, \theta \ \text{and} \ \vdash \neg \theta, \Delta_2,$$
where
$$\mathrm{Var}(\theta) \subseteq \mathrm{Var}(\Delta_1, \langle n \rangle \neg \varphi_0, \varphi_0) = \mathrm{Var}(\Gamma_1)$$
and
$$\mathrm{Var}(\theta) \subseteq \mathrm{Var}(\Delta_2) = \mathrm{Var}(\Gamma_2).$$

The equalities hold because of the components Φ_4^i. Since $\bigvee \Phi_4^i$ is equivalent to $\bot$ and can be dropped from a disjunction, we obviously have
$$\vdash \Phi_1^1, \Phi_2^1, \Phi_3^1, \langle n \rangle \theta, \theta, \langle n \rangle \neg \varphi_0, \varphi_0$$
and hence
$$\vdash \Phi_1^1, \Phi_2^1, \{\langle n \rangle \psi : \psi \in \Gamma_1\}, \langle n \rangle \theta, [n]\varphi_0,$$
by Lemma 1. All the formulas in this sequent except for $\langle n \rangle \theta$ belong to Γ_1, hence $\vdash \Gamma_1, \langle n \rangle \theta$.

On the other hand, from $\vdash \neg \theta, \Delta_2$ we similarly obtain
$$\vdash \Phi_1^2, \Phi_2^2, \{\langle n \rangle \psi : \psi \in \Gamma_2\}, [n]\neg \theta$$
and hence $\vdash \Gamma_2, \neg \langle n \rangle \theta$. It follows that $\langle n \rangle \theta$ interpolates $(\Gamma_1; \Gamma_2)$, which is impossible.

Thus, $(\Delta_1, \langle n \rangle \neg \varphi_0, \varphi_0 \,;\, \Delta_2)$ is inseparable and can be extended to a maximal inseparable pair $(\Delta_1'; \Delta_2')$ such that $\mathrm{Var}(\Delta_i') = \mathrm{Var}(\Delta_i) = \mathrm{Var}(\Gamma_i)$ for $i = 1, 2$. We observe that $(\Gamma_1; \Gamma_2) R_n(\Delta_1'; \Delta_2')$. Indeed, Conditions 1, 2 and 4 are obviously satisfied. Also, if $\langle m \rangle \psi \in \Gamma_i$ and $m < n$, then $\langle m \rangle \psi \in \Delta_i \subseteq \Delta_i'$. On the other hand, if $m < n$ and $\langle m \rangle \psi \in \Delta_i'$ then $\mathrm{Var}(\langle m \rangle \psi) \subseteq \mathrm{Var}(\Gamma_i)$ and hence either $\langle m \rangle \psi \in \Gamma_i$ or $\neg \langle m \rangle \psi \in \Gamma_i$, by Lemma 3 (iii). Yet, $[m]\neg \psi \in \Gamma_i$ implies

$[m]\neg\psi \in \Delta_i$, whence Δ'_i contains both $\langle m \rangle \psi$ and its negation contradicting Lemma 3 (iv). Thus, we conclude $\langle m \rangle \psi \in \Gamma_i$, as required.

Since $\varphi_0 \in \Delta'_1$, by the induction hypothesis we obtain $(\Delta'_1; \Delta'_2) \not\vdash \varphi_0$. Hence, $(\Gamma_1; \Gamma_2) \not\vdash [m]\varphi_0$. ∎

From the previous lemma we obtain a proof of Theorem 2 in a standard way. Assume $(\Gamma; \Delta)$ is inseparable. Extend $\Gamma \cup \Delta$ to a finite adequate set Φ and build the corresponding model $\mathcal{W}$. Let x be any maximal inseparable pair of sequents over Φ containing $(\Gamma; \Delta)$. By Lemma 5, $\mathcal{W}, x \not\vDash \bigvee(\Gamma \cup \Delta)$. ∎

COROLLARY 6 (Craig interpolation for **J**). *If* $\mathbf{J} \vdash \varphi \to \psi$, *then there is a formula* θ *such that* $\mathrm{Var}(\theta) \subseteq \mathrm{Var}(\varphi) \cap \mathrm{Var}(\psi)$ *and*

$$\mathbf{J} \vdash \varphi \to \theta \quad \text{and} \quad \mathbf{J} \vdash \theta \to \psi.$$

COROLLARY 7. *Craig interpolation property holds for* **GLP**.

Proof. If $\mathbf{GLP} \vdash \varphi \to \psi$ then $\mathbf{J} \vdash M^+(\varphi \to \psi) \to (\varphi \to \psi)$ by (Red). Since every subformula $[i]\xi$ of $\varphi \to \psi$ belongs either to φ or to ψ, we have

$$\mathbf{J} \vdash M^+(\varphi) \land \varphi \to (M^+(\psi) \to \psi).$$

Let θ be the corresponding interpolant. Then obviously $\mathbf{GLP} \vdash \varphi \to \theta$, $\mathbf{GLP} \vdash \theta \to \psi$, and $\mathrm{Var}(\theta) \subseteq \mathrm{Var}(\varphi) \cap \mathrm{Var}(\psi)$. ∎

Open questions:

1. Can we also obtain in **J** an interpolant θ satisfying an additional condition $\mathrm{Op}(\theta) \subseteq \mathrm{Op}(\varphi) \cap \mathrm{Op}(\psi)$, that is, if modalities occurring in θ occur both in φ and in ψ? The given proof of Theorem 2 only implies that $\mathrm{Op}(\theta)$ is contained in $\mathrm{Op}(\varphi) \cup \mathrm{Op}(\psi)$. The stronger interpolation theorem obviously fails for **GLP**, as the example $[0]p \to [1]p$ shows.

2. Does **J** satisfy uniform interpolation? Lyndon interpolation?

3. Does the sequential inference rule formulated in Lemma 1 provide a complete cut-free sequent calculus for **J**, taken together with a standard Tait-style axiomatization of propositional logic?

Concerning the last question, by a modification of the given proof of Theorem 2 one can easily prove a weaker (analytic) form of cut-elimination theorem.

Consider the standard Tait-style sequent calculus rules for propositional logic with cut (see [12]) together with the inference rule

$$\frac{\Delta, \Diamond_{\geq n}\Gamma, \Gamma, \langle n \rangle \neg \varphi, \varphi}{\Delta, \langle n \rangle \Gamma, [n]\varphi}, \qquad (J)$$

where Δ consists of formulas of the form $\langle m \rangle \psi$ and $[m]\psi$, for $m < n$. By Lemma 1, this rule is admissible in **J**. On the other hand, all modal axioms of **J** translated into the Tait-style language are easily derivable from (J). Hence, (J) provides a sequent-style axiomatization of **J**.

Fix an adequate set of formulas Φ. Write $\vdash_\Phi \Gamma$ if Γ has a proof in which all cut-formulas have the form $[m]\psi$ (or $\langle m\rangle\neg\psi$) such that $[m]\psi \in \Phi$.

THEOREM 8. *Let Φ be the smallest adequate set containing Γ. Then, Γ is provable in $\mathbf{J}$ iff $\vdash_\Phi \Gamma$.*

Proof (sketch). Assume $\nvdash_\Phi \Gamma$. We consider sequents contained in Φ and provability with the cut-rule restricted to Φ. We build a $\mathbf{J}$-model $\mathcal{W}$ falsifying Γ essentially in the same way as we did in Theorem 2. Let W denote the set of all maximal sequents $\Delta \subseteq \Phi$ such that $\nvdash_\Phi \Delta$. Define $\Delta R_n \Sigma$ iff the following conditions hold:

1. For any $\langle n\rangle\varphi \in \Delta$, $\varphi \in \Sigma$ and
$$\forall k \geq n \ (k \in \mathrm{Op}(\Phi) \Rightarrow \langle k\rangle\varphi \in \Sigma);$$

2. For any $m < n$,
$$\langle m\rangle\varphi \in \Delta \iff \langle m\rangle\varphi \in \Sigma;$$

3. There is a $\langle n\rangle\varphi \in \Sigma$ such that $\langle n\rangle\varphi \notin \Delta$.

Further, let $\Delta \Vdash p$ iff $p \notin \Delta$. As in Theorem 2, it is easy to verify that $\Delta \nVdash \varphi$ whenever $\varphi \in \Delta$. Hence, $\mathcal{W} \nvDash \Gamma$. ∎

3 Fixed points

As a standard corollary of interpolation we obtain Beth definability property for $\mathbf{J}$ and $\mathbf{GLP}$.

COROLLARY 9 (Beth definability for $\mathbf{J}$). *If q does not occur in $\varphi(p)$ and*
$$\mathbf{J} \vdash \varphi(p) \wedge \varphi(q) \to (p \leftrightarrow q),$$
then there is a ψ such that $\mathrm{Var}(\psi) = \mathrm{Var}(\varphi(p)) \setminus \{p\}$ and
$$\mathbf{J} \vdash \varphi(p) \to (p \leftrightarrow \psi).$$

Proof. Let ψ be the interpolant of the implication
$$\mathbf{J} \vdash \varphi(p) \wedge p \to (\varphi(q) \to q).$$
∎

A similar property obviously holds for $\mathbf{GLP}$.

We obtain the fixed point property for $\mathbf{J}$ and $\mathbf{GLP}$ using a method of Smoryński and Bernardi (cf. [6]). First, we prove the so-called Bernardi lemma for $\mathbf{J}$.

LEMMA 10. *Suppose q does not occur in $\varphi(p)$ and p only occurs in $\varphi(p)$ within the scope of a modality. Then $\mathbf{J}$ proves the following formula B_φ:*
$$\Box^+(p \leftrightarrow \varphi(p)) \wedge \Box^+(q \leftrightarrow \varphi(q)) \to (p \leftrightarrow q).$$

Proof. We show that B_φ is valid in all finite J-models $\mathcal{W}$. With every $x \in \mathcal{W}$ we associate a sequence of numbers

$$D(x) := \langle d_0(x), d_1(x), \ldots, d_n(x) \rangle,$$

where $d_i(x)$ denotes the depth of x in $\mathcal{W}$ w.r.t. relation R_i inductively defined by

$$d_i(x) := \sup\{d_i(y) + 1 : xR_i y\},$$

and n is the maximal number such that R_n is non-empty on $\mathcal{W}$. We consider a lexicographic ordering of such sequences and make the following observation.

OBSERVATION 11. *For all $x, y \in \mathcal{W}$ and any k, if $xR_k y$ then $D(x) < D(y)$.*

Proof. Suppose $xR_k y$. For each $i < k$ we have $d_i(x) = d_i(y)$, since by (I) the same points z are R_i-accessible from x and from y. Also, obviously $d_k(x) > d_k(y)$, hence the result. ■

Suppose $\mathcal{W}$ is given and $\mathcal{W} \nvDash B_\varphi$. By considering a suitable generated submodel we may assume that

$$\mathcal{W} \vDash p \leftrightarrow \varphi(p), q \leftrightarrow \varphi(q) \qquad (**)$$

and $\mathcal{W} \nvDash p \leftrightarrow q$. Select $x \in \mathcal{W}$ such that p and q have have different evaluations at x and $D(x)$ is the minimal possible. By $(**)$ we have that $\varphi(p)$ and $\varphi(q)$ have different evaluations at x. Since p only occurs within the scope of modality in $\varphi(p)$, $\varphi(p)$ is a boolean combination of formulas of the form $[k]\psi(p)$ and variables different from p, q. Hence, there must exist a subformula $[k]\psi(p)$ of $\varphi(p)$ such that $[k]\psi(p)$ and $[k]\psi(q)$ have different evaluations at x. It follows that for some y such that $xR_k y$ the formulas $\psi(p)$ and $\psi(q)$ have different evaluations at y.

Let $\mathcal{W}_y$ denote the submodel of $\mathcal{W}$ generated by y. For each $z \in \mathcal{W}_y$ one has $xR_i z$, for some i. (If $yR_m z$ and $m < k$ then $xR_m z$ by (I), and if $m \geq k$ then $xR_k z$ by (J).) Hence, for all $z \in \mathcal{W}_y$, $D(z) < D(x)$. Therefore, by the choice of x, $\mathcal{W}_y \vDash p \leftrightarrow q$. It follows that for all subformulas $\theta(p)$ of $\varphi(p)$, $\mathcal{W}_y \vDash \theta(p) \leftrightarrow \theta(q)$. In particular,

$$\mathcal{W}, y \Vdash \psi(p) \leftrightarrow \psi(q),$$

a contradiction. ■

COROLLARY 12 (Fixed points in **J**). *Suppose q does not occur in $\varphi(p)$ and p only occurs in $\varphi(p)$ within the scope of a modality. Then there is a ψ (a fixed point of $\varphi(p)$) such that $\mathrm{Var}(\psi) = \mathrm{Var}(\varphi(p)) \setminus \{p\}$ and $\mathbf{J} \vdash \psi \leftrightarrow \varphi(\psi)$. Moreover, any two fixed points of $\varphi(p)$ are provably equivalent in **J**.*

Proof. Apply Beth definability property for the formula $\Box^+(p \leftrightarrow \varphi(p))$. Then we obtain a formula ψ such that

$$\mathbf{J} \vdash \Box^+(p \leftrightarrow \varphi(p)) \to (p \leftrightarrow \psi).$$

We show that ψ is the required fixed point.

LEMMA 13. $\mathbf{J} \vdash \Box^+(p \leftrightarrow \psi) \to (p \leftrightarrow \varphi(p))$.

Proof. Consider a finite **J**-model $\mathcal{W}$ and a node $x \in \mathcal{W}$ with the minimal $D(x)$ such that $\mathcal{W}, x \Vdash \Box^+(p \leftrightarrow \psi)$ and $\mathcal{W}, x \nVdash p \leftrightarrow \varphi(p)$. As before, we obviously have $\mathcal{W}_x \vDash p \leftrightarrow \psi$. Let p' be a fresh variable evaluated as follows: $\mathcal{W}, y \Vdash p'$ iff $\mathcal{W}, y \Vdash p$, for all $y \neq x$, and $\mathcal{W}, x \Vdash p'$ iff $\mathcal{W}, x \nVdash p$.

If $y \in \mathcal{W}_x$ and $y \neq x$ then $\mathcal{W}, y \Vdash p' \leftrightarrow \varphi(p')$, since p' and p have the same evaluation above x and $D(x)$ was chosen minimally. Since p occurs within the scope of a modality in $\varphi(p)$ we have $\mathcal{W}, x \Vdash \varphi(p)$ iff $\mathcal{W}, x \Vdash \varphi(p')$. Therefore, $\mathcal{W}, x \Vdash p' \leftrightarrow \varphi(p')$, since p' and p have opposite evaluations at x.

We conclude that $\mathcal{W}, x \Vdash \Box^+(p' \leftrightarrow \varphi(p'))$ and by the choice of ψ we must have $\mathcal{W}, x \Vdash p' \leftrightarrow \psi$. This implies $\mathcal{W}, x \Vdash p \leftrightarrow \psi \leftrightarrow p'$, quod non. ∎

As an immediate corollary of this lemma (substituting ψ for p) we obtain

$$\mathbf{J} \vdash \psi \leftrightarrow \varphi(\psi).$$

If ψ_1 and ψ_2 are two fixed points of $\varphi(p)$, then obviously

$$\mathbf{J} \vdash \Box^+(\psi_i \leftrightarrow \varphi(\psi_i)), \text{ for } i = 1, 2.$$

Hence, by Bernardi's lemma $\mathbf{J} \vdash \psi_1 \leftrightarrow \psi_2$. ∎

COROLLARY 14. *The fixed-point property holds for* **GLP**.

Proof. Given a formula $\varphi(p)$ in which p only occurs within the scope of a modality, we obtain a ψ such that $\mathbf{J} \vdash \psi \leftrightarrow \varphi(\psi)$. Obviously, the same equivalence also holds in a stronger system **GLP**.

To show the uniqueness, assume $\mathbf{GLP} \vdash \psi_1 \leftrightarrow \varphi(\psi_1)$, for another formula ψ_1. Denoting $\theta := \psi_1 \leftrightarrow \varphi(\psi_1)$ we obtain by (*Red*):

$$\mathbf{J} \vdash M^+(\theta) \to (\psi_1 \leftrightarrow \varphi(\psi_1)).$$

It follows that

$$\mathbf{J} \vdash \Box^+ M^+(\theta) \to \Box^+(\psi_1 \leftrightarrow \varphi(\psi_1)).$$

Since we also have $\mathbf{J} \vdash \Box^+(\psi \leftrightarrow \varphi(\psi))$, this implies

$$\mathbf{J} \vdash \Box^+ M^+(\theta) \to (\psi_1 \leftrightarrow \psi),$$

by Bernardi's lemma. Taking into account that $\mathbf{GLP} \vdash \Box^+ M^+(\theta)$, for any formula θ, this implies $\mathbf{GLP} \vdash \psi \leftrightarrow \psi_1$. ∎

Open question: Can the syntactic proofs of the fixed-point property for the Gödel-Löb logic (see e.g. [14, 11]) be adapted to the cases of **J** and **GLP**?

BIBLIOGRAPHY

[1] L.D. Beklemishev. Provability algebras and proof-theoretic ordinals, I. *Annals of Pure and Applied Logic*, 128:103–123, 2004.
[2] L.D. Beklemishev. Reflection principles and provability algebras in formal arithmetic. *Uspekhi Matematicheskikh Nauk*, 60(2):3–78, 2005 (Russian). English translation in: *Russian Mathematical Surveys*, 60(2): 197–268, 2005.
[3] L.D. Beklemishev. The Worm principle. In Z. Chadzitakis, P. Koepke, and W. Pohlers, editors, *Logic Colloquium '02, Lecture Notes in Logic, v. 27*, pages 75–95. AK Peters, 2006.

[4] L.D. Beklemishev. Kripke semantics for provability logic GLP. *Annals of Pure and Applied Logic*, 161:756–774, 2010. Preprint: Logic Group Preprint Series 260, University of Utrecht, November 2007. http://preprints.phil.uu.nl/lgps/.
[5] L.D. Beklemishev, J. Joosten, and M. Vervoort. A finitary treatment of the closed fragment of Japaridze's provability logic. *Journal of Logic and Computation*, 15(4):447–463, 2005.
[6] G. Boolos. *The Logic of Provability*. Cambridge University Press, Cambridge, 1993.
[7] L. Carlucci. Worms, gaps and hydras. *Mathematical Logic Quarterly*, 51(4):342–350, 2005.
[8] K.N. Ignatiev. The closed fragment of Dzhaparidze's polymodal logic and the logic of Σ_1-conservativity. ITLI Prepublication Series X–92–02, University of Amsterdam, 1992.
[9] K.N. Ignatiev. On strong provability predicates and the associated modal logics. *The Journal of Symbolic Logic*, 58:249–290, 1993.
[10] G.K. Japaridze. *The modal logical means of investigation of provability*. Thesis in Philosophy, Moscow, 1986 (Russian).
[11] P. Lindström. Provability logic – a short introduction. *Theoria*, 62(1-2):19–61, 1996.
[12] H. Schwichtenberg. Some applications of cut-elimination. In J. Barwise, editor, *Handbook of Mathematical Logic*, pages 867–896. North Holland, Amsterdam, 1977.
[13] I. Shapirovsky. PSPACE-decidability of Japaridze's polymodal logic. In C. Areces and R. Goldblatt, editors, *Advances in Modal Logic, v. 7*, pages 289–304. King's College Publications, 2008.
[14] C. Smoryński. *Self-Reference and Modal Logic*. Springer-Verlag, Berlin, 1985.

K4.Grz and hereditarily irresolvable spaces

GURAM BEZHANISHVILI, LEO ESAKIA, DAVID GABELAIA[1]

ABSTRACT. We show that if we interpret modal diamond ◇ as the derived set operator of a topological space, then **K4.Grz** is exactly the modal logic of hereditarily irresolvable spaces. We also introduce preordinal spaces and show that **K4.Grz** is in fact the modal logic of these spaces.

2000 Mathematics Subject Classification: 03B45
Key words: modal logic, topological semantics, derived set operator, hereditarily irresolvable spaces

1 Introduction

It is a classical result in topology, known as the Cantor-Bendixson theorem, that each topological space X can be decomposed into the union of two disjoint subspaces F and U, where F is closed and dense-in-itself and U is open and scattered.[2]

In [14] Hewitt gave an alternative decomposition of a topological space. We recall that a subset A of a topological space X is *dense* if the closure of A is X, that X is *resolvable* if X can be written as the union of two disjoint dense subspaces of X, and that X is *irresolvable* if X is not resolvable. It is easy to see that each resolvable space is dense-in-itself. Hewitt [14] showed that although in general the converse is not true, it is true for large classes of topological spaces.

We call a space X *hereditarily irresolvable* or simply *HI* if each nonempty subspace of X is irresolvable. An alternative to the Cantor-Bendixson decomposition was given by Hewitt [14] who showed that each space X can be decomposed into the union of two disjoint subspaces F and U, where F is closed and resolvable and U is open and HI.

The same way there is a close connection between dense-in-itself and resolvable spaces, there is a close connection between scattered and HI-spaces. It is not difficult to observe that each scattered space is HI. On the other hand, as shown in [4], the converse is not true in general, but it is true for large classes of topological spaces.

---

[1] The second and third authors were partially supported by the Georgian National Science Foundation Grant GNSF/ST06/3-017

[2] We recall that a point x of a topological space X is *isolated* if $\{x\}$ is an open subset of X, that X is *dense-in-itself* if X has no isolated points, and that X is *scattered* if each nonempty subspace of X has an isolated point.

For a topological space X, we denote by d the derived set operator of X; for $A \subseteq X$ and $x \in X$ we recall that $x \in d(A)$ iff for each open neighborhood U of x we have $A \cap (U - \{x\}) \neq \emptyset$. It was shown in [9] that if we interpret modal diamond $\diamond$ as the derived set operator of a topological space, then the modal logic of scattered spaces (that is, the set of modal formulas valid in all scattered spaces under the above interpretation) is the well-known modal logic **GL** of Gödel and Löb, which is obtained from the basic modal logic **K** by adding Löb's axiom $\square(\square p \to p) \to \square p$.

Typical examples of scattered spaces are ordinals in the interval topology, where the interval topology on an ordinal γ is the topology generated by the intervals $(\alpha, \beta) = \{\lambda \in \gamma : \alpha < \lambda < \beta \}$, $(-\infty, \alpha) = \{\lambda \in \gamma : \lambda < \alpha \}$, and $(\alpha, +\infty) = \{\lambda \in \gamma : \alpha < \lambda \}$, with $\alpha, \beta \in \gamma$. In [1] (and independently in [6]) the result of [9] was sharpened by showing that **GL** is the logic of all ordinals; in fact, **GL** is the logic of any ordinal $\alpha \geq \omega^\omega$.

It is the goal of this note to axiomatize the modal logic of HI-spaces. Since the class of HI-spaces properly contains the class of scattered spaces, the modal logic of HI-spaces is contained in the modal logic of scattered spaces. We show that the modal logic of HI-spaces is axiomatized by adding the well-known Grzegorczyk axiom **grz** $= \square(\square(p \to \square p) \to p) \to \square p$ to the modal logic **K4** $=$ **K** $+ \square p \to \square\square p$. The resulting system **K4.Grz** has been investigated in the literature [2, 10, 11, 15]. In particular, [2] shows that **K4.Grz** has the finite model property, while [10] and [11] establish a close connection between **K4.Grz**, **S4.Grz** $=$ **S4** $+$ **grz**, and the modalized Heyting calculus **mHC**.

In this note we show that **K4.Grz** is the modal logic of HI-spaces. This is an analogue of the result of [9] that **GL** is the modal logic of scattered spaces. We also introduce pre-ordinal spaces and prove that **K4.Grz** is in fact the modal logic of these spaces. This is an analogue of the result of [1, 6] that **GL** is the modal logic of ordinals. Parts of the paper are based on the PhD thesis of the third author [13].

2 Topological completeness of K4.Grz

It is the goal of this section to show that **K4.Grz** is the modal logic of HI-spaces. As our first task, we prove that **K4.Grz** defines the class of HI-spaces.

As usual, for a topological space X, we denote by cl the closure operator of X. For $A \subseteq X$ and $x \in X$ we recall that $x \in \text{cl}(A)$ iff for each open neighborhood U of x we have $A \cap U \neq \emptyset$. It is well known and easy to see that $\text{cl}(A) = A \cup d(A)$ for each $A \subseteq X$.

LEMMA 1. *Let X be a topological space. Then $X \models$ **grz** iff X is an HI-space.*

Proof. It is easy to verify that $X \models$ **grz** iff $d(A) \subseteq d(A - d(d(A) - A))$ for each $A \subseteq X$; this, by [10, Lem. 3] (see also [11, Main Lemma]), is equivalent to $\text{cl}(A) \subseteq \text{cl}(A - \text{cl}(\text{cl}(A) - A))$ for each $A \subseteq X$; which, by [4, Thm. 2.4] (see also [7, Thm. 2.1]) is equivalent to X being an HI-space. ∎

Our next task is to show that HI-spaces are enough to refute non-theorems of **K4.Grz**, thus establishing that **K4.Grz** is the modal logic of HI-spaces. To do so, we will need to recall relational semantics of **K4.Grz**.

Let $\langle P, \leq \rangle$ be a poset. For $L \subseteq P$ we call $\langle L, \leq \rangle$ a *linear order* if for each $p, q \in L$ we have $p \leq q$ or $q \leq p$. We call $\langle P, < \rangle$ a *strict partial order* if $<$ is irreflexive, transitive, and antisymmetric; that is, the reflexive closure of $<$ is a partial order. Also, for $L \subseteq P$, we call $\langle L, < \rangle$ a *strict linear order* if for each distinct $p, q \in L$ we have $p < q$ or $q < p$; that is, the reflexive closure of $<$ is a linear order.

By a *frame* we mean a pair $\mathfrak{F} = \langle W, R \rangle$, where W is a nonempty set and R is a transitive relation on W. We call a frame $\mathfrak{F}$ *rooted* if there exists $r \in W$—called a *root* of $\mathfrak{F}$—such that rRw for all $w \in W$ distinct from r.

DEFINITION 2. We call a frame $\mathfrak{F} = \langle W, R \rangle$ an *unreflexive partial order* if it is antisymmetric. Let $V \subseteq W$. We call $\langle V, R \rangle$ an *unreflexive linear order* if for each distinct $w, v \in W$ we have wRv or vRw.

Equivalently, R is an unreflexive partial order iff its reflexive closure is a partial order, and R is an unreflexive linear order iff its reflexive closure is a linear order.

DEFINITION 3. A frame $\mathfrak{F} = \langle W, R \rangle$ is a *transitive tree* or simply a *tree* if $\mathfrak{F}$ is rooted and for each $w \in W$ the set $R^{-1}(w) = \{v \in W \mid vRw\}$ is a strict linear order (with respect to R).

If we 'decorate' a tree with some reflexive 'loops', we get an unreflexive tree.

DEFINITION 4. We call a frame $\mathfrak{F} = \langle W, R \rangle$ an *unreflexive tree* or simply a *u-tree* if $\mathfrak{F}$ is a rooted unreflexive partial order and $R^{-1}(w) = \{v \in W \mid vRw\}$ is an unreflexive linear order for each $w \in W$.

As follows from [2] (see also [13]), **K4.Grz** is determined by the class of finite u-trees. Since this result will be instrumental in determining that **K4.Grz** is the modal logic of HI-spaces, we state it as a theorem.

THEOREM 5. **K4.Grz** *is determined by the class of finite u-trees. In other words, a modal formula φ is provable in* **K4.Grz** *iff φ is valid in all finite u-trees.*

Let X be a topological space and $\mathfrak{F} = \langle W, R \rangle$ a frame. Recall that $U \subseteq W$ is an *R-upset* of $\mathfrak{F}$ if $w \in U$ and wRv imply $v \in U$. Then the collection τ_R of R-upsets of $\mathfrak{F}$ forms a topology on W, known as the *Alexandroff topology* (in which the intersection of any family of open sets is again open). Let $f : X \to W$ be a map. We recall [3, Def. 2.6] that f is a *d-morphism* if $f : X \to \langle W, \tau_R \rangle$ is continuous and open, $f^{-1}(w)$ is a discrete subspace of X whenever w is irreflexive, and $f^{-1}(w)$ is a dense-in-itself subspace of X whenever w is reflexive. A key feature of d-morphisms is that if $f : X \to W$ is an onto d-morphism and φ is refuted on $\mathfrak{F}$, then φ is also refuted on X [3, Cor. 2.9].

Our strategy in establishing that **K4.Grz** is the modal logic of HI-spaces will be as follows. Let **K4.Grz** $\not\vdash \varphi$. By Theorem 5, there exists a finite u-tree $\mathfrak{T} = \langle W, R \rangle$ such that $\mathfrak{T} \not\models \varphi$. We will build an HI-space X based on $\mathfrak{T}$ in such a way that there is an onto d-morphism $f : X \to W$. This will imply that $X \not\models \varphi$, which will complete the proof.

In building X, one of our basic building blocks will be a dense-in-itself HI-space. The simplest such space is *El'kin's space* introduced in [8]. Following

[4], we call a space X a *filtral space* if the set $\tau - \{\emptyset\}$ of nonempty open subsets of X is a filter. We will denote $\tau - \{\emptyset\}$ by ∇.

DEFINITION 6. *El'kin's space* is a countable filtral space $E = \langle E, \tau \rangle$ such that $\nabla = \tau - \{\emptyset\}$ is a free ultrafilter.

Clearly each subset of E is either open or closed (such spaces are usually called *door spaces*). Moreover, E is a dense-in-itself, T_1, connected space. Furthermore, it follows from [4, Prop. 2.7] that the set of dense subsets of E is exactly ∇ and that E is an HI-space.

Let $\mathfrak{T} = \langle W, R \rangle$ be a u-tree. We associate with $\mathfrak{T}$ a topological space X by placing a copy of El'kin's space E instead of each reflexive point of $\mathfrak{T}$. Formally, for each $w \in W$ let E_w be a homeomorphic copy of El'kin's space such that E_w and E_v are disjoint whenever $w \neq v$. Take the disjoint union $X = \bigsqcup_{w \in W} X_w$, where $X_w = E_w$ if wRw and $X_w = \{w\}$ otherwise; now define a topology τ on X by declaring U open iff the following two conditions are satisfied:

1. $U \cap X_w \neq \emptyset$ implies $U \cap X_w \in \nabla_w$;

2. $U \cap X_w \neq \emptyset$, wRv, and $v \neq w$ imply $U \cap X_v = X_v$.

Here ∇_w is an ultrafilter over X_w, which is free iff $X_w = E_w$.

To make this definition and subsequent arguments more transparent, we will employ the following notation:

For a subset $A \subseteq X$ let $A_w = A \cap X_w$. We call A_w the *w-trace* of A. Also let $I_A = \{w \in W \mid A_w \neq \emptyset\}$. We call I_A the *index set* of A. Clearly $A = \bigcup_{w \in I_A} A_w$.

In this terminology condition (1) means that a w-trace of an open set is open in the space $\langle X_w, \nabla_w \cup \{\emptyset\} \rangle$ and condition (2) means that the index set of an open set always constitutes an upset of $\mathfrak{T}$. In fact, condition (2) says a bit more; it allows the trace U_w to be a proper subset of X_w only if w is a minimum (with respect to R) of the index set I_U.

LEMMA 7. *The τ defined above is indeed a topology.*

Proof. That $\emptyset, X \in \tau$ is obvious. Suppose $\{U_i \mid i \in I\} \subseteq \tau$. We show that $\bigcup U_i \in \tau$. Let $(\bigcup U_i) \cap X_w \neq \emptyset$. Then $U_j \cap X_w \neq \emptyset$ for some $j \in I$. Therefore, by condition (1) for U_j we have $U_j \cap X_w \in \nabla_w$, which implies that $(\bigcup U_i) \cap X_w \in \nabla_w$. Thus, condition (1) is satisfied for $\bigcup U_i$. Next let $(\bigcup U_i) \cap X_w \neq \emptyset$, wRv, and $w \neq v$. Then $U_j \cap X_w \neq \emptyset$ for some $j \in I$, and using (2) for $U_j \in \tau$ we obtain $U_j \cap X_v = X_v$. It follows that $(\bigcup U_i) \cap X_v = X_v$. Consequently, condition (2) is also satisfied for $\bigcup U_i$, and so $\bigcup U_i \in \tau$.

Now suppose that $U, V \in \tau$. We show that $U \cap V \in \tau$. Let $(U \cap V) \cap X_w \neq \emptyset$. Then $U \cap X_w \neq \emptyset$ and $V \cap X_w \neq \emptyset$. Therefore, by (1) for $U, V \in \tau$, we have $U \cap X_w \in \nabla_w$ and $V \cap X_w \in \nabla_w$. But then $(U \cap V) \cap X_w = (U \cap X_w) \cap (V \cap X_w) \in \nabla_w$ since ∇_w is a filter. Thus, condition (1) is satisfied for $U \cap V$. Next let $(U \cap V) \cap X_w \neq \emptyset$, wRv, and $w \neq v$. Since $U \cap X_w \neq \emptyset$ and $V \cap X_w \neq \emptyset$, using (2) for $U, V \in \tau$ we obtain $U \cap X_v = X_v$ and $V \cap X_v = X_v$. It follows that $(U \cap V) \cap X_v = X_v$. Consequently, condition (2) is satisfied for $U \cap V$, and so $U \cap V \in \tau$. ∎

The next lemma appeared as Theorem 4.22 in [13], but the proof contained a gap. Below we give a corrected version of the proof.

LEMMA 8. *The topological space $\langle X, \tau \rangle$ defined above is an HI-space.*

Proof. We show that each nonempty subspace Y of X is an irresolvable space. Let $A, B \subseteq Y$ be dense subsets of Y. It is sufficient to show that $A \cap B \neq \emptyset$. Since $\mathfrak{T} = \langle W, R \rangle$ is a finite u-tree, each subset of W has a maximal point. Let w be a maximal point of I_Y. Consider $A_w = A \cap X_w$ and $B_w = B \cap X_w$. Clearly $A_w, B_w \subseteq Y_w \neq \emptyset$. First we show that $A_w, B_w \neq \emptyset$. If $A_w = \emptyset$, then consider any point $x \in Y_w$. Since w is a maximal point of I_Y, then $V = X_w \cup \bigcup \{X_v \mid wRv\}$ is an open neighborhood of x which misses A. This contradicts the density of A in Y. Thus, $A_w \neq \emptyset$. A similar argument shows that $B_w \neq \emptyset$.

Next we show that $A_w \cap B_w \neq \emptyset$. Suppose that $A_w \cap B_w = \emptyset$. Then we claim that $A_w \in \nabla_w$. If not, then we take any $x \in B_w \subseteq X_w - A_w$ and consider $U = (X_w - A_w) \cup \bigcup \{X_v \mid wRv \text{ and } w \neq v\}$. Since $A_w \notin \nabla_w$ and ∇_w is an ultrafilter, $X_w - A_w \in \nabla_w$. Moreover, $I_U = \{w\} \cup R(w)$ is an upset of $\mathfrak{T}$ and for each v with wRv and $w \neq v$ we have $U_v = X_v$. Therefore, U is an open neighborhood of x. Note that $A_w \cap U = \emptyset$. Also, as w is maximal in I_Y (and thus in I_A), then $A_v = \emptyset$ for each v with wRv and $w \neq v$. Thus, $A \cap U = \emptyset$, which means that $x \notin \text{cl}(A)$. As $x \in B_w \subseteq Y$, this contradicts the density of A in Y. Consequently, $A_w \in \nabla_w$. A similar argument shows that $B_w \in \nabla_w$. Since ∇_w is a filter, we obtain $A_w \cap B_w \neq \emptyset$, a contradiction.

Thus, $A_w \cap B_w \neq \emptyset$, so $A \cap B \neq \emptyset$, and so Y is an irresolvable space. Consequently, X is an HI-space. ∎

Now we define $f : X \to \mathfrak{F}$ by $f(x) = w$ iff $x \in X_w$; that is, f sends an element of X to its index.

LEMMA 9. *The map f defined above is an onto d-morphism.*

Proof. That f is onto is obvious; that $f : X \to \langle W, \tau_R \rangle$ is continuous and open follows easily from conditions (1) and (2). It is also clear that if w is irreflexive, then $f^{-1}(w) = \{w\}$ is a discrete subspace of X. It is left to be shown that if w is reflexive, then $f^{-1}(w)$ is dense-in-itself. But $f^{-1}(w) = E_w$. We show that $E_w \subseteq d(E_w)$. Let $x \in E_w$ and U be an open neighborhood of x in X. By condition (1), $U_w = U \cap E_w \in \nabla_w$. Since ∇_w is a free ultrafilter, U_w is infinite. Therefore, $E_w \cap (U_w - \{x\}) \neq \emptyset$. Thus, $x \in d(E_w)$. ∎

Now we are in a position to prove our first main result.

THEOREM 10. **K4.Grz** *is the modal logic of the class of HI-spaces.*

Proof. It is sufficient to show that if $\mathbf{K4.Grz} \not\vdash \varphi$, then there exists an HI-space X such that $X \not\models \varphi$. By Theorem 5, $\mathbf{K4.Grz} \not\vdash \varphi$ implies that there exists a finite u-tree $\mathfrak{T} = \langle W, R \rangle$ such that $\mathfrak{T} \not\models \varphi$. By Lemmas 8 and 9, there exists an HI-space X and an onto d-morphism $f : X \to W$. By [3, Cor. 2.9], $X \not\models \varphi$. ∎

3 Pre-ordinal spaces and K4.Grz

In this section we sharpen Theorem 10 by introducing a special class of HI-spaces, we call *pre-ordinal spaces*, and showing that **K4.Grz** is in fact the modal logic of these spaces. Our result is reminiscent of ordinal completeness of **GL** [1, 6]. Pre-ordinal spaces are constructed from ordinals by placing copies of El'kin's space instead of designated points of ordinals as follows.

Let γ be an ordinal and let $A \subseteq \gamma$. We associate with γ a topological space X by placing a copy of El'kin's space E instead of each $\alpha \in A$. Formally, for each $\alpha \in \gamma$ let E_α be a homeomorphic copy of El'kin's space such that E_α and E_β are disjoint whenever $\alpha \neq \beta$. Take the disjoint union $X = \bigsqcup_{\alpha \in \gamma} X_\alpha$, where $X_\alpha = E_\alpha$ if $\alpha \in A$ and $X_\alpha = \{\alpha\}$ otherwise. To make the next definition and subsequent arguments more transparent, we will employ the following notation:

For a subset $Y \subseteq X$ let $Y_\alpha = Y \cap X_\alpha$. We call Y_α the α-*trace* of Y. Also let $I_Y = \{\alpha \in \gamma \mid Y_\alpha \neq \emptyset\}$. We call I_Y the *index set* of Y. Clearly $Y = \bigcup_{\alpha \in I_Y} Y_\alpha$.

Now we define a topology τ on X by declaring U open iff the following two conditions are satisfied:

1. $U_\alpha \in \tau_\alpha$ for each $\alpha \in \gamma$;

2. I_U is open in γ.

Here $\tau_\alpha = \nabla_\alpha \cup \{\emptyset\}$ and ∇_α is an ultrafilter over X_α, which is free iff $\alpha \in A$.

In this terminology condition (1) means that an α-trace of an open set is open in the space $\langle X_\alpha, \tau_\alpha \rangle$ and condition (2) means that the index set of an open set always constitutes an open set of γ.

LEMMA 11. *The τ defined above is indeed a topology.*

Proof. That $\emptyset, X \in \tau$ is obvious. Suppose $\{U_i \mid i \in I\} \subseteq \tau$. Then $(\bigcup U_i) \cap X_\alpha = \bigcup(U_i \cap X_\alpha)$. Since $U_i \cap X_\alpha \in \tau_\alpha$ for each α, also $(\bigcup U_i) \cap X_\alpha = \bigcup(U_i \cap X_\alpha) \in \tau_\alpha$. Therefore, condition (1) is satisfied for $\bigcup U_i$. Next, note that $I_{\bigcup U_i} = \bigcup I_{U_i}$. Since each I_{U_i} is open in γ, so is $I_{\bigcup U_i} = \bigcup I_{U_i}$. Consequently, condition (2) is also satisfied for $\bigcup U_i$, and so $\bigcup U_i \in \tau$.

Now suppose that $U, V \in \tau$. Then $(U \cap V) \cap X_\alpha = (U \cap X_\alpha) \cap (V \cap X_\alpha)$. Since $U \cap X_\alpha, V \cap X_\alpha \in \tau_\alpha$, so is $(U \cap V) \cap X_\alpha = (U \cap X_\alpha) \cap (V \cap X_\alpha)$. Therefore, condition (1) is satisfied for $U \cap V$. Next we show that $I_{U \cap V}$ is open in γ. Let $\beta \in I_{U \cap V}$. If β is an isolated point, then there is an open neighborhood $\{\beta\}$ of β contained in $I_{U \cap V}$. If not, then as I_U and I_V are open and $\beta \in I_{U \cap V} \subseteq I_U, I_V$, there exist $\beta_1, \beta_2 < \beta$ such that $(\beta_1, \beta] \subseteq I_U$ and $(\beta_2, \beta] \subseteq I_V$. We may assume without loss of generality that $\beta_1 < \beta_2$. Therefore, $(\beta_2, \beta] \subseteq I_U \cap I_V$. Thus, for each $\delta \in (\beta_2, \beta]$, we have $U_\delta, V_\delta \in \nabla_\delta$. Since ∇_δ is a filter, $U_\delta \cap V_\delta \in \nabla_\delta$. It follows that $(U \cap V)_\delta = U_\delta \cap V_\delta \in \nabla_\delta$, and so $\delta \in I_{U \cap V}$. Thus, $(\beta_2, \beta] \subseteq I_{U \cap V}$ and so $I_{U \cap V}$ is open in γ. Consequently, condition (2) is satisfied for $U \cap V$, and so $U \cap V \in \tau$. ■

LEMMA 12. *The topological space $\langle X, \tau \rangle$ defined above is an HI-space.*

Proof. We show that each nonempty $Y \subseteq X$ is an irresolvable space. Let $B, C \subseteq Y$ be dense subsets of Y. It is sufficient to show that $B \cap C \neq \emptyset$. Since γ is a scattered space, each subspace of γ has an isolated point. Let β be an isolated point of I_Y. Clearly $Y_\beta \neq \emptyset$. Consider B_β and C_β. First we show that $B_\beta, C_\beta \neq \emptyset$. If $B_\beta = \emptyset$, then consider any point $x \in Y_\beta$. Since β is an isolated point of I_Y, there exists an open subset U of γ such that $U \cap I_Y = \{\beta\}$. Therefore, $V = \bigcup \{X_\delta \mid \delta \in U\}$ is an open neighborhood of x which misses B. This contradicts the density of B in Y. Thus, $B_\beta \neq \emptyset$. A similar argument shows that $C_\beta \neq \emptyset$.

Next we show that $B_\beta \cap C_\beta \neq \emptyset$. Suppose that $B_\beta \cap C_\beta = \emptyset$. Then we claim that $B_\beta \in \nabla_\beta$. If not, then we take any $x \in C_\beta \subseteq X_\beta - B_\beta$ and consider $V = (X_\beta - B_\beta) \cup \bigcup \{X_\delta \mid \delta \in U$ and $\delta \neq \beta\}$. Since $B_\beta \notin \nabla_\beta$ and ∇_β is an ultrafilter, $X_\beta - B_\beta \in \nabla_\beta$. Therefore, V is an open neighborhood of x. Note that $B_\beta \cap V = \emptyset$. Thus, $B \cap V = \emptyset$, which means that $x \notin \operatorname{cl}(B)$. As $x \in C_\beta \subseteq Y$, this contradicts the density of B in Y. Consequently, $B_\beta \in \nabla_\beta$. A similar argument shows that $C_\beta \in \nabla_\beta$. As ∇_β is a filter, we obtain $B_\beta \cap C_\beta \neq \emptyset$, a contradiction.

Thus, $B_\beta \cap C_\beta \neq \emptyset$, so $B \cap C \neq \emptyset$, and so Y is an irresolvable space. Consequently, X is an HI-space. ∎

DEFINITION 13. We call the HI-spaces constructed above *pre-ordinal spaces*.

We show that **K4.Grz** is in fact the modal logic of the class of pre-ordinal spaces. We will show this by utilizing the fact that **GL** is the modal logic of all ordinals [1, 6]. We refer to [5, Sec. 3] for a simplified proof of this result. Let **K4.Grz** $\nvdash \varphi$. It is sufficient to find a pre-ordinal space X such that $X \not\models \varphi$. By Theorem 5, there exists a finite u-tree $\mathfrak{T} = \langle W, R \rangle$ such that $\mathfrak{T} \not\models \varphi$. Let $\mathfrak{T}^-$ be the tree obtained from $\mathfrak{T}$ by deleting all the reflexive arrows; that is $\mathfrak{T}^- = \langle W, R^- \rangle$, where R^- is obtained from R by subtracting the diagonal. By [5, Lem. 3.4], there exits $n \in \omega$ and an onto d-morphism $g : \omega^n + 1 \to \mathfrak{T}^-$. Let V denote the set of reflexive points of $\mathfrak{T}$; that is $V = \{w \in W \mid wRw\}$. We set $A = g^{-1}(V) \subseteq \omega^n + 1$ and take the pre-ordinal space X constructed from $\omega^n + 1$ by placing a copy of El'kin's space instead of each $\alpha \in A$. Define $f : X \to \mathfrak{T}$ by $f(x) = w$ iff $x \in X_\alpha$ and $g(\alpha) = w$.

LEMMA 14. $f : X \to \mathfrak{T}$ *is an onto d-morphism.*

Proof. Note that for each $A \subseteq X$ and $w \in W$ we have $f(A) = g(I_A)$ and $f^{-1}(w) = \bigcup \{X_\alpha \mid \alpha \in g^{-1}(w)\}$. Now since g is onto, continuous, and open, the definition of X implies that f is onto, continuous, and open. If w is a reflexive point of $\mathfrak{T}$, then by the construction we have $f^{-1}(w) = \bigcup \{E_\alpha \mid \alpha \in g^{-1}(w)\}$, which is obviously dense-in-itself. Finally, if w is an irreflexive point of $\mathfrak{T}$, then by the construction we have $f^{-1}(w) = g^{-1}(w)$, which is a discrete subspace of X because $g^{-1}(w)$ is a discrete subspace of $\omega^n + 1$. Consequently, f is an onto d-morphism. ∎

Now we are in a position to prove our second main result.

THEOREM 15. **K4.Grz** *is the modal logic of the class of pre-ordinal spaces.*

Proof. Let $\mathbf{K4.Grz} \not\vdash \varphi$. By Theorem 5, there exists a finite u-tree $\mathfrak{T} = \langle W, R \rangle$ such that $\mathfrak{T} \not\models \varphi$. By Lemmas 12 and 14, there exists a pre-ordinal space X and an onto d-morphism $f : X \to W$. By [3, Cor. 2.9], $X \not\models \varphi$. ∎

REMARK 16. It is easy to see that each pre-ordinal space X constructed from a finite u-tree is countable since $\omega^n + 1$ and each of the E_α's are countable. It follows that $\mathbf{K4.Grz}$ is already the modal logic of countable pre-ordinal spaces. Moreover, by enumerating all finite u-trees and taking the topological sum of the corresponding pre-ordinal spaces, we obtain a single countable pre-ordinal space which generates $\mathbf{K4.Grz}$. In fact, we can build this pre-ordinal space based on ω^ω, which is reminiscent of the corresponding result that $\mathbf{GL}$ is the modal logic of ω^ω [1, 6].

REMARK 17. It was observed in [12] that $\mathbf{K4.Grz}$ admits a provability-like interpretation. An important extension of $\mathbf{K4.Grz}$ is obtained by postulating $\mathbf{g} = \neg\Box\bot \to \neg\Box\neg\Box\bot$, which is a modal version of Gödel's Second Incompleteness Theorem. By [13, Thm. 4.14], $\mathbf{G} = \mathbf{K4.Grz} + \mathbf{g}$ is complete with respect to the class of finite u-trees with irreflexive top. Topologically speaking, $\mathbf{G}$ defines the class of weakly scattered HI-spaces [13, Thm. 4.29], where we recall that a space X is *weakly scattered* if the set of isolated points of X is dense in X. It is an easy consequence of Lemmas 8 and 9 that $\mathbf{G}$ is the modal logic of weakly scattered HI-spaces. Moreover, it follows from Lemmas 12 and 14 that $\mathbf{G}$ is in fact the modal logic of (countable) weakly scattered pre-ordinal spaces. Note that the pre-ordinal space X constructed from an ordinal γ and $A \subseteq \gamma$ is weakly scattered iff A contains only limit points of γ.

BIBLIOGRAPHY

[1] M. Abashidze. Ordinal completeness of the Gödel-Löb modal system. In *Intensional logics and the logical structure of theories (Telavi, 1985)*, pages 49–73. "Metsniereba", Tbilisi, 1988. (In Russian).

[2] M. Amerbauer. Cut-free tableau calculi for some propositional normal modal logics. *Studia Logica*, 57(2-3):359–372, 1996.

[3] G. Bezhanishvili, L. Esakia, and D. Gabelaia. Some results on modal axiomatization and definability for topological spaces. *Studia Logica*, 81(3):325–355, 2005.

[4] G. Bezhanishvili, R. Mines, and P. J. Morandi. Scattered, Hausdorff-reducible, and hereditarily irresolvable spaces. *Topology Appl.*, 132(3):291–306, 2003.

[5] G. Bezhanishvili and P. J. Morandi. Scattered and hereditarily irresolvable spaces in modal logic. *Archive for Mathematical Logic*, 49(3):343–365, 2010.

[6] A. Blass. Infinitary combinatorics and modal logic. *J. Symbolic Logic*, 55(2):761–778, 1990.

[7] B. ten Cate, D. Gabelaia, and D. Sustretov. Modal languages for topology: Expressivity and definability. *Annals of Pure and Applied Logic*, 159(1-2):146–170, 2009.

[8] A. El'kin. Ultrafilters and irresolvable spaces. *Vestnik Moskov. Univ. Ser. I Mat. Meh*, 24(5):51–56, 1969.

[9] L. Esakia. Diagonal constructions, Löb's formula and Cantor's scattered spaces. In *Studies in logic and semantics*, pages 128–143. "Metsniereba", Tbilisi, 1981. (In Russian).

[10] L. Esakia. The modal version of Gödel's Second Incompleteness Theorem and McKinsey's system. In *Logical Investigations*, volume 9, pages 292–300. Moscow, Nauka, 2002. (In Russian).

[11] L. Esakia. The modalized Heyting calculus: a conservative modal extension of the intuitionistic logic. *J. Appl. Non-Classical Logics*, 16(3-4):349–366, 2006.

[12] L. Esakia. Around provability logic. *Annals of Pure and Applied Logic*, 161(2):174–184, 2009.

[13] D. Gabelaia. *Topological Semantics and Two-Dimensional Combinations of Modal Logics*. PhD thesis, King's College, London, 2004.

[14] E. Hewitt. A problem of set-theoretic topology. *Duke Mathematical Journal*, 10:309–333, 1943.

[15] T. Litak. The non-reflexive counterpart of Grz. *Bull. Sect. Logic Univ. Łódź*, 36(3-4):195–208, 2007.

Unprovable Ramsey-type statements reformulated to talk about primes

ANDREY BOVYKIN

ABSTRACT. Let us say that two finite sets of natural numbers are primality-isomorphic if there is a difference-preserving primality-preserving and nonprimality-preserving bijection between them. Let Φ be the statement "every infinite set $B \subseteq \mathbb{N}$ has an infinite subset A such that for any $x < y < z$ in A, the interval $\left\{\dfrac{x+y}{2}, \ldots, \dfrac{3x+y}{2}\right\}$ is primality-isomorphic to the interval $\left\{\dfrac{x+z}{2}, \ldots, \dfrac{3x+z}{2}\right\}$". We prove in RCA$_0$, that the Hardy-Littlewood k-tuple Conjecture implies that Φ is equivalent to the Regressive Ramsey Theorem for pairs, a statement that axiomatises ACA$_0$, and so in particular implies all theorems of Peano Arithmetic.

Let Ψ be the statement "for every infinite set $B \subseteq \mathbb{N}$ there is an infinite $A \subseteq B$ such that for any $k < m < n$ in A, $p_m \equiv p_n \mod p_k$", where p_n is the nth prime. We show, using $I\Sigma_1$-provability of an effective version of Dirichlet's theorem on primes in arithmetical progressions, that Ψ is again equivalent to the Regressive Ramsey Theorem for pairs and thus implies all theorems of Peano Arithmetic.

Finally, for every $n \geq 1$, let $P(n)$ be the statement "for all $m > n$, there is N such that for every polynomial $p(x_1, x_2, \ldots, x_n)$ with integer coefficients, there is a set $H \subseteq \{0, 1, 2, \ldots, N-1\}$ of size at least m such that $|H| > \min H$ and the values of the polynomial p are prime on all n-element subsets of H or composite on all n-element subsets of H". For every $n \geq 2$, the statement $P(n)$ is equivalent to the Paris-Harrington Principle in dimension n, and hence is $I\Sigma_{n-1}$-unprovable. In particular the statement "for all n, $P(n)$ holds" is equivalent to the Paris-Harrington Principle and hence is not provable in Peano Arithmetic.[1]

In this note we show how to use prime constellations, residues modulo a prime number and primality and non-primality of polynomials in place of colours as in Ramsey theory, to formulate some simple and attractive strong (unprovable) statements about prime numbers. In all examples in this note, our strong statements are mere reformulations of the Regressive Ramsey theorem, the Kanamori-McAloon Principle and the Paris-Harrington Principle.

The background material on Unprovability Theory is in [2]. We use standard names for arithmetical theories of this part of the spectrum of arithmetical strength. The theory $I\Delta_0 + \exp$, also denoted $I\Delta_0(\exp)$ (and its variation EA as well as the conservative second-order version EFA) is the theory where usual

---
[1]This note was written on the occasion of the 70th birthday of Grigori Mints.

concrete mathematics (that does not use genuine infinitary methods and does not talk about functions that grow faster than finite towers of exponents) takes place. See [1] for a discussion of this theory and its strength. A stronger theory $I\Sigma_1$, the one-quantifier induction arithmetic (and its variations, including primitive recursive arithmetic PRA and the conservative second-order extension RCA_0) is often identified with the intuitive concept of "finitary reasoning" or "all possible elementary methods in mathematics" but is actually somewhat stronger than the informal perception of "elementary methods". The theory $I\Sigma_2$, the two-quantifier induction arithmetic, is an extension of $I\Sigma_1$ and is believed to be able to incorporate the rest of concrete mathematical proofs of first-order arithmetical theorems from the past, including all "non-elementary methods", such as theorems of complex analysis etc. The theory PA, first-order Peano Arithmetic (and its second-order conservative version ACA_0) is very strong and is often identified with "finite" or "separable" mathematics, the last outpost of the finite before truly infinitary methods kick in. All definitions and discussion of these theories can be found in S. Simpson's book [11], which is the standard reference.

The classical number-theoretic theorems and conjectures we mention in this note can be found in most introductory number theory textbooks. I was occasionally using the textbook [10]. Let me formulate the three important unsolved problems in number theory that will often be mentioned below.

A finite sequence of zeros, ones and stars is called a constellation. We say that a constellation is realised in $\mathbb{N}$ if there is an order-preserving, difference-preserving function that maps zeros into composite numbers, ones into prime numbers and stars into any numbers. (By difference-preserving we mean that two symbols of a constellation whose locations differ by n are mapped into two numbers whose difference is n.) We say that a constellation is allowable if it satisfies the following straightforward condition: for every prime number p, the set of places where 1 occurs in our constellation does not cover all residues modulo p. The Hardy-Littlewood k-tuple Conjecture says that every allowable constellation is realised infinitely-often (and even gives the asymptotic number of occurrences of each constellation below x). Only two easy cases of the Hardy-Littlewood Conjecture are known to hold: the case of an arbitrarily long string of zeros and the case of an arbitrary string of zeros and a single one.

The Buniakovsky Conjecture [4] says that for every irreducible polynomial $p(x)$ with integer coefficients, if $p(x)$ does not have local obstruction (i.e., there is no prime number that divides $p(x)$ for all x) then $p(x)$ takes infinitely-many prime values.

Hypothesis H is the ultimate generalisation of the Buniakovsky Conjecture to finite collections of polynomials: for any finite collection or irreducible polynomials $p_1(x), p_2(x), \ldots, p_n(x)$, if there is no prime number that divides $\prod_{i=1}^{m} p_i(x)$ for all x, then on some infinite set of arguments x, the polynomials $p_i(x)$ are simultaneously prime.

1 Primality-isomorphic intervals of natural numbers

We say that two intervals of natural numbers are primality-isomorphic if there is an order-preserving, primality-preserving and nonprimality-preserving bijection between them. For an interval $[m,n]$, we define $isotype([m,n])$ as the sequence of zeros and ones of length $n-m+1$ such that for every $i < n-m+1$, the ith entry is 0 if $i+m$ is composite and 1 if $i+m$ is prime.

THEOREM 1. *Let Φ be the statement "every infinite set B has an infinite subset A such that for any $x < y < z$ in A, the interval $\left\{\dfrac{x+y}{2}, \ldots, \dfrac{3x+y}{2}\right\}$ is primality-isomorphic to the interval $\left\{\dfrac{x+z}{2}, \ldots, \dfrac{3x+z}{2}\right\}$". Then RCA_0 proves that, assuming the Hardy-Littlewood conjecture, Φ is equivalent to the Regressive Ramsey Theorem for pairs, thus implying all theorems of PA.*

For every set X, $[X]^n$ denotes the set of all n-element subsets of X. We define a function f of n arguments $x_1 < x_2 < \ldots < x_n$ to be regressive if $f(x_1, x_2, \ldots, x_n) \leq x_1$ and 2^x-regressive if $f(x_1, x_2, \ldots, x_n) \leq 2^{x_1}$. We say that a set H is f-min-homogeneous if for all $x_1 < x_2 < \ldots < x_n$ and $x_1 < y_2 < \ldots < y_n$ in H, $f(x_1, x_2, \ldots, x_n) = f(x_1, y_2, \ldots, y_n)$. RegRT^2 is the statement "for every regressive $f\colon [\mathbb{N}]^2 \to \mathbb{N}$, there exists an infinite f-min-homogeneous set". $\mathrm{RegRT}^2(2^x)$ is the statement "for every 2^x-regressive $f\colon [\mathbb{N}]^2 \to \mathbb{N}$, there exists an infinite f-min-homogeneous set". It is known that RegRT^2 is equivalent to the infinite Ramsey Theorem for triples and two colours RT^3_2 ("for any infinite set $B \subseteq \mathbb{N}$ and any function $f\colon [B]^3 \to 2$, there exists an infinite set such that f is constant on its 3-element subsets"). It is also known that RT^3_2 axiomatises ACA_0, and thus implies all theorems of Peano Aritmetic [11].

Let us first show that Φ follows from $\mathrm{RegRT}^2(2^x)$. Consider any infinite set B and a colouring $iso\colon [B]^2 \to \mathbb{N}$ defined as follows: $iso(x,y) = isotype\left(\left\{\dfrac{x+y}{2}, \ldots, \dfrac{3x+y}{2}\right\}\right)$. Since there are fewer than 2^x possible isomorphism types, the function is 2^x-regressive. A min-homogeneous infinite subset of B is as needed.

We used a seemingly stronger version of $\mathrm{RegRT}^2(2^x)$ by applying the principle to an arbitrary set B, not to $\mathbb{N}$, which could theoretically turn out to be strictly stronger, for example false. Let us show that it is still equivalent to RT^3_2. Consider any colouring $f\colon [B]^2 \to \mathbb{N}$ with $f(x,y) \leq 2^x$. Put

$$g(x,y,z) = \begin{cases} 0 & \text{if } f(x,y) = f(x,z) \\ 1 & \text{otherwise} \end{cases}$$

Using RT^3_2, choose an infinite homogeneous set and notice that the colour is 0, so we are done.

Before we go into the proof of Theorem 1, we need the following lemma.

LEMMA 2. *Given numbers n and r and an allowable constellation C there is a number s such that there are at least n allowable constellations of the form*

where each I is a string of zeros and ones of length s.

The lemma is provable in $I\Delta_0 + \exp$.

Proof. (Proof of Lemma 2.) Let ℓ be the length of the sequence C. For every prime number $p_i \leq \ell + 2$, fix some residue a_i modulo p_i that occurs in C and define the sets

$$P_i = \{m \mid m \geq \ell + r \text{ and } m = k \cdot p_i + a_i \text{ for some } k \in \mathbb{N}\}.$$

By well-known properties of residues, the set $\bigcap_{p_i \leq \ell+2} P_i$ is infinite (and recurring with period $\prod_{p_i \leq \ell+2} p_i$). Set

$$b_1 = \min \bigcap_{p_i \leq \ell+2} P_i.$$

Clearly, for all primes $p_i \leq \ell + 2$, b_1 realises only existing residues modulo p_i and for every prime $p > \ell + 2$, b_1 does not complete the set of all residues modulo p (since we left two spare places when wrote $\ell + 2$).

Do the same process to define b_i for all $i \leq n$. Suppose the number b_i has been built. For every $p_i \leq b_{n-1} + 2$, fix a residue a_i occurring so far in C or as b_k for some $k = 1, 2, \ldots, i-1$ and define again

$$P_i = \{m \mid m > b_{n-1}, \ m = k \cdot p_i + a_i \text{ for some } k \in \omega\}.$$

Set $b_i = \min \bigcap_{p_j \leq b_{i-1}+2} P_j$ and define

$$C' = C \underbrace{* * * * * *}_{r} 000 \ldots 01_{b_1} 0 \ldots 001_{b_n},$$

where new ones stand in the places $b_1, b_2, \ldots, b_n$ and the rest are zeros. Notice that C' is an allowable constellation, and so is any constellation obtained from C' by substituting some of b_i's ones by zeros. Therefore we have built 2^n-many allowable constellations that continue C. The number $s = b_n - r - \ell$ can be estimated, since at each stage there is a rough exponential bound $b_i \leq b_{i-1} + \prod_{p_i \leq b_{i-1}+2} p_i$, so the construction can be conducted within $I\Delta_0 + \exp$. ∎

Now let us prove Theorem 1.

Proof. (Proof of Theorem 1). Consider an arbitrary regressive colouring $g\colon [\mathbb{N}]^2 \to \mathbb{N}$ and build a set $B_g \subseteq \mathbb{N}$ with its nth element denoted by b_n such that for any $n < m < k$ in $\mathbb{N}$, if $iso(b_n, b_m) = iso(b_n, b_k)$ then $g(n, m) = g(n, k)$, so the application of Φ to B_g will pick out the desired min-homogeneous set for g.

Let us define a sequence of points $\langle b_n \mid n = 0, 1, 2 \ldots \rangle$ and a certain auxiliary tree T. Set b_0 to be arbitrary and fix an allowable interval constellation I_0 on $[0, b_0]$. Let the root of T be I_0.

Define the point b_1 as follows. Apply the Hardy-Littlewood Conjecture to I_0 to find (using Lemma 2) the first realisation $[a, a + b_0]$ such that, defining $b_1 = 2a - b_0$, we have: there are at least two allowable constellations of the form

$$I_0 \underbrace{* * * * * * * *}_{c_0} I,$$

where I is of length $b_1 + 1$ and $c_0 = \dfrac{b_1 - 3b_0}{2}$.

Consider the set

$$[0, b_0] \underbrace{* * * * * *}_{c_0} \left[\dfrac{b_1 - b_0}{2}, 3b_1 - b_0 2\right]$$

and fix two allowable constellations on this set: $I_0 ******I_{00}$ and $I_0 ******I_{01}$. These two constellations form the second level in T, and are the two immediate successors of the root I_0.

Let us find a point b_2 such that

1. $b_2 > 17b_1$;

2. $iso(b_0, b_2) = I_0$, $iso(b_1, b_2) = I_{0,g(1,2)}$;

3. both I_{00} and I_{01} have three different (and different from each others') continuations, i.e. there are three constellations of the form

$$I_0 \underbrace{* * * * * * *}_{c_0} I_{00} \underbrace{* * * * * * *}_{c_1} I$$

and another three allowable constellations of the form

$$I_0 \underbrace{* * * * * * *}_{c_0} I_{01} \underbrace{* * * * * * *}_{c_1} I,$$

where I is of length $b_2 + 1$ and $c_1 = \dfrac{b_2 - 3b_1}{2}$.

The three continuations of $I_0 * * * * * I_{00}$ will be called I_{000}, I_{001} and I_{002} and the three continuations of $I_0 * * * * * I_{01}$ will be called I_{010}, I_{011} and I_{012}. It is important that we chose b_2 so that all these six constellations of length $b_2 + 1$ are different.

Now the general case. Suppose $b_0, b_1, \ldots, b_{n-1}$ have been defined, as well as the first $(n - 1)$ levels of the tree T.

Find a point b_n such that

1. $b_n > 17b_{n-1}$;

2. $iso(b_0, b_n) = I_0$;
   $iso(b_1, b_n) = I_{0,g(1,n)}$;
   $iso(b_2, b_n) = I_{0,g(1,n),g(2,n)}$;
   $\vdots$
   $iso(b_{n-1}, b_n) = I_{0,g(1,n),\ldots,g(n-1,n)}$;

3. for every branch in T_{n-1}, i.e. for every constellation of the form

$$C = \underbrace{I_0 * * * * * *}_{c_0} \underbrace{I_{0k_1} * * * *}_{c_1} I_{0k_1 k_2} * \ldots * * I_{0k_1 k_2 \ldots k_{n-1}},$$

where $k_i \in \{0, 1, \ldots, i\}$, there are $(n+1)$ different allowable continuations of the form

$$C \underbrace{* * * * * * * *}_{c_{n-1}} I,$$

where I is of length $b_n + 1$ and $c_{n-1} = \dfrac{b_n - 3b_{n-1}}{2}$.

Define B to be the set of all b_n for $n \in \mathbb{N}$. (Notice that it is well possible that $g(n,m) = g(n,k)$ but $iso(b_n, b_m) \neq iso(b_n, b_k)$ but it does not concern us since we are only interested in the inverse of this relation.)

Now, apply Φ to B and extract the set A as in Φ. Now notice that the set

$$\{m \in \mathbb{N} \mid b_m \in A\}$$

is min-homogeneous for g. ∎

There is another way to think about the proof of the strength of Φ, namely in terms of choosing an infinite branch through our tree T (which we eventually do when we extract A). So we could think not in terms of $\Phi \leftrightarrow \mathrm{RegRT}^2$ but in terms of equivalence with full König's Lemma (which is equivalent to ACA_0) [7].

Let φ_2 be the statement "for all m, there is N such that for any set $a_1 < a_2 < \cdots < a_N$, there is $H \subseteq N$ such that whenever $i < j < k$ and are in H then the sets $iso(a_i, a_j)$ and $iso(a_i, a_k)$ are primality-isomorphic".

COROLLARY 3. $I\Sigma_1$ *proves that the Hardy-Littlewood conjecture implies* $\varphi_2 \to \mathrm{KM}^2$. *Thus φ_2 and the Hardy-Littlewood conjecture cannot be both provable in $I\Sigma_1$.*

Proof. The proof is identical to the proof of Theorem 1 above. ∎

It is now possible to formulate some other related unprovable statements about polynomials and primes if instead of the Hardy-Littlewood k-tuple Conjecture we use the Buniakovsky Conjecture or Hypothesis H. Since each of those unprovable statements uses exactly the same idea (realisation of constellations corresponding to colours) and a very similar proof, we shall stop here now.

Also, it is possible to generalise the statement φ_2 to the statement φ_n, equivalent to KM^n, using the function $iso\left(x_1, \dfrac{x_2 + x_3 + \ldots + x_n}{n-1}\right)$.

2 Basic congruences and Dirichlet's theorem

Let p_n be the nth prime. Consider the statement Ψ: "for every infinite set $B \subseteq \mathbb{N}$, there is an infinite subset $A \subseteq B$ such that for any $k < m < n$ in A, $p_m \equiv p_n \mod p_k$".

THEOREM 4. RCA_0 *proves that Ψ is equivalent to* RegRT^2, *and hence implies all of* ACA_0.

Proof. $\mathrm{RegRT}^2 \to \Psi$ is easy. Set $f \colon [B]^2 \to \mathbb{N}$ to be defined as follows: $f(x, y)$ is the residue of p_y modulo p_x. Clearly f is x^2-regressive (and even $x \ln x$-regressive), so choose an infinite f-min-homogeneous set A and notice that this set is as needed in Ψ.

$\Psi \to \mathrm{RegRT}^2$. Given a regressive colouring $g \colon [\mathbb{N} \smallsetminus \{0\}]^2 \to \mathbb{N} \smallsetminus \{0\}$, build a set $B_g \subseteq \mathbb{N}$ consisting of primes such that for any $m < n$ in $\mathbb{N}$,

$$b_m \quad \mathrm{mod}\ b_n = g(m, n).$$

Set $b_1 = 2$, $b_2 = 3$. Clearly, $g(1, n) = 1$ for all $n > 1$, so for all prime numbers p, we have $p \bmod b_1 = g(1, n)$ for all $n \in \mathbb{N}$. Suppose for $n \geq 2$ we have chosen prime numbers $b_1, b_2, \ldots, b_n$. Find a number $a \in \{1, 2, \ldots, b_1 b_2 \ldots b_n - 1\}$ such that

$$a \quad \mathrm{mod}\ b_1 = g(1, n+1) = 1$$

$$a \quad \mathrm{mod}\ b_2 = g(2, n+1)$$

$$\vdots$$

$$a \quad \mathrm{mod}\ b_n = g(n, n+1).$$

This number exists because all b_i are prime. Notice also that since $g(i, j) \neq 0$, a is not divisible by any b_i. Now, every member of the arithmetic progression

$$b_1 b_2 \ldots b_n \cdot k + a$$

satisfies the same set of congruences modulo b_i ($i = 1, 2, \ldots, n$) as a, so, we can use Dirichlet's theorem and set b_{n+1} to the first prime member of this arithmetic progression for some $k \geq 1$.

$I\Sigma_1$-provability of Dirichlet's theorem can be found in Cegielski [5]. See also discussion and some of the history of the question in Avigad [1][2].

Therefore our proof of the equivalence $\Psi \leftrightarrow \mathrm{RegRT}^2$ is clearly being conducted in RCA_0.

Define $B_g = \{n \in \mathbb{N} \mid p_n \in B\}$. Apply Ψ to B_g to get a subset $A \subseteq B_g$ such that for $k < m < n$ in A, $p_m \equiv p_n \bmod p_k$. Notice that A is the g-min-homogeneous set we were looking for. ∎

COROLLARY 5. *The statement ψ_2 defined as "for all n there is N such that for every set B of size N, there is a subset $A \subseteq B$ of size n such that for all $m < k < \ell$ in A, $p_k \equiv p_\ell \bmod p_m$" is not provable in $I\Sigma_1$.*

Proof. The proof repeats the proof of Theorem 4 above. ∎

---

[2] A related question concerns provability of the Prime Number Theorem. $I\Sigma_1$-provability of the Prime Number Theorem, was done by Grigori Mints already in 1975, see [9]. (A $I\Delta_0(\exp)$-provability proof can also be found in [6].)

It is possible to transform ψ_2 into an equivalent statement in Π_2^0 form by substituting the quantifier "for every finite set B of size N" by a bounded quantifier with an explicit upper bound $f(N)$ such that the set $\{b_1, b_2, \ldots, b_N\}$ is stated to be chosen from $\{0, 1, \ldots, f(N)\}$.

As with many strong Π_2^1 statements, both Φ and Ψ can be approximated by their "densities" in the sense of J. Paris. The resulting first-order statements are equivalent to 1-consistency of PA (and thus are much stronger than φ_2 or ψ_2) and talk in a certain iterative way about prime numbers. But those statements no longer look particularly interesting, so we omit them here. Another way to gain more strength than ψ_2, but still end up with interesting assertions (while staying in the language of first-order arithmetic) is to imitate KM^n in the same way as in the proof of Theorem 4, by multiple applications of Dirichlet's theorem. This is quite straightforward and we also omit it.

3 Primality and non-primality of polynomials

THEOREM 6. *For every $n \geq 1$, let $P(n)$ be the statement "for all $m > n$, there is N such that for every polynomial $p(x_1, x_2, \ldots, x_n)$ with integer coefficients, there is $H \subseteq \{0, 1, 2, \ldots, N-1\}$ of size at least m such that $|H| > \min H$ and p is prime on all n-element subsets of H or composite on all n-element subsets of H".*

For every $n \geq 2$, the statement $P(n)$ is equivalent to PH$_2^n$*, and hence is $I\Sigma_{n-1}$-unprovable. In particular the statement "for all n, $P(n)$ holds" is equivalent to* PH *and thus is not provable in Peano Arithmetic.*

Notice that since there are polynomials all of whose positive values are prime [8] and polynomials all of whose positive values are composite, it is important to mention both "primality" and "non-primality" cases here, to stay consistent.

We routinely think of each polynomial as a cut-off function from $\mathbb{N}$ to $\mathbb{N}$, setting $p(x_1, x_2, \ldots, x_n) = 0$ if the value turns out to be negative. Let us also mention what we mean by "primality on a set H". We mean that for all $x_1 < x_2 < \ldots < x_n$ in H, the number $p(x_1, x_2, \ldots, x_n)$ is prime. Similarly for "being composite on a set H".

Proof. It is clear that for every n, PH$_2^n$ implies $P(n)$ and that PH$_2$ implies "for all n, $P(n)$ holds". So let us now prove the opposite direction.

Consider an arbitrary m and find the number N as provided by the principle $P(n)$ for m. Consider an arbitrary colouring $f\colon [N]^n \to 2$. Let us build a polynomial $p_f(x_1, \ldots, x_n)$ with integer coefficients such that for all $x_1 < x_2 < \ldots < x_n < N$,

$$p_f(x_1, x_2, \ldots, x_n) \text{ is prime} \iff f(x_1, x_2, \ldots, x_n) = 1.$$

This is not difficult because we have finitely-many such functions f but an infinite supply of various polynomials we can use to imitate f by their primality or non-primality.

First let us define for every $k_1 < k_2 < \ldots < k_n$, an auxiliary polynomial

$$g_{k_1 k_2 \ldots k_n}(x_1, x_2, \ldots, x_n) =$$

$$\prod_{i_1<N, i_1\neq k_1}(x_1-i_1)\cdot \prod_{i_2<N, i_2\neq k_2}(x_2-i_2)\cdot \ldots \cdot \prod_{i_n<N, i_n\neq k_n}(x_n-i_n).$$

Clearly, for $x_1 = k_1, \ldots, x_n = k_n$, $g_{k_1 k_2 \ldots k_n}(x_1, x_2, \ldots, x_n) \neq 0$, but for all other arguments $x_1, \ldots, x_n < N$, $g_{k_1 k_2 \ldots k_n}(x_1, x_2, \ldots, x_n) = 0$.

Now consider C_N^n-many infinite sequences:

$$1 + i \cdot g_{k_1 k_2 \ldots k_n}(k_1, k_2, \ldots, k_n)$$

that is, one sequence for each n-element subset $k_1 < k_2 < \ldots < k_n$ of $\{0, 1, 2, \ldots, N-1\}$.

Each of these sequences has infinitely-many composite values and, by Dirichlet's theorem, infinitely-many prime values, and we have a primitive recursive bound on when the first composite value and the first prime value is guaranteed.

For each $k_1 < k_2 < \ldots < k_n < N$, find and fix the natural number $M_{k_1 k_2 \ldots k_n}$ such that:

- if $f(k_1, k_2, \ldots, k_n) = 0$ then $1 + M_{k_1 k_2 \ldots k_n} \cdot g_{k_1 k_2 \ldots k_n}(k_1, k_2, \ldots, k_n)$ is composite;

- if $f(k_1, k_2, \ldots, k_n) = 1$ then $1 + M_{k_1 k_2 \ldots k_n} \cdot g_{k_1 k_2 \ldots k_n}(k_1, k_2, \ldots, k_n)$ is prime.

Now, set

$$p_f(x_1, x_2, \ldots, x_n) = 1 + \sum_{k_1 < k_2 < \ldots < k_n < N} M_{k_1 k_2 \ldots k_n} \cdot g_{k_1 k_2 \ldots k_n}(x_1, x_2, \ldots, x_n).$$

Clearly, this polynomial is as needed.

Hence the set $H \subseteq \{0, 1, 2, \ldots, N-1\}$ of constant primality or non-primality for p_f is the homogeneous set for f needed in the Paris-Harrington Principle, so $P(n)$ implies PH_2^n. The proof of this implication has been carried out in $I\Sigma_1$ because the explicit bounds in Dirichlet's theorem are proved in $I\Sigma_1$. ∎

At this moment we may want to formulate the following statement A (an easy consequence of the Infinite Ramsey Theorem): "for every polynomial p of several variables, there is an infinite set H on which primality or non-primality of p is constant". It would be tempting to try a compactness argument to show that A implies "for all n, $P(n)$ holds" thus implying PH. However, it does not work (the infinite branch is not defined by a polynomial). There is more to say about this statement and several other interesting statements concerning prime values of polynomials. This is work in progress and will appear in due course in [3].

Although all unprovability proofs in this note are very simple, the main ideas (constellations of primes, residues modulo a prime and primality or non-primality as colours) appear as ingredients inside more serious arguments in the big project [3]. We extracted and isolated the unprovable statements Φ, φ_2, Ψ and $\forall n\ P(n)$ of the current paper since they are simple, compact and might have some independent interest.

BIBLIOGRAPHY

[1] Avigad, J. (2003). Number theory and elementary arithmetic. *Philosophia Mathematica* (3), vol. 11, pp. 257–284.
[2] Bovykin, A. (2008). Brief introduction to unprovability. *Logic Colloquium 2006*. Lecture Notes in Logic, pp. 38–64.
[3] Bovykin, A. (2009). Unprovable statements about prime values of polynomials. Preprint. Work in progress.
[4] Buniakovsky V. (1857). Nouveaux théorèmes relatifs à la distinction des nombres premiers et à la décomposition des entiers en facteurs. *Memoirs of St. Petersburg Academy of Sciences.*, **6**, pp. 305–329.
[5] Cegielski, P. (1992). Le théorème de Dirichlet est finitiste. Technical Report 92.40. Laboratiore Informatique Théoretique et Programmation. Institut Blaise Pascal, Paris.
[6] Cornaros, C., Dimitracopoulos, C. (1994). The prime number theorem and fragments of PA, *Archive for mathematical logic*, 33, pp. 265–281.
[7] Clote, P., McAloon, K. (1983). Two further combinatorial theorems equivalent to the 1-consistency of Peano Arithmetic. *Journal of Symbolic Logic*, 48, pp. 1090–1104.
[8] Matiyasevich, Yu. (1992). Hilbert's tenth problem. MIT Press.
[9] Mints, G. (1976). What can be done in PRA. *Zapiski nauchnyh seminarov LOMI*, vol. 60, pp. 93–102.
[10] Pollack, P. (2004). Not Always Buried Deep. Selections from analytic and combinatorial number theory. Book manuscript. Available online.
[11] Simpson, S. (2009). Subsystems of Second-Order Arithmetic. Second Edition. Association for Symbolic Logic.

Maximum Satisfiability and Subexponential Time

EVGENY DANTSIN

ABSTRACT. It is known that the Boolean satisfiability problem for formulas in 3-CNF (3-SAT) can be solved in time asymptotically less than $\mathcal{O}(2^n)$ where n is the number of variables in the input formula. A challenging open question is whether there exists a constant $c > 0$ such that no algorithm solves 3-SAT in time $\mathcal{O}(2^{cn})$, or 3-SAT can be solved in subexponential time, i.e., in time $\mathcal{O}(2^{\varepsilon n})$ for an arbitrarily small $\varepsilon > 0$. Impagliazzo, Paturi, and Zane in their seminal paper [7] obtained deep results that shed more light on this intriguing question. The present paper extends this line of research to the case of maximum satisfiability. In particular, it is shown that for any $k \geq 2$, the following equivalence holds: MAX k-SAT has subexponential complexity if and only if 3-SAT has subexponential complexity.

1 Introduction

In the Boolean satisfiability problem (SAT) we are given a Boolean formula in conjunctive normal form and we are asked whether the formula is satisfiable. The problem is **NP**-complete; all known algorithms for SAT take exponential time in the worst case. However, there may be a significant difference in the efficiency between exponential-time algorithms, for example between an algorithm running in time $\mathcal{O}(2^n)$ and an algorithm running in time $\mathcal{O}(2^{n/100})$. How large is the exponent for SAT?

The Exponential-Time Hypothesis. Given a formula F with n variables, a straightforward way to test satisfiability of F is to enumerate all 2^n truth assignments. The exponent n in this bound can be lowered for the restriction of SAT to formulas in which every clause has at most k variables; this restriction is denoted by k-SAT and it remains **NP**-complete for $k \geq 3$. Namely, it was shown in [3, 10, 13] that for any k, there is an algorithm that solves k-SAT in time $\mathcal{O}(2^{c_k n})$ where $c_k < 1$ is a constant depending on k. For example, 3-SAT can be solved by a randomized algorithm in time $\mathcal{O}(2^{0.404n})$ [11] and by a deterministic algorithm in time $\mathcal{O}(2^{0.559n})$ [1]. How far can we lower the constants c_k?

It is a challenging open question which of the following two possibilities holds:

- 3-SAT has a threshold for the exponent, i.e., there exists a positive constant c such that no algorithm solves 3-SAT in time $\mathcal{O}(2^{cn})$;

- 3-SAT can be solved in subexponential time, i.e., for any $\varepsilon > 0$, there is an algorithm that solves 3-SAT in time $\mathcal{O}(2^{\varepsilon n})$.

The first possibility is known as the *Exponential-Time Hypothesis* (*ETH* for short) defined by Impagliazzo and Paturi in [6]. We currently cannot prove or disprove ETH (even assuming $\mathbf{P} \neq \mathbf{NP}$), but we could learn more by analyzing implications of ETH and implications of its negation. A number of such implications were obtained in [2, 6], see also a survey in [4]. Here is just one of them, an interesting implication of ETH. For any $k \geq 3$, let s_k denote the infimum of all $\delta \geq 0$ such that k-SAT can be solved in time $\mathcal{O}(2^{\delta n})$. Then, assuming that ETH holds, the sequence $s_3, s_4, \ldots$ is strictly increasing infinitely often [6].

Subexponential complexity. It is natural to relate ETH to the complexity of other problems. For example, does ETH imply that some other problems cannot be solved in subexponential time? Or, how is a possible threshold for the exponent in SAT related to possible thresholds for other problems? To analyze such questions, we need an appropriate definition of the class of problems that have subexponential complexity and we need an appropriate notion of reducibility that preserves subexponential complexity.

The complexity class **SE** (subexponential time) was defined in [7] as the set of pairs $(\mathcal{A}, p)$ where $\mathcal{A}$ is a problem and p is a complexity parameter of $\mathcal{A}$ such that $\mathcal{A}$ can be solved in subexponential time with respect to p (see Section 2 for a precise definition). It is important to specify a parameter because subexponential complexity with respect to one parameter does not necessarily imply subexponential complexity with respect to others. Using this notation and denoting the number of variables in a formula by n, ETH can be written as

$$(3\text{-SAT}, n) \notin \mathbf{SE}.$$

Impagliazzo, Paturi, and Zane defined and studied reductions that preserve subexponential complexity [7]. Using such reductions, they obtained deep results concerning ETH, the class **SNP**, and lower bounds for depth-3 circuits. Let us mention only those of their results and techniques that are directly related to the present paper.

For any $k \geq 3$, we have

$$(k\text{-SAT}, n) \in \mathbf{SE} \quad \Leftrightarrow \quad (k\text{-SAT}, m) \in \mathbf{SE}$$

where n and m are respectively the number of variables and the number of clauses. One part of this equivalence is obvious (subexponential complexity with respect to n implies subexponential complexity with respect to m), but the other part is very nontrivial. This part of the equivalence was proved using the so-called *sparsification lemma*, a powerful tool for subexponential reducibility. The lemma says that for an arbitrarily small $\varepsilon > 0$, any formula F with n variables reduces to the disjunction of $2^{\varepsilon n}$ "sparse" formulas, i.e., formulas where every variable occurs a constant number of times.

Another result to mention is the completeness of 3-SAT under subexponential reducibility: for any $k > 3$, both $(k\text{-SAT}, n)$ and $(k\text{-SAT}, m)$ are reducible to 3-SAT with parameter m (and hence also with parameter n).

Main results. This paper shows the relationship between subexponential complexity of k-SAT and subexponential complexity of the maximum k-satisfiability problem (MAX k-SAT). The main results can be summarized as follows.

1. It is shown in Section 3 that MAX k-SAT with parameter m is reducible to $(k+1)$-SAT with parameter n. It follows from this reduction that for any $k \geq 2$,

$$(\text{MAX } k\text{-SAT}, m) \notin \mathbf{SE} \iff \text{ETH}.$$

In particular, if MAX 2-SAT has subexponential complexity with respect to m then for any $k > 2$, MAX k-SAT also has subexponential complexity with respect to m.

2. It is shown in Section 4 that the weighted version of MAX k-SAT is reducible to the unweighted version. Hence, for any $k \geq 2$,

$$(\text{WEIGHTED MAX } k\text{-SAT}, m) \in \mathbf{SE} \iff (\text{MAX } k\text{-SAT}, m) \in \mathbf{SE}.$$

3. It is shown in Section 5 that ETH can be expressed in terms of k-SAT and MAX k-SAT restricted to sparse formulas. Namely, ETH holds if and only if the restriction of (3-SAT, m) to formulas in which every variable has at most 3 occurrences is not in $\mathbf{SE}$. Also, ETH holds if and only if the restriction of (MAX 2-SAT, m) to formulas in which every variable has at most 12 occurrences is not in $\mathbf{SE}$.

Organization of the paper. Definitions and notation are given in Section 2. The main results are proved in Sections 3–5. An open question concerning subexponential complexity of MAX k-SAT is discussed in Section 6.

2 Preliminaries

Boolean satisfiability. By a *variable* we mean a Boolean variable that takes truth values true or false (identified with 1 and 0 respectively). A *literal* is a variable or its negation; complementary literals are denoted by a and $\bar{a}$. A *clause* is a disjunction of literals that does not contain a variable together with its negation. It is common to identify a clause $a_1 \vee \ldots \vee a_k$ with a set of its literals and to write this clause as $\{a_1, \ldots, a_k\}$. The *length* of a clause is the number of its literals. A *CNF formula* (or simply *formula*) is a conjunction of clauses; CNF stands for conjunctive normal form. A k-*CNF formula* is a formula where the length of every clause is at most k. A formula $C_1 \wedge \ldots \wedge C_m$ is identified with a set of its clauses and it is written as $\{C_1, \ldots, C_m\}$. A *truth assignment* (or simply *assignment*) is a mapping from a set of variables to {true, false}. of literals: if a variable v is mapped to true then $v \in A$; if v is mapped to false then $\neg v \in A$. An assignment A_1 is an *extension* of an assignment A_2 if $A_1 \supseteq A_2$. An assignment A *satisfies* a clause C if at least one literal in C is true under A.

The *satisfiability problem* (denoted by SAT) is stated as follows: given a formula F, is there an assignment that satisfies F? The *maximum satisfiability problem* (MAX SAT) is usually defined as an optimization problem, but in this paper, we consider its decision version defined as follows. An instance of MAX SAT is a pair $\langle F, s \rangle$ where F is a formula and s is an integer. The problem is: given such an instance, is there an assignment that satisfies at least s clauses in F? The k-*satisfiability problem* (k-SAT) and the *maximum*

k-*satisfiability problem* (MAX k-SAT) are the restrictions of respectively SAT and MAX SAT to k-CNF formulas. Some other variants of MAX SAT are also considered in this paper, for example the weighted version of MAX SAT (defined in Section 4) and MAX SAT restricted to formulas in which every variable occurs a constant number of times (defined in Section 5).

Complexity parameters. Let $\mathcal{A}$ be a problem (language) in the class **NP**. Since $\mathcal{A} \in \mathbf{NP}$, there exists a polynomial-time verifier V for $\mathcal{A}$. An instance x is a "yes" instance of $\mathcal{A}$ if and only if V accepts $\langle x, y \rangle$ where y is a *certificate* for x. Let p be a polynomial-time computable function that bounds shortest certificates, i.e., for any "yes" instance x, there is a certificate y with $|y| \leq p(x)$. Any such function p is called a *complexity parameter* of $\mathcal{A}$. We write $(\mathcal{A}, p)$ to denote problem $\mathcal{A}$ with specified complexity parameter p.

In this paper, we consider SAT and its variants with two complexity parameters: the number n of variables and the number m of clauses in the input formula.

Subexponential reductions. The class **SE** (stands for "subexponential time") is defined as follows [7]: a problem $\mathcal{A}$ with complexity parameter p belongs to **SE** if for any $\varepsilon > 0$, the problem $\mathcal{A}$ can be solved by a deterministic algorithm in time $|x|^{\mathcal{O}(1)} \cdot 2^{\varepsilon p(x)}$, where x is an instance of A and $|x|$ denotes the size of x.

The *Exponential Time Hypothesis* (ETH) is the following statement [6]:

$$(3\text{-SAT}, n) \notin \mathbf{SE}.$$

Two types of reductions that preserve subexponential complexity were defined in [7]: strong many-one reductions and SERFs. Let $(\mathcal{A}, p)$ and $(\mathcal{B}, q)$ be two problems with specified parameters. A many-one reduction r from $\mathcal{A}$ to $\mathcal{B}$ is called a *strong many-one reduction* from $(\mathcal{A}, p)$ to $(\mathcal{B}, q)$ if for any instance x of $\mathcal{A}$, we have

$$q(r(x)) = \mathcal{O}(p(x)).$$

Suppose that for any $\varepsilon > 0$, there is a Turing reduction $T_\varepsilon^\mathcal{B}$ from $\mathcal{A}$ to $\mathcal{B}$ such that the following holds:

- For any $\varepsilon > 0$, the reduction $T_\varepsilon^\mathcal{B}$ runs in time $|A|^{\mathcal{O}(1)} \cdot 2^{\varepsilon p(A)}$ where A is an instance of $\mathcal{A}$.

- When running on A, each reduction $T_\varepsilon^\mathcal{B}$ is allowed to query its oracle only for instances B such that $q(B) = \mathcal{O}(p(A))$ and $|B| = |A|^{\mathcal{O}(1)}$.

Any such collection $\{T_\varepsilon^\mathcal{B}\}_{\varepsilon > 0}$ is called a *subexponential reduction family* (*SERF* for short) that reduces $(\mathcal{A}, p)$ to $(\mathcal{B}, q)$. If such a collection exists, we say that $(\mathcal{A}, p)$ is *SERF-reducible* to $(\mathcal{B}, q)$.

Notation fixed throughout the paper.

- n denotes the number of variable in a given formula;

- m denotes the number of clauses in a given formula.

3 Maximum k-satisfiability

This section shows that for any $k \geq 2$, the following equivalence holds:

(1) (MAX k-SAT, m) $\notin$ **SE** $\Leftrightarrow$ ETH.

LEMMA 1. *For any $k \geq 2$, there exists a strong many-one reduction from* MAX k-SAT *with parameter m to $(k+1)$-SAT with parameter n.*

Proof. Let $k \geq 2$ be fixed. We construct a many-one reduction that transforms an instance $\langle F, s \rangle$ of MAX k-SAT into a $(k+1)$-CNF formula φ such that

- φ is satisfiable if and only if F has an assignment that satisfies at least s clauses in F;

- the number of variables in φ is $\mathcal{O}(m)$ where m is the number of clauses in F.

Let $x_1, \ldots, x_n$ and $C_1, \ldots, C_m$ be respectively the variables and clauses of the input formula F. The formula φ is the conjunction of three formulas α, β, and γ, where α is just the input formula F and the formulas β and γ are defined below. To define β, we introduce new variables $y_1, \ldots, y_m$ corresponding to the clauses $C_1, \ldots, C_m$. For each clause C_i, if C_i is the disjunction of literals $a_1, \ldots, a_k$, we form the equivalence

$$y_i \equiv a_1 \vee \ldots \vee a_k.$$

and write it as a $(k+1)$-CNF formula:

(2) $(\bar{y}_i \vee a_1 \vee \ldots \vee a_k) \wedge (y_i \vee \bar{a}_1) \wedge \ldots \wedge (y_i \vee \bar{a}_k)$

Now we define β to be the conjunction of the formulas (2) for all clauses $C_1, \ldots, C_m$. Obviously, the input formula F has an assignment that satisfies at least s clauses if and only if $\alpha \wedge \beta$ has a satisfying assignment in which at least s of the variables $y_1, \ldots, y_m$ have the true value. Note that $\alpha \wedge \beta$ has $n + m$ variables.

The formula γ encodes the most non-trivial task of the reduction: the count of the number of variables y_i set to true and the comparison of this number with s. In fact, γ encodes a bounded fan-in circuit that is the composition of two circuits denoted by COUNT and COMPARISON. The COUNT circuit has m input gates and $\lceil \log_2(m+1) \rceil$ output gates. The circuit computes the number of ones assigned to the input variables and it outputs this number in binary representation. The COMPARISON circuit has $2\lceil \log_2(m+1) \rceil$ input gates that represent two $\lceil \log_2(m+1) \rceil$-bit numbers (in binary). It outputs 1 if the first of these numbers is greater than or equal to the second number. In the composition of COUNT and COMPARISON, the sum computed by COUNT is the first input number for COMPARISON. The second input number is s from the instance $\langle F, s \rangle$.

How many gates are needed to implement COUNT and COMPARISON? The counting circuit can be implemented using the iterated-addition technique (see for example [14, Section 1.3]), which gives a circuit of size $\mathcal{O}(m)$. The comparison of two numbers in binary can also be done with a circuit of linear size. Therefore, the composition of COUNT and COMPARISON is a bounded fan-in circuit of size $\mathcal{O}(m)$. This circuit can be encoded by a 3-CNF formula in the standard way used in the reduction from CIRCUIT SAT to SAT, see for example [9, Section 8.1].

Thus, φ is a $(k+1)$-CNF formula with $\mathcal{O}(n+m)$ variables. Since n does not exceed km, the number of variables in φ is in fact $\mathcal{O}(m)$. It is easy to see that this formula is satisfiable if and only if F has an assignment that satisfies at least s clauses in F. It is also easy to see that the construction of φ from $\langle F, s \rangle$ can be done in polynomial time. ∎

THEOREM 2. *For any $k \geq 2$, if (MAX k-SAT, m) $\notin$ SE then ETH holds.*

Proof. For any $k \geq 2$, Lemma 1 gives a strong many-one reduction from MAX k-SAT with parameter m to $(k+1)$-SAT with parameter n. It is shown in [7, Corollary 2] that the sparsification procedure provides a SERF that reduces $(k+1)$-SAT with parameter n to 3-SAT with parameter m. For any $k \geq 3$, there is a strong many-one reduction from $(k$-SAT, $m)$ to $(3$-SAT, $m)$, see for example [7, Lemma 10]. Since n is at most $3m$, this reduction is also a reduction to $(3$-SAT, $n)$. Thus, (MAX k-SAT, m) is SERF-reducible to $(3$-SAT, $n)$ and, therefore, if $(3$-SAT, $n) \in$ **SE** then (MAX k-SAT, $m) \in$ **SE**. The claim is just the contrapositive form of this implication. ∎

REMARK 3 (another way to prove the reducibility in Lemma 1). In Lemma 1 above, the reducibility is proved by giving an explicit transformation of an instance of MAX k-SAT into the corresponding instance of $(k+1)$-SAT. This reducibility could also be proved in an indirect way, using a method briefly sketched in [7]. This method is to show that MAX k-SAT with complexity parameter m belongs to the class called *Size-Constrained* **SNP** and defined as follows. The class **SNP** consists of parameterized problems that are expressible in second-order logic by formulas of the form

$$\exists R_1 \ldots \exists R_p \forall x_1 \ldots \forall x_q \varphi$$

where $\exists R_1, \ldots, \exists R_p$ are second-order quantifiers, $\forall x_1, \ldots, \forall x_q$ are first-order quantifiers, and φ is a quantifier-free formula, see [7, 8] for details. Size-Constrained **SNP** is an extension of **SNP** where second-order existential quantifiers are allowed to be *size-constrained*. Such a quantifier has one of the following forms: either $\exists R, |R| = s$, or $\exists R, |R| < s$, or $\exists R, |R| > s$.

Using circuits for counting and comparing, it can be proved that if problem $\mathcal{A}$ with parameter p is in Size-Constrained **SNP** then there is an integer $k \geq 3$ such that $(\mathcal{A}, p)$ is strongly reducible to k-SAT with complexity parameter n [7, Theorem 5]. Therefore, if (MAX k-SAT, m) belongs to Size-Constrained **SNP** then there is an integer k' such that (MAX k-SAT, m) is strongly reducible to $(k'$-SAT, $n)$.

Now let us move on to the other part of equivalence (1).

THEOREM 4. *For any $k \geq 2$, ETH implies (MAX k-SAT, m) $\notin$ SE.*

Proof. Obviously, it suffices to prove that ETH implies (MAX 2-SAT, m) $\notin$ **SE**. Since ETH is equivalent to (3-SAT, m) $\notin$ **SE**, we can prove the claim by giving a strong many-one reduction from (3-SAT, m) to (MAX 2-SAT, m).

Proofs of the **NP**-completeness of MAX 2-SAT usually refer to the many-one reduction from 3-SAT proposed in [5], see for example [9, Theorem 9.2]. However, this reduction is in fact a reduction to the weighted version of MAX k-SAT. Since we consider the unweighted version (so far), we have to modify the usual reduction as follows. Let F be a 3-CNF formula. Without loss of generality, we can assume that every clause in F contains exactly three literals: use the unit clause elimination rule and then replace each clause $a \vee b$ by two clauses $a \vee b \vee x$ and $a \vee b \vee \overline{x}$ where x is a new variable. Now for each clause $a \vee b \vee c$, let us introduce seven new variables $y_1, \ldots, y_7$ and form 16 new clauses:

$$a \vee y_1, \quad \overline{y}_1, \quad b \vee y_2, \quad \overline{y}_2, \quad c \vee y_3, \quad \overline{y}_3,$$
$$\overline{a} \vee y_4, \quad \overline{b} \vee \overline{y}_4, \quad \overline{a} \vee y_5, \quad \overline{b} \vee \overline{y}_5, \quad \overline{b} \vee y_6, \quad \overline{b} \vee \overline{y}_6,$$
$$a \vee y_7, \quad b \vee y_7, \quad c \vee y_7, \quad \overline{y}_7.$$

It is straightforward to check that the new clauses have the following two properties:

- any assignment that satisfies $a \vee b \vee c$ can be extended to satisfy 13 of the new clauses and cannot be extended to satisfy more than 13 new clauses;

- any assignment that falsifies $a \vee b \vee c$ can be extended to satisfy 12 of the new clauses and cannot be extended to satisfy more than 12 new clauses.

Let φ be a formula obtained from F as above; the number of variables in φ is $n + 7m$ and the number of clauses is $16m$. Clearly, F is satisfiable if and only if there is an assignment A to the variables of φ such that A satisfies $13m$ clauses in φ. This reduction is strong because the number of clauses in φ is linear in the number of clauses in F. ∎

4 Weighted version

According to our definitions in Section 2, a formula is a set of clauses. Now we define a *weighted formula* to be a multiset of clauses; the *weight* of a clause is the number of occurrences of this clause in the multiset. The weighted version of MAX SAT is denoted by WEIGHTED MAX SAT. As before, we use m to denote the number of (possibly repeated) clauses in a weighted formula, i.e., m is the number of members in the corresponding multiset. An instance of WEIGHTED MAX SAT is a pair $\langle F, s \rangle$ where F is a weighted formula and s is an integer. The problem is: given such an instance, is there an assignment that satisfies at least s elements in the multiset F? Similar to the unweighted case, WEIGHTED MAX k-SAT is the restriction of this problem to instances with clauses of length at most k.

THEOREM 5. *For any $k \geq 2$, the following equivalence holds:*

(3) (WEIGHTED MAX k-SAT, m) $\in$ **SE** $\Leftrightarrow$ (MAX k-SAT, m) $\in$ **SE**.

Proof. Since WEIGHTED MAX SAT is a generalization of MAX SAT, one part of equivalence (3) is obvious: if WEIGHTED MAX k-SAT can be solved by an $\mathcal{O}(2^{\varepsilon m})$ algorithm then MAX k-SAT can be solved by the same algorithm. To prove the other part, we give a strong many-one reduction from WEIGHTED MAX k-SAT to MAX k-SAT.

For any instance $\langle F, s \rangle$ of WEIGHTED MAX k-SAT, let us define an instance $\langle F', s' \rangle$ of MAX k-SAT such that $\langle F, s \rangle$ is a "yes" instance if and only if $\langle F', s' \rangle$ is a "yes" instance. Let $C_1, \ldots, C_m$ be the members of the multiset F. We obtain F' from F by replacing each clause C_i by two new clauses as follows.

- Case 1: the clause C_i consists of a single literal a. Introduce a new variable x_i and replace the clause a by two clauses $a \vee x_i$ and $\overline{x}_i$.

- Case 2: the clause C_i is a disjunction $a_1 \vee \ldots \vee a_p$ where $p > 1$. Partition $\{a_1, \ldots, a_p\}$ into any two nonempty subsets

$$\{a_1, \ldots, a_p\} = \{a_1, \ldots, a_q\} \cup \{a_{q+1}, \ldots, a_p\}.$$

Introduce a new variable x_i and replace the clause $a_1 \vee \ldots \vee a_p$ by two clauses

(4) $\quad a_1 \vee \ldots \vee a_q \vee x_i,$

(5) $\quad a_{q+1} \vee \ldots \vee a_p \vee \overline{x}_i.$

Thus, F is transformed into a k-CNF formula F' that contains $2m$ pairwise distinct clauses (since all introduced variables $x_1, \ldots, x_m$ are distinct).

Let A be any assignment to the variables of F. Suppose A satisfies exactly r members in F. Then the following holds:

- The assignment A can be extended by assigning truth values to the variables $x_1, \ldots, x_m$ so that the extended assignment A' satisfies exactly $r + m$ clauses in F'. Indeed, suppose that A satisfies a clause $a_1 \vee \ldots \vee a_p$ and its possible copies in F. Then a value for the corresponding variable x_i can be chosen so that both clauses (4) and (5) are satisfied (similarly for the case $p = 1$). Therefore, A' satisfies at least $2r$ clauses in F'. In addition to these satisfied clauses, A' satisfies $m - r$ more clauses. This is because if A falsifies $a_1 \vee \ldots \vee a_p$ then any extension of A satisfy exactly one of clauses (4) and (5).

- No extension of A can satisfy more than $r + m$ clauses. This because any extension of A cannot satisfy more than all $2r$ clauses obtained from r satisfied members of F and it cannot satisfy more than $m - r$ clauses out of all $2(m - r)$ clauses obtained from $m - r$ falsified members of F.

Now it is clear that $\langle F, s \rangle$ is a "yes" instance of WEIGHTED MAX k-SAT if and only if $\langle F', s + m \rangle$ is a "yes" instance of MAX k-SAT. Since the transformation above can be performed in polynomial time and it only doubles the number of clauses, this transformation is a strong many-one reduction from WEIGHTED MAX k-SAT to MAX k-SAT for any $k \geq 2$. ∎

COROLLARY 6. *For any $k \geq 2$, the following equivalence holds:*

(6) (WEIGHTED MAX k-SAT, m) $\notin$ **SE** $\Leftrightarrow$ ETH.

Proof. Immediately follows from Theorems 2, 4, and 5. ∎

5 Restriction to sparse formulas

Let F be a formula and x be a variable occurring in F. The number of occurrences of x in F is called the *frequency* of x in F. We call F a (k, f)-*formula* if every clause has at most k literals and the frequency of every variable is at most f. It is common to refer to (k, f)-formulas, where f is a constant, as to *sparse* formulas. We write (k, f)-SAT and MAX (k, f)-SAT to denote respectively the restrictions of k-SAT and MAX k-SAT to (k, f)-formulas.

It is well known that restrictions of k-SAT to sparse formulas remain **NP**-complete. For example, $(3, 3)$-SAT is still **NP**-complete. To see why, let us transform a 3-CNF formula F into the following $(3, 3)$-formula F'. For each variable x that has $p \geq 3$ occurrences in F, we replace all occurrences of x by new variables $x_1, \ldots, x_p$ and add p new clauses

(7) $\overline{x}_1 \vee x_2, \quad \overline{x}_2 \vee x_3, \quad \ldots, \quad \overline{x}_p \vee x_1,$

that express the requirement that all $x_1, \ldots, x_p$ must have the same truth value. Clearly, the resulting formula F' is satisfiable if and only if F is satisfiable.

It is also known the **NP**-completeness of MAX k-SAT is preserved under restriction to sparse formulas. For example, MAX $(2, 3)$-SAT remains **NP**-complete [12].

Can we express ETH in terms of sparse formulas? The **NP**-completeness of the restrictions above does not automatically imply that there are reductions that preserve (up to a linear increase) parameters m or n. Such reductions are constructed in the following theorem.

THEOREM 7. *The following two equivalences hold:*

$$\text{ETH} \Leftrightarrow ((3,3)\text{-SAT}, m) \notin \mathbf{SE}; \qquad (8)$$

$$\text{ETH} \Leftrightarrow (\text{MAX } (2,12)\text{-SAT}, m) \notin \mathbf{SE}. \qquad (9)$$

Proof. It is easy to prove the $\Leftarrow$-parts of equivalences (8) and (9). Suppose that $((3,3)\text{-SAT}, m) \notin \mathbf{SE}$. Then $(3\text{-SAT}, m)$ is not in **SE** either, and therefore ETH holds (since $n \leq 3m$). Similarly, if $(\text{MAX } (2,12)\text{-SAT}, m)$ is not in **SE** then $(\text{MAX } 2\text{-SAT}, m)$ is not in **SE** and hence, by Theorem 2, ETH holds.

To prove the $\Rightarrow$-part of equivalence (8), let us consider the standard reduction from 3-SAT to $(3,3)$-SAT sketched above: replace all occurrences of x by new variables $x_1, \ldots, x_p$ and add clauses (7). Since the resulting formula F' has $\mathcal{O}(m)$ clauses where m is the number of clauses in F, this reduction is a strong many-one reduction from $(3\text{-SAT}, m)$ to $((3,3)\text{-SAT}, m)$. Therefore, if $((3,3)\text{-SAT}, m) \in \mathbf{SE}$ then $(3\text{-SAT}, m) \in \mathbf{SE}$. As shown in [7], if $(3\text{-SAT}, m) \in \mathbf{SE}$ then $(3\text{-SAT}, n) \in \mathbf{SE}$. Thus, supposing that $((3,3)\text{-SAT}, m)$ is in **SE**, we can conclude that ETH does not hold.

Now it remains to prove $\Rightarrow$-part of equivalence (9). A straightforward way is to combine two reductions:

- the reduction from 3-SAT to $(3,3)$-SAT (see above);
- the reduction from $(3,3)$-SAT to MAX 2-SAT (using the transformation of 3-CNF formulas into 2-CNF formulas described in the proof of Theorem 4).

The composition of them will give us a strong many-one reduction from 3-SAT with parameter m to MAX $(2,12)$-SAT with parameter m. Let us describe this reduction in more details.

Let F be a 3-CNF formula with m clauses. First, we transform it into a $(3,3)$-formula F' as above. If this formula F' contains clauses of length 1, we get rid of them applying the unit clause elimination rule. Then we get rid of clauses of length 2 as follows: for each clause $a \vee b$, we introduce a new variable x and replace $a \vee b$ by two clauses $a \vee b \vee x$ and $a \vee b \vee \overline{x}$. Let F'' denote the resulting formula. Obviously, we have the following:

- F'' consists of $\mathcal{O}(m)$ clauses of length 3;
- every variable in F'' appears in at most 3 clauses;
- F'' is satisfiable if and only if F is satisfiable.

The second step is to transform F'' into a 2-CNF formula φ using the same method as in the proof of Theorem 4. For each clause $a \vee b \vee c$ in F'', we introduce seven new variables $y_1, \ldots, y_7$ and replace $a \vee b \vee c$ by 16 new clauses:

$$
\begin{array}{lll}
a \vee y_1, \ \overline{y}_1, & b \vee y_2, \ \overline{y}_2, & c \vee y_3, \ \overline{y}_3, \\
\overline{a} \vee y_4, \ b \vee \overline{y}_4, & \overline{a} \vee y_5, \ b \vee \overline{y}_5, & \overline{b} \vee y_6, \ b \vee \overline{y}_6, \\
a \vee y_7, \ b \vee y_7, & c \vee y_7, \ \overline{y}_7. &
\end{array}
$$

The resulting formula φ has the following properties:

- φ consists of $16m''$ clauses of length at most 2, where m'' is the number of clauses in F'';
- every variable in φ appears in at most 12 clauses (any variable has at most four occurrences in a group of 16 clauses, and it may appear in at most three such groups);
- φ has an assignment that satisfies at least $13m''$ clauses if and only if F'' is satisfiable (see the proof of Theorem 4).

Thus, the transformation of F into φ is a strong many-one reduction from 3-SAT with parameter m to MAX $(2,12)$-SAT with parameter m. Hence we have

$$(\text{3-SAT}, m) \notin \mathbf{SE} \quad \Rightarrow \quad (\text{MAX } (2,12)\text{-SAT}, m) \notin \mathbf{SE}.$$

Since (3-SAT, n) is SERF-reducible to (3-SAT, m) [7], we have another implication:

$$(\text{3-SAT}, n) \notin \mathbf{SE} \quad \Rightarrow \quad (\text{3-SAT}, m) \notin \mathbf{SE}$$

where the antecedent is ETH. Therefore, if ETH holds then MAX $(2,12)$-SAT with parameter m is not in $\mathbf{SE}$. ∎

6 Open question

One of the most important results proved in [7] is the following implication:

(10) $(k\text{-SAT}, m) \in \mathbf{SE} \Rightarrow (k\text{-SAT}, n) \in \mathbf{SE}$.

Does a similar implication hold for MAX k-SAT:

(11) $(\text{MAX } k\text{-SAT}, m) \in \mathbf{SE} \Rightarrow (\text{MAX } k\text{-SAT}, n) \in \mathbf{SE}$?

The proof of (10) is based on the sparsification procedure that reduces satisfiability of a k-CNF formula over n variables to satisfiability of the disjunction of at most $2^{\varepsilon n}$ sparse k-CNF formulas, where ε can be arbitrarily small. A possible approach to answering (11) could be to extend the sparsification procedure to the case of maximum satisfiability. It seems that the most difficult point for such an extension is the subsumption step (the branching step can be easily extended to the case of weighted maximum satisfiability).

BIBLIOGRAPHY

[1] T. Brueggemann and W. Kern. An improved local search algorithm for 3-SAT. *Theoretical Computer Science*, 329(1–3):303–313, December 2004.

[2] C. Calabro, R. Impagliazzo, and R. Paturi. A duality between clause width and clause density for SAT. In *Proceedings of the 21st Annual IEEE Conference on Computational Complexity, CCC 2006*, pages 252–260. IEEE Computer Society, 2006.

[3] E. Dantsin, A. Goerdt, E. A. Hirsch, R. Kannan, J. Kleinberg, C. Papadimitriou, P. Raghavan, and U. Schöning. A deterministic $(2 - 2/(k + 1))^n$ algorithm for k-SAT based on local search. *Theoretical Computer Science*, 289(1):69–83, October 2002.

[4] E. Dantsin and E. A. Hirsch. Worst-case upper bounds. In *Handbook of Satisfiability*, chapter 12, pages 403–424. IOS Press, 2009.

[5] M. R. Garey, D. S. Johnson, and L. J. Stockmeyer. Some simplified NP-complete graph problems. *Theoretical Computer Science*, 1(3):237–267, 1976.

[6] R. Impagliazzo and R. Paturi. On the complexity of k-SAT. *Journal of Computer and System Sciences*, 62(2):367–375, 2001.

[7] R. Impagliazzo, R. Paturi, and F. Zane. Which problems have strongly exponential complexity. *Journal of Computer and System Sciences*, 63(4):512–530, 2001.

[8] Ph. G. Kolaitis and M. Y. Vardi. The decision problem for the probabilities of higher-order properties. In *Proceedings of the 19th Annual ACM Symposium on Theory of Computing, STOC 1987*, pages 425–435. ACM, 1987.

[9] C. H. Papadimitriou. *Computational Complexity*. Addison-Wesley, 1994.

[10] R. Paturi, P. Pudlák, M. E. Saks, and F. Zane. An improved exponential-time algorithm for k-SAT. *Journal of the ACM*, 52(3):337–364, May 2005.

[11] D. Rolf. Improved bound for the PPSZ/Schöning algorithm for 3-SAT. *Journal on Satisfiability, Boolean Modeling and Computation*, 1:111–122, November 2006.

[12] V. Raman, B. Ravikumar, and S. Srinivasa Rao. A simplified NP-complete MAXSAT problem. *Information Processing Letters*, 65(1):1–6, 1998.

[13] U. Schöning. A probabilistic algorithm for k-SAT based on limited local search and restart. *Algorithmica*, 32(4):615–623, 2002.

[14] H. Vollmer. *Introduction to Circuit Complexity: A Uniform Approach*. Springer, 1999.

On constructive semantics of natural language

ALEXANDER DIKOVSKY

1 Introduction

The traditional approach to formal semantics of Natural Language is analytical in the sense that it represents the *hearer's stance*, in order to be uniform with 'Speaker's stance'. However, to interpret the discourse, the hearer resolves complex extralinguistic problems, such as scene analysis, events sequencing, propositional attitude analysis, co-reference resolution, etc. using complex extralinguistic ontological knowledge. So from this stance it is very difficult, if possible, to separate the linguistic faculty in the strict sense from the rational agent's faculties. This partly explains why the traditional formal NL semantics are defined in terms of truth in models. Even the underspecified NL semantics (cf. [2, 3, 15]) only postpone the truth analysis by separating non-compositional noun phrase scoping analysis and predication-argument analysis. On the contrary, the *Speaker's stance* is strictly linguistic. It consists in introducing objects, stating facts about them and relating them to the objects previously introduced. The speaker's task is to *express*, not to *interpret*. Let us see an example. *Currently, insurers can increase premiums by (levying surcharges if they determine (a driver)$_{\Downarrow_x}$ is more than 50 percent to blame for a collision)$_{\Downarrow_e}$. (Such penalties)$_{\Downarrow_p\ (e\ \in\ p)}$ often cost $0_{\Uparrow_x}$ hundreds of dollars annually for up to six years. (About half of (the 50,000 cases disputed each year)$_{\Downarrow_c\ (c\ \sim\ p)})_{\Downarrow_{c_h}}$ part$\widetilde{_{0.5}}(c_h, c)$ are overturned by the appeals board. (Those drivers)$_{\Downarrow_d\ \text{of}\ -\ \text{concern}(d, c_h)}$ are issued refunds.* [The Boston Globe, March 2, 2009].

Here are tagged the constituents describing entities and events related between them within this discourse. Suppose that $\Downarrow_x$ in *(a driver)$_{\Downarrow_x}$* means something like: "a new semantic object x will identify the entity denoted by the selected occurrence of *a driver* in the discourse", that $\Uparrow_x$ is the object identified by x and that x $\sim$ y means that x and y identify the same entity. Then e identifies the act of *levying*, which is a special case of the *penalties* p that cost much to the *drivers* x (elided in the text). Further, c are the *disputed cases*, c $\sim$ p states that c identifies the same object as p and c_h identifies *about half of them..overturned...* Finally, d are the *drivers* concerned with the cases c_h.

From the Speaker's stance these facts and referential relations are *given*. The Speaker, as a linguistic agent, should only express them. At that, "to express" means two tasks: the first being to implement the facts and the relations by sentences, the other being to relate them through co-reference with the entities and the relations introduced in the preceding discourse (i.e. to construct/update the discourse model). Of course, model updates may introduce inconsistencies: a fact $p(a)$ in the model for the preceding discourse will contradict the new fact $\neg p(b)$ if the co-reference $a \sim b$ is also added. But the task of interpretation of the discourse, in particular of checking consistency of the new model with respect to the models of the preceding propositions, is the task of a rational agent which represents a part of Hearer's stance.

This restriction of formal Speaker's stance NL semantics to models' *construction / updates* permits one to significantly simplify both, the basic features of the language of semantical expressions, and the logical means needed for the definition of such semantics. In particular, there is no more need for object variables: all entities and events may be denoted by *constants* (cf. the conventional dynamic logical discourse semantics, such as DRT [8, 14], where the objects are denoted by reference variables because an update consists in finding an object assignment making new facts compatible with the original model). The other implicit simplification is that the models for this semantics are *finite*. Indeed, on the one hand, they contain only the facts and the co-reference relations of concern for the objects "mentioned" in finite discourses and, on the other hand, the Open World Assumption is no more needed when consistency is omitted. We avoid using the term "model" for such possibly inconsistent finite relational structures and use the term *context* instead. By the way, the consistency check becomes a polynomial time procedure. In this way, we arrive at a formal semantics which is *constructive* in the sense that the semantical operator applied to a succeeding expression in the sequence of semantical expressions defining the meaning of a discourse *updates* the finite contexts of the preceding expressions. One may hope that due to the Speaker's stance restrictions this constructive semantics is monotonic. On the other hand, the language of this semantics should possess features permitting us to explain different kinds of determiners, quantifier words, plurality markers, etc. and the semantics should systematically compute new facts, objects and relations between them from these markers. So the new semantics should also be constructive in the logical sense of the term.

Our general goal is to define the constructive Speaker's stance semantics in a completely compositional way and to show that it is monotonic and can be implemented in polynomial time. In the restricted limits of this paper we cannot elaborate upon the dynamic aspect of this semantics so we dwell on its language and its definition in a given context (i.e. on its *static semantics*).

2 Discourse Plans

Semantic expressions of our semantics, called *discourse plans* (DP), have previously been explained in our papers [5, 4]. Below, a discourse is seen as a sequence of DP. The complete syntax of DP is defined together with their static semantics. In this section, we introduce, comment and illustrate their main features.

Predication modulo diatheses. Following the tradition going back at least to Aristotle the semantics of adjectives and of nouns is expressed through properties. Following the standard representation in first order logic, the predication, i.e. the semantics of verbs is expressed through predicates. For instance, the meaning of *John opened the door with the new key* might be represented by $\exists x \exists y (key(x) \wedge new(x) \wedge door(y) \wedge open(j,y,x))$. It is not as simple to explain the meaning of adverbs (cf. *John easily opened the door with the new key*) because to do this requires higher order predicates. But the fundamental problem is elsewhere. As early as in 1879, G. Frege [6] remarked that in the sentences *Bei Platae siegten die Griechen über die Perser* (*By Plataea the Greeks vanquished the Persians*) and *By Platae wurden die Perser von den Griechen besiegt* (*By Plataea the Persians were vanquished by the Greeks*) the verb *siegen* (*vanquish*) in active voice (*siegten*) and in passive voice (*wurden .. besiegt*) express the same predication. It may seem that the problem of semantic representation of the predication invariant of voice variations is purely technical, because it is resolved by the permutation of arguments (in Lambda-notation: $\lambda xy.siegen$ represents *siegten* and $\lambda yx.siegen$ represents *wurden .. besiegt*). This might work more or less, if it were not for two important points: (1) the right interpretation of nouns in argument positions of verbs depends on a preposition/preposition absence rather than on the position number ($siegen(0{:}x,0{:}y)$ is equivalent to $siegen(0{:}y,\text{VON}{:}x)$); (2) in both voice forms the first argument is *topical* (the case in point) and the second is *focal* (i.e. related with the topical one through the predication of the verb). In linguistic theories of verbal semantics, in the place of language dependent features, such as prepositions or cases, so called *thematic roles* (or just *roles*), e.g., SUBJECT, OBJECT, AGENT, INSTRUMENT are used (see, e.g., [18] for more details). So the adequate functional type of a verb form should not only specify the right argument and value types, but also specify the arguments' roles. For instance, the type of the active form of *siegen* is $t_1 = (S|\mathbf{A}^n O|\mathbf{P}^n \to s)$ and the type of its passive form is $t_2 = (\text{OBJ}^n \text{AGT}^n \to s)$, where $S|\mathbf{A}$ is the general role "SUBJECT or AGENT", $O|\mathbf{P}$ is "OBJECT or PATIENT", OBJ is OBJECT, AGT is AGENT, $\mathbf{n}$ is the type of nominal meanings and $\mathbf{s}$ is the type of sentential meanings. Assigning to the verb *siegen* type t_1, one states that applying the meaning of this verb to the meaning of a noun in the $S|\mathbf{A}$-position and to the meaning of a noun in the $O|\mathbf{P}$-position we obtain a sentential meaning (similar for t_2). In modern linguistic semantical theories such featured types are called *diatheses* (see [12, 13] for discussion and details). In particular, t_1 is the canonical active diathesis of *siegen*, whereas t_2 is its passive diathesis. Now, the abovementioned Frege's remark applied to these diatheses means that the adequate semantics of *siegen* must be invariant with respect to specific diathesis choices. Needless to say, different diatheses of the same verb are differently implemented by the Speaker (in linguistic terms: *have different surface forms*). At the same time, they specify different points of view of the Speaker on the same event. In particular, in the cited example of Frege, the active diathesis t_1 represents the view with Greeks as the topic and Persians as the focus, whereas the passive diathesis represents the inverse view of the same event. The point (2) above may serve to provide a systematic representation of the set of a verb's diatheses as derivatives of one of them, chosen as canonical. Viz., one may identify verbal positions by the corresponding view and derive every diathesis from the canonical one by assigning to every argument a new view and eventually a new role. This transformation is called *diathetic shift*. In the place of "view", we will use a more specific term *communicative rank* (or just *rank*). E.g., denoting by $\vec{\mathsf{T}}$ the rank of the topical argument, by $\odot$ the rank of the focal argument and by $\oplus$ that of the third (*background*) argument, if any, we can describe the derivation of *siegen* to the passive form as the diathetic shift $\text{OBJ} \hookleftarrow \text{O}|\mathbf{P}_{\vec{\mathsf{T}}}$, $\text{AGT} \hookleftarrow \text{S}|\mathbf{A}_{\odot}$, which means that the $\text{O}|\mathbf{P}$-argument (focal in the canonical diathesis) is moved through assignment of the new rank $\vec{\mathsf{T}}$ to the topical (first) position and obtains the new role OBJ, whereas the $\text{S}|\mathbf{A}$-argument moves to the (second) focal position and obtains the new role AGT. In these terms, the semantics of a verb must be invariant with respect to diathetic shifts. The next example shows that this semantics is not trivial. In the sentence *John opened the door with the new key* the verb *open* has the canonical diathesis $t_3 = (\text{S}|\mathbf{A}^{\mathbf{n_a}} \text{O}|\mathbf{P}^{\mathbf{n}} \text{INS}^{\mathbf{n}} \to \mathbf{s_{eff}})$ in which $\mathbf{n_a}$, the type of animated nominals, is a kind of nominal types, INS is the instrumental role whose canonical rank $\oplus$ corresponds to the third argument, and $\mathbf{s_{eff}}$, the sentential type of caused effect, is a kind of sentential types. This verb has the diathesis $t_4 = (\text{AGT}^{\mathbf{n}} \text{OBJ}^{\mathbf{n}} \to \mathbf{s_{eff}})$ in the sentence *The new key opened the door*. Comparing t_3 and t_4, we see that the $\text{S}|\mathbf{A}$-argument is absent in t_4, i.e. it is deleted by the corresponding diathetic shift. In order to delete an argument, we will assign to it the *peripheral* rank $\ominus$. So t_4 is derived from t_3 using the diathetic shift $\emptyset \hookleftarrow \text{S}|\mathbf{A}_{\ominus}$, $\text{OBJ} \hookleftarrow \text{O}|\mathbf{P}_{\odot}$, $\text{AGT} \hookleftarrow \text{INS}_{\vec{\mathsf{T}}}$. In general, such diathetic shifts are caused at the surface level by change of voice, nominalization, conversion to infinitive, etc.

DP as feature trees. Now we can state that DP are feature typed terms (generally called *feature trees*), i.e. functional terms in which the arguments of a functor $f^{\mathbf{t}}$ are identified in its type $\mathbf{t}$ by pairwise different sorts of arguments. The abovementioned roles are such sorts. Here is how feature trees are defined.

DEFINITION 1. Let $\mathbf{S}$ be a set of *sorts* and $\mathbf{T}$ be a set of *primitive types* with a partial *genericity* order $\preceq$ on it (intuitively, $\mathbf{u} \preceq \mathbf{v}$ means $\mathbf{u}$ is a kind of $\mathbf{v}$). The expressions $\mathbf{t} = (S_1^{\mathbf{u_1}} \ldots S_n^{\mathbf{u_n}} \to \mathbf{v})$, where $n \geq 0$, $\mathbf{v}, \mathbf{u_1}, \ldots, \mathbf{u_n} \in \mathbf{T}$ and $S_1, \ldots, S_n$ are pairwise different sorts, are *functional* (or *composite*) types. If $n = 0$, then $\mathbf{t} = \mathbf{v}$. Let $\mathbf{F} = \{f^{\mathbf{t}}/n \mid n \geq 0\}$ be a set of typed functors such that the type $\mathbf{t}$ has n argument subtypes (one functor may have several types). Finally, let D be a set of labels we will call *determiners*.
(i) Every expression $d\ f^{\mathbf{v'}}/0$ is a (*determined*) feature tree (*f-tree*) of type $\mathbf{v}$, if $\mathbf{v} \succeq \mathbf{v'}$ and $d \in D$.
(ii) If $t_1^{\mathbf{u'_1}}, \ldots, t_n^{\mathbf{u'_n}}$ are f-trees of primitive types $\mathbf{u'_1}, \ldots, \mathbf{u'_n}$, $d \in D$, $f^{\mathbf{t}}/n \in \mathbf{F}$, $\mathbf{t} = (S_1^{\mathbf{u_1}} \ldots S_n^{\mathbf{u_n}} \to \mathbf{v'})$, $\mathbf{v'} \preceq \mathbf{v}$ and $\mathbf{u'_1} \preceq \mathbf{u_1}, \ldots, \mathbf{u'_n} \preceq \mathbf{u_n}$, then the term $d\ f(S_1 : t_1, \ldots, S_n : t_n)$ is an f-tree of type $\mathbf{v}$. □

In order to distinguish the sentences, the verbs, the nouns, the adjectives / adverbs from the semantical objects realizing them, we will use for the latter different terms: *sententials, verbals, nominals, attributors*. We will use the term *semanteme* for DP constants representing words. The next example shows some types and semantemes of these types used in DP.

EXAMPLE 2. **Some nominal types and nominals.** $\mathbf{n}$ (*nominals*), $\mathbf{n_a} \preceq \mathbf{n}$ (*animated nominals*), $\mathbf{n_{count}} \preceq \mathbf{n}$ (*countable nominals*), $\mathbf{u_{ncount}} \preceq \mathbf{n}$ (*uncountable nominals*) are examples of nominal types. $\mathbf{nct} = (\text{STATE}^{\mathbf{a_{grad}}} \to$

n_{ncount}) (cf. *hot milk*), **ctr** = (CONTENTS$^{n_{ncount}}$ FULLNESS$^{a_{grad}}$QUANT$^{a_{card}}$ → $n_{container}$) (cf. *two full glasses of beer*) are compound nominal types. MILK^{nct}, SAND^{nct} are nominals of type **nct**. GLASS^{ctr}, PACK^{ctr} are nominals of type **ctr**.

Some attributor types: **a** (*attributors*), $\mathbf{a_{grad}} \preceq \mathbf{a}$ (*gradable attributors*, cf. RED$^{a_{grad}}$, FAST$^{a_{grad}}$), $\mathbf{a_{degr}} \preceq \mathbf{a}$ (*degree attributors*, cf. VERY$^{a_{degr}}$, A_BIT$^{a_{degr}}$), $\mathbf{a_{ord}} \preceq \mathbf{a}$ (*ordinal attributors*, cf. FIRST$^{a_{ord}}$) $\mathbf{a_{card}} \preceq \mathbf{a}$ (*cardinal attributors*, cf. FIVE$^{a_{card}}$, MANY$^{a_{card}}$), $\mathbf{a_{prec}} \preceq \mathbf{a}$ (*precision attributors*, cf. ABOUT$^{a_{prec}}$, NEARLY$^{a_{prec}}$).

Verbals have types ($\phi \to \mathbf{s'}$), where $\mathbf{s'} \preceq \mathbf{s}$ and $\mathbf{s}$ is the *sentential type*. The sorts in ϕ identifying their arguments are divided into *roles* R and *attributes* A. The roles identify the *core* arguments, the attributes identify *circumstantials* and *propositional parameters* denoted PP (see examples below). If a verbal V has several types: $types(V) = \{\mathbf{t_0}, \ldots, \mathbf{t_p}\}$, we call them *diatheses* of V. One of the diatheses, $\mathbf{t_0}$, is selected as *canonical*. Other diatheses are the result of transformations of $\mathbf{t_0}$, called *diathetic shifts*. The diathetic shifts concern only core arguments. Below, representing diatheses, we will denote the roles by R and the attributes by A.

DEFINITION 3. Let $\mathbf{t_0} = (R_1^{\mathbf{u_1}} \ldots R_n^{\mathbf{u_n}} A_1^{\mathbf{v_1}} \ldots A_m^{\mathbf{v_m}} \to \mathbf{v})$ be the canonical diathesis of V and $\mathbf{t_i} = ((R')_1^{\mathbf{u'_1}} \ldots (R')_k^{\mathbf{u'_k}} A_1^{\mathbf{v_1}} \ldots A_m^{\mathbf{v_m}} \to \mathbf{v'}) \in types(V)$ be some other diathesis of V. Then $D_i = (\mathbf{t_i}, \mathbf{d}_i)$ is a *diathetic shift* of $\mathbf{t_0}$ if $\mathbf{d}_i : \{1, \ldots, k\} \stackrel{1-1}{\to} \{i_1 \ldots, i_k\}$, for $1 \leq i_1 < \ldots < i_k \leq n$, is a bijection preserving types: $\mathbf{u'_j} = \mathbf{u}_{i_j}$, $1 \leq j \leq k$. We call this bijection *argument shift* and denote it by $\mathbf{d}_i : k \stackrel{1-1}{\mapsto} n$. $V[\mathbf{d}_i]^{\mathbf{t_i}} =_{df} V^{\mathbf{t_i}}$ is a *derivative* of the canonical form $V^{\mathbf{t_0}}$ through D_i.

EXAMPLE 4. For the verbal OPEN, the argument shift $\{1 \mapsto 2, 2 \mapsto 1, 3 \mapsto 3\}$ corresponds to its diathesis of passive as in the sentence *The door was opened with the key by John's girl-friend* and $\{3 \mapsto 1, 2 \mapsto 2\}$ corresponds to its argument alternation diathesis as in the sentence *The key easily opened the door*.

In the definition of static semantics of DP we will use this representation of verbal derivatives through argument shift and resulting diathesis. Clearly, it can be extracted from the representation of diathetic shifts through rank assignments we showed above. We use the latter representation in examples of DP. In particular, in Fig. 1 we show a DP of the sentence *The new key easily opened the door*. In this figure, the DP is presented in a graphical form where solid lines labeled with roles link verbals to their core arguments and dashed lines labeled with attributes link semantemes to their attributor type arguments: circumstantials for verbals, qualifiers for nominals (e.g., $NEW^{a_{grad}}$ represents the value of attribute STATE of KEY^n). There is also a group of propositional parameters. We don't include them in verbal types because these attributes are common to all verbals. In particular, among the PP-attributes shown in Fig. 1 there are PSTATUS (declarative in Fig. 1), SIGN (*positive*), EVEXTENT, a generalized aspect (*pointwise* • in Fig. 1, *continuous interval* (__)

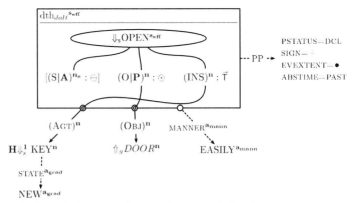

Figure 1. DP of *The new key easily opened the door*

in Fig. 2) and time parameter ABSTIME (PAST in Fig. 1, PRES in Fig. 2). There are other PP we don't show in this figure (cf. RELTIME (*relative time*) ($\|t$), simultaneous with t, in Fig. 2).

Diathetic shifts are represented as assignments to core arguments of new roles and communicative ranks: the canonical assignments are shown inside the box, the new assignments determining the type of the verbal derivative are shown outside the box.

Expressions $\Downarrow_s$, $\mathbf{H}\Downarrow_x^1$, $\Uparrow_y$ are *determiners*. Intuitively, they create or allow access to objects o which have a set value $\|o\|$ called *extension* of o. The most general form of a determiner defined below is $\mathbf{D}_x = \mathbf{QC}_x u$, where x is a *global reference*, u is a *local reference*, $\mathbf{C}_x$ is either one of expressions $\Downarrow_x^k$, $\Downarrow_{xry}^k$ (*objectification operators*), or $\Uparrow_x$ (*access operator*), and $\mathbf{Q}$ is one of two *modes of access*: $\mathbf{I}$ (*individual*) and $\mathbf{H}$ (*holistic*). Intuitively, the objectification operators $\Downarrow_x^k$, $\Downarrow_{xry}^k$ create a new object o with extension $\|o\|$ whose cardinality is bounded by k (a number or ω) and bind the global reference x and the local reference u with this object. Besides this, the latter operator relates the object x with a previously created object y through binary relation $\mathbf{r}$: xry. Global and local references differ in scopes: u is visible only in the subplan (i.e. subterm) determined by $\mathbf{D}_x$, whereas x is visible in the discourse starting from this subplan. The access operator $\Uparrow_x$ applies to a global reference x previously bound with an object o by a determiner $\mathbf{D}_x$. Its value depends on the mode of access $\mathbf{Q}$ in $\mathbf{D}_x$: if $\mathbf{Q} = \mathbf{H}$ (holistic access) then the value is $\{o\}$, otherwise it is $\|o\|$ (individual access).

The last feature of DP not yet mentioned is the operator ι whose intuitive reading is "*such object .. that ..*" and which is used to express the meaning of relative and comparative clauses. In Fig. 2 we show the DP of a variant of "donkey sentences" borrowed from [9], which uses this operator. In this example, we see the use of *absolute* individual access determiners (cf. $(\mathbf{I}\Downarrow_{x_f} u_f)\text{FARMER}^{na}$), i.e. those which create objects without relating them with other objects. In Fig. 3 is shown a DP using *relativized* determiners, which create objects and relate them with previously created objects through the inclusion relation $\overset{?}{\in}$ (cf. $(\mathbf{H}\Downarrow^1_{y_f \overset{?}{\in} x_f} u'_f)(\text{Bob}^{na})_{sh})$. Intuitively, this

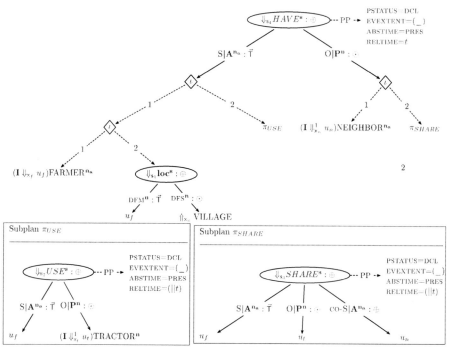

Figure 2. A DP of *Every farmer in the village, who uses a tractor, has a neighbor with whom he shares it.*

determiner states that the object o_B, a realization of the *shifter* constant [1] $(Bob^{na})_{sh}$, belongs to the extension of the object o_f, a realization of the semanteme FARMER^{na} in DP in Fig. 2.

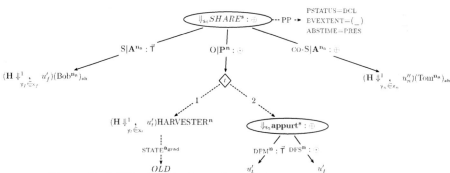

Figure 3. A DP of *Bob shares his old harvester with Tom.*

Before we proceed to the definition of the semantics of DP, we will note two of their special properties which shouldn't pass unnoticed.

Determiners vs. quantifiers. We see that in the DP in our examples the determiners are used in the place of quantifiers. Great difficulties created by

---
[1]The term proposed by R. Jakobson for the semantemes, such as proper names, whose meaning is completely determined by the context.

quantifiers for compositional semantics definitions, after several decades of debates, have led the semanticians to the pragmatic solution to leave out from semantics the choice of quantifier scope, i.e. to underspecify semantics. The DP determiners avoid this problem due to the joint use of local and global references. The scope of the former, similar to that of quantifiers, is limited by the determined subplan, whereas the scope of the latter extends to the rest of the discourse. Importantly, the *dynamic* semantics of determiners (not included in this paper) is defined so that at the moment, when in a DP $\pi_2 = \mathbf{D}_y \pi'_2$, determined by $\mathbf{D}_y = (\mathbf{Q} \Downarrow^k_{y \in x}, u_2)$, the object y is added to the extension of the object x created in a preceding DP $\pi_1 = \mathbf{D}_x \pi'_1$, determined by $\mathbf{D}_x = (\mathbf{I} \Downarrow^l_x u_1)$, the local reference u_1 of $\mathbf{D}_x$ becomes bound by $\{y\}$. In this way, in the new context, y becomes available in DP π_1.

Attributes vs. properties. As we have noted above, in conventional logical semantics the meaning of the adjectives is expressed using properties of entities. This choice creates semantical problems. One of them was the subject of long and substantial discussions in NL semantics. For instance, in the preceding sentence, the adjectives *long* and *substantial* are related to two different components of the meaning of *discussions*, one *temporal* (a physical characteristics of a process), another *cognitive*. Cf. another example: *his favorite book stood first on the shelf*. Here *first* is related to the position of the *book*, whereas *favorite* is related to its literary qualities. This phenomenon called *copredication* has led to the emergence of formal lexical semantics [16, 1]. It creates a problem of compositional representation of meaning of nouns, providing a correspondence between components of this meaning and the meaning of adjectives modifying the nouns. In DP semantics, the adjectives are interpreted not as properties, but as values of *attributes*. One can see that in DP the subplans representing adjective phrases are in the attribute argument positions of nominals (see the DP in Fig. 4) and the subplans representing adverb phrases are in attribute argument positions of verbals.

Figure 4. A DP of *about three nearly full glasses of hot milk*.

Below we define the semantics of the attributes as functions on objects. This decision resolves the problem of copredication. Viz., there is no need to stratify the meaning of nominals because different functional attributes may express different components of a primitive nominal meaning: STATE(MILK) = COLD, ORIGIN(MILK) = COW'S. Moreover, in this way the semantic function-argument dependency has the same direction as the syntactic dependency (from noun to adjective), not the opposite, as in conventional logical semantics.

3 Fundamentals of DP Semantics

Dynamic and *static* DP semantics are defined in a subset of set theory extended with specific constants: $LC = \{L_W \mid W$ is a semanteme$\}$ (*lexical class constants*), $R_\mathbf{g}$ (*global object references*), $R_\mathbf{l}$ (*local object references*), $\mathbf{O^t}$ (*objects of type* $\mathbf{t}$, $\mathbf{O}$ is the union of all $\mathbf{O^t}$), attribute names and some other constants.
Contexts. Both semantics are relative to finite relational structures: the former to *dynamic contexts* (*d-contexts*), the latter to *static contexts* (*s-contexts*).

DEFINITION 5. A *d-context* is a finite structure $\Sigma = (D, I)$, where D is a finite collection of sets and I is a finite function from D into D with four particular restrictions[2]: $\gamma_\Sigma = I{\restriction}R_\mathbf{g}$ (*global assignment*), $\lambda_\Sigma = I{\restriction}R_\mathbf{l}$ (*local assignment*), $\theta_\Sigma = I{\restriction}\mathbf{O^n}$ (*nominal objects' evaluation*) and $\hbar_\Sigma = I{\restriction}LC$ (*horizon line* of Σ). $\sigma = \langle \gamma_\Sigma, \lambda_\Sigma, \theta_\Sigma, \hbar_\Sigma \rangle$ is the *s-context corresponding to* Σ.

$\gamma_\Sigma(\mathsf{x}) = o$ means that the global reference x is bound with the object o, $\lambda_\Sigma(u) = s$ means that the local reference u is bound with the set s, $\theta_\Sigma(o) = s$ means that the nominal type object o has d-extension $|o|^\Sigma = s$ and $\hbar_\Sigma(L_W) = s$ means that s is the part of the d-extension of L_W accessible in Σ.

Lexical semantics. DP semantics rests upon a set of lexical axioms. The axioms relate a set of functions (*attributes*) with every lexical class constant L_W representing a DP semanteme W. For space reasons, we replace the axioms by simplified postulates. The constant $\bot \notin \mathbf{O}$ used in them is an "*uncertain value*".

For $\mathbf{u} \in \mathbf{T}$, $LEX(\mathbf{u})$ denotes the set of all DP semantemes of types $(\phi \to \mathbf{u})$ or $\mathbf{u}$. We suppose that every semanteme W has a unique set code $W^\star$.

POSTULATE 6. $L_\mathbf{u} \subseteq \mathbf{O^u}$. $\mathbf{u} \preceq \mathbf{v}$ iff $L_\mathbf{u} \subseteq L_\mathbf{v}$. $L_W \subseteq L_\mathbf{u}$ for $W \in LEX(\mathbf{u})$.

POSTULATE 7. The extension $\|o\|$ of an attributor type object $o \in \mathbf{O^u}$, $\mathbf{u} \preceq \mathbf{a}$, is a semanteme code: $\|o\| \in \{W^\star \mid W \in LEX(\mathbf{u})\}$.

POSTULATE 8. For all semantemes $W \in LEX(\mathbf{u})$, the classes L_W have the same attributes. The set of these attributes is denoted $Att(\mathbf{u})$.

POSTULATE 9. For every attribute $\mathrm{A} \in Att(\mathbf{u})$, there is an attributor type $\mathbf{v} \preceq \mathbf{a}$ such that: $\mathrm{A} : \mathbf{O^u} \to (\mathbf{O^v} \cup \{\bot\})$, if $Att(\mathbf{v}) \neq \emptyset$, and $\mathrm{A} : \mathbf{O^u} \to (\{W^\star \mid W \in LEX(\mathbf{v})\} \cup \{\bot\})$, if $Att(\mathbf{v}) = \emptyset$. This function (denoted $\mathrm{A}^\mathbf{v}$) is finite in the sense that $\{o \mid \mathrm{A}(o) \neq \bot\}$ is a finite set. $\mathbf{v}$ is its *value type*. For attributor type objects $o \in \mathbf{O^u}$, $\mathbf{u} \preceq \mathbf{a}$, and for their attributes $\mathrm{A}^\mathbf{v} \in Att(\mathbf{u})$, $o.\mathrm{A} =_{df} \|\mathrm{A}(o)\|$ if $\mathrm{A}(o) \neq \bot$.

If $Att(\mathbf{u}) = \{\mathrm{A}_1^{\mathbf{v}_1}, \ldots, \mathrm{A}_m^{\mathbf{v}_m}\}$, then the types in $DT(\mathbf{u}) =_{df} \{\mathbf{v}_1, \ldots, \mathbf{v}_m\}$ are *immediately dependent* on $\mathbf{u}$ (denoted $\mathbf{v}_i \triangleleft^{imm} \mathbf{u}$). Let $\trianglelefteq$ (*lexical dependency*) be the reflexive-transitive closure of $\triangleleft^{imm}$. So $\mathbf{v} \trianglelefteq \mathbf{u}$ implies $\mathbf{v} \preceq \mathbf{a}$.

POSTULATE 10. $\trianglelefteq$ is a partial order.

There are also axioms defining a hierarchy of primitive types. We only cite

---
[2] $f{\restriction}S = \{\langle x_1, x_2 \rangle \in f \mid x_1 \in S\}$.

two of their consequences and give several examples.
1. There are three $\preceq$-maximal (i.e. most general) types: **s**, **n** and **a**.
2. $\mathbf{a_{degr}}, \mathbf{a_{prec}}$ are $\trianglelefteq$-minimal attributor types: $DT(\mathbf{a_{degr}}) = DT(\mathbf{a_{prec}}) = \emptyset$.

EXAMPLE 11. The types in Example 2 have the following properties.
$DT(\mathbf{n_{ncount}}) = \{\mathbf{a_{grad}}\}$. $\mathbf{a_{grad}} \in DT(\mathbf{n}) \cap DT(\mathbf{s})$ and $DT(\mathbf{a_{grad}}) = \{\mathbf{a_{degr}}\}$. $\mathbf{a_{ord}} \in DT(\mathbf{n_{count}})$ and $DT(\mathbf{a_{ord}}) = \{\mathbf{a_{prec}}\}$. $\mathbf{a_{card}} \in DT(\mathbf{n_{meas}}) \cap DT(\mathbf{n_{count}})$ and $DT(\mathbf{a_{card}}) = \{\mathbf{a_{prec}}\}$.

One can see that due to these axioms with every attributor DP is uniquely related a set of constraints on values and extensions of attributes.

EXAMPLE 12. The semanteme GLASS in the DP in Fig. 4 has one core CONTENTS-argument and two attributes: $Att(\mathbf{n_{container}}) = \{\text{FULLNESS}^{\mathbf{a_{grad}}}, \text{QUANT}^{\mathbf{a_{card}}}\}$. Respectively, the system of constraints for GLASS is defined as $AC(\text{FULLNESS}^{\mathbf{a_{grad}}} = \pi_1, \text{QUANT}^{\mathbf{a_{card}}} = \pi_2) = AC(\text{FULLNESS}^{\mathbf{a_{grad}}} = \pi_1) \cup AC(\text{QUANT}^{\mathbf{a_{card}}} = \pi_2)$, where π_1 and π_2 are the two attributor subplans of this DP: FULLNESS-branch and QUANT-branch. The components are computed recursively: $AC(\text{DEGREE} = \text{NEARLY}) = \{\text{DEGREE}(o_f) = NEARLY^\star, \|o_f\| = FULL^\star\}$ for $o_f = \gamma(\mathbf{z}_f)$, $AC(\text{FULLNESS}^{\mathbf{a_{grad}}} = \pi_1) = \{\text{FULLNESS}(o) = o_f\} \cup AC(\text{DEGREE} = \text{NEARLY})$ for $o = \gamma(\mathbf{x})$. Similar for $AC(\text{QUANT}^{\mathbf{a_{card}}} = \pi_2)$.

Shifted product. DP semantics reduces all verbal's derivatives to its canonical form. For this it uses a special product allowing to relate their arguments.

DEFINITION 13. Let $n > 0$, $\mathbf{d}: k \stackrel{1-1}{\mapsto} n$ be an argument shift and $s_1, \ldots, s_n$ be a sequence of sets. The *shifted product* of this sequence (under shift $\mathbf{d}$) is:
$$\prod_{1 \leq i \leq k}^{\mathbf{d}} s_i =_{df} M_1 \times \ldots \times M_n,$$
where $M_i = s_{\mathbf{d}^{-1}(i)}$ for $i \in range(\mathbf{d})$ and $M_i = \{\bot\}$ otherwise.

4 Static Semantics

In this section we define in parallel the syntax[3] and the static semantics of DP. The correspondence between d- and s-contexts being inessential for this semantics, we fix an s-context $\sigma = \langle \Gamma, \Lambda, \Theta, H \rangle$ in which, for every DP π, will be defined its *s-extension* $\|\pi\|^\sigma$. So Γ is a global assignment, Λ is a local assignment, Θ is a nominal objects' evaluation and H is a horizon line. As we shall see, every composite subplan π of a DP is uniquely identified by a global reference x introduced by a determiner: $\pi = \mathbf{D}_\mathsf{x}\pi'$. The static semantics $\|\pi\|^\sigma$ will be defined through the extension $\|\Gamma(\mathsf{x})\|^\sigma$ of the object $\Gamma(\mathsf{x})$.

I. Primitives.
I.1. Lexical classes. For a non-attributor semanteme W, $\|L_W\|^\sigma = H(L_W)$.
I.2. Null plans (intuitively, corresponding to existentially bound arguments). For a null nominal plan $\pi =\Downarrow_\mathsf{x} 0^\mathbf{n}$ (**Ex:** *Testamentary succession*$_{\text{OBJ}:\Downarrow_\mathsf{x} 0^\mathbf{n}}$ *goes to Mary*),
$\|\pi\|^\sigma = \|\Gamma(\mathsf{x})\|^\sigma$, where $\|\Gamma(\mathsf{x})\|^\sigma = \{\bot\}$.
For a null attributor plan $\pi =\Downarrow_\mathsf{x} 0^\mathbf{a}$ (**Ex:** *happy*$_{\text{DEGREE}:0^{\mathbf{a_{degr}}}}$ *as goblin*),
$\|\pi\|^\sigma = \|\Gamma(\mathsf{x})\|^\sigma$, where $\|\Gamma(\mathsf{x})\|^\sigma = \bot$.

---
[3]Because of space limits we omit the rules of visibility of references.

I.3. Shifter plans. Let $\pi = \Downarrow_x (K^{n'})_{sh}$, where $(K^{n'})_{sh}$ is a nominal shifter constant of type n' (e.g., $(\mathbf{speaker}^{n_a})_{sh}$, $(John^{n_a})_{sh}$). Then:
$\|\pi\|^\sigma = \|\Gamma(x)\|^\sigma$, where $\|\Gamma(x)\|^\sigma = \{((K)_{sh})^*\}$.

I.4. Reference plans. Let u^t and x^t be local and global references.
$\|\pi\|^\sigma = \|\Lambda(u)\|^\sigma$ for $\pi = u$.
$\|\pi\|^\sigma = \|\pi_0\|^\sigma$ for $\pi = \Uparrow_{x^t}$ and $\pi_0 = \mathbf{D}_x \pi'$, $\mathbf{D}_x$ being a determiner.

I.5. Primitive attributor plans. $\pi = W$, where $W \in LEX(\mathbf{v})$ and $\mathbf{v} \preceq \mathbf{a}$ is a $\trianglelefteq$-minimal attributor type (e.g. $\mathbf{a}_{\mathbf{degr}}$, $\mathbf{a}_{\mathbf{prec}}$), are the only nonreferenced DP. For such DP, $\|\pi\|^\sigma = W^*$.

II. Compound DP.
Sentential plans.

II.1. Unit sentential plans. Let $\pi = \Downarrow_x V[\mathbf{d}](R_1 : \pi_1, \ldots, R_k : \pi_k, A_1 : \pi'_1, \ldots, A_m : \pi'_m)$ be a sentential DP in which $\pi'_i = \Downarrow_{x_i} \pi''_i$, $1 \leq i \leq m$, are composite attributor DP. Then:
$\|\pi\|^\sigma = \|\Gamma(x)\|^\sigma$,
$\|\Gamma(x)\|^\sigma = \prod\limits_{1 \leq i \leq k}^{\mathbf{d}} \|\pi_i\|^\sigma$.
$\Gamma(x) \in \|L_\mathbf{V}\|^\sigma$,
$A_i(\Gamma(x)) = \Gamma(x_i)$ and $\Gamma(x).A_i = \|\pi'_i\|^\sigma$, $1 \leq i \leq m$.

II.2. Coordinated sentential plans. Let $\pi = \Downarrow_x \mathcal{C}^{(n)}(\pi_1, \ldots, \pi_n)$, where $n > 1$ and $\pi_i = \Downarrow_{x_i} \pi'_i$ are unit sentential DP of sentential types $\mathbf{s_i}$, $1 \leq i \leq n$. Then:
$\|\pi\|^\sigma = \|\Gamma(x)\|^\sigma$, where $\|\Gamma(x)\|^\sigma = \langle \Gamma(x_1), \ldots, \Gamma(x_n) \rangle$,
$\|\Gamma(x_i)\|^\sigma = \|\pi'_i\|^\sigma$, $1 \leq i \leq n$.

Nominal plans.

II.3. Absolute unit determined nominal plans. Let $\pi = \mathbf{D}_x \hat{\pi}$, where $\mathbf{D}_x = (\mathbf{Q} \Downarrow^\mathbf{k}_x u)$ is a determiner in which $\mathbf{Q} \in \{\mathbf{H}, \mathbf{I}\}$, $x^{n'}$ is a global reference, u is a local reference, $\mathbf{k}$ is a number or ω, N^t is a nominal of type $\mathbf{t} = (\mathbf{s}_1^{n_1} \ldots \mathbf{s}_k^{n_k} \mathbf{A}_1^{v_1} \ldots \mathbf{A}_m^{v_m} \rightarrow \mathbf{n'})$, $\hat{\pi} = N^t(S_1 : \pi_1, \ldots, S_k : \pi_k, A_1 : \pi'_1, \ldots, A_m : \pi'_m)$ is a determinerless nominal DP, where $\mathbf{n'} \preceq \mathbf{n}$, $\pi_i = \mathbf{D}_{x_i} \hat{\pi}_i$, $1 \leq i \leq k$, are core argument nominal DP and $\pi'_j = \Downarrow_{y_j} \hat{\pi}'_j$, $1 \leq j \leq m$, are composite attributor DP (see the DP in Fig. 4 and Example 12). Then:
$\|\pi\|^\sigma = \{\Gamma(x)\}$, if $\mathbf{Q} = \mathbf{H}$, and $\|\pi\|^\sigma = \|\Gamma(x)\|^\sigma$, if $\mathbf{Q} = \mathbf{I}$,
$\|\Gamma(x)\|^\sigma = \Theta(\Gamma(x))$, $\bot \in \|\Gamma(x)\|^\sigma$ and $card(\|\Gamma(x)\|^\sigma) \leq \mathbf{k}$,
$\Gamma(x) \in \|L_N\|^\sigma$,
$S_i(\Gamma(x)) = \Gamma(x_i)$, $1 \leq i \leq k$.
$A_j(\Gamma(x)) = \Gamma(y_j)$ and $\Gamma(x).A_j = \|\pi'_j\|^\sigma$, $1 \leq j \leq m$.

II.4. Relativized unit determined nominal plans. Let $\pi = \mathbf{D}_x \pi_1$, where $\mathbf{D}_x = (\mathbf{Q} \Downarrow^\mathbf{k}_{xry} u)$ is a determiner in which $\mathbf{Q} \subset \{\mathbf{H}, \mathbf{I}\}$, $\mathbf{r} \in \{\stackrel{?}{\in}, \sim, \subset, /, \ldots\}$, π_1 is a determinerless nominal plan and y is a global object reference identifying in the preceding discourse a nominal plan $\mathbf{D}_y \pi_0$ with determiner $\mathbf{D}_y = (\mathbf{Q}_0 \Downarrow^{\mathbf{k_0}}_y u_0)$ (see the DP in Fig. 2). Then $\|\pi\|^\sigma$ is defined as in the preceding case. Besides this, the following $\mathbf{r}$-conditions also hold:
$\|\mathbf{r}\|^\sigma(\Gamma(x), \Gamma(y))$ if $\mathbf{r} \in \{\sim, \subset, /, \ldots\}$,
$\Gamma(x) \in \|\Gamma(y)\|^\sigma$, $\Lambda(u_0) = \{\Gamma(x)\}$ and $card(\|\Gamma(y)\|^\sigma) \leq \mathbf{k_0}$ if $\mathbf{r} = \stackrel{?}{\in}$.

II.5. Relative determined nominal plans. Let $\pi = \iota_R(\pi_0 \mid \hat{\pi}_0)$, where

$\pi_0 = \mathbf{D}_x \pi_0'$ is a unit determined nominal plan, u is the local reference in $\mathbf{D}_x$, R is a role and $\hat{\pi}_0 = \Downarrow_y V[\mathbf{d}](\text{R}_1 : \hat{\pi}_1, \ldots, \text{R}_i : u, \ldots, \text{R}_k : \hat{\pi}_k, \text{A}_1 : \hat{\pi}_1', \ldots, \text{A}_m : \hat{\pi}_m')$ is a sentential plan such that $\text{R}_i = \text{R}$. Let also $I_\text{R}^\sigma(\pi_0) = \{x \mid (\exists y_1, \ldots, y_n)(\langle y_1, \ldots, y_n \rangle \in \|\Gamma(\mathbf{y})\|^\sigma \ \& \ x = y_{d^{-1}(i)})\}$. Then:
$\|\Gamma(\mathbf{x})\|^\sigma = \|\pi_0\|^\sigma \cap I_\text{R}^\sigma(\hat{\pi}_0)$ and
$\|\pi\|^\sigma = \{\Gamma(\mathbf{x})\}$, if $\mathbf{Q} = \mathbf{H}$, and $\|\pi\|^\sigma = \|\Gamma(\mathbf{x})\|^\sigma$, if $\mathbf{Q} = \mathbf{I}$.
If π_0 is a relativized unit determined nominal plan (i.e. $\mathbf{D}_x = \mathbf{Q} \Downarrow_{xry}^k u$), then the r-conditions also hold.
Ex: Relative and comparative clauses.

II.6. Aggregate nominal plans. Let $\pi = \mathbf{D}_x \mathcal{A}(\pi_1, \ldots, \pi_n)$, where $\mathbf{D}_x = \mathbf{Q} \Downarrow_x^k u$ and $\pi_i = \mathbf{D}_{x_i} \pi_i'$, $1 \leq i \leq n$, are determined nominal DP. Then:
$\|\Gamma(\mathbf{x})\|^\sigma = \Theta(\Gamma(\mathbf{x}))$ and $\Gamma(\mathbf{x}_1), \ldots, \Gamma(\mathbf{x}_n) \in \|\Gamma(\mathbf{x})\|^\sigma$ if $\mathbf{k} = \omega$,
$\|\Gamma(\mathbf{x})\|^\sigma = \{\Gamma(\mathbf{x}_1), \ldots, \Gamma(\mathbf{x}_n)\}$ if $\mathbf{k} = n$,
$\|\pi\|^\sigma = \{\Gamma(\mathbf{x})\}$, if $\mathbf{Q} = \mathbf{H}$, and $\|\pi\|^\sigma = \|\Gamma(\mathbf{x})\|^\sigma$, if $\mathbf{Q} = \mathbf{I}$.
Ex: *(Students$_{\Downarrow_{x_1}}$ and professors$_{\Downarrow_{x_2}}$)$_{\Downarrow_x}$ went on strike.*

Attributor plans.

II.7. Lexicalized attributor plans. Let $\pi = \Downarrow_x W^\mathbf{t}(\text{A}_1 : \pi_1, \ldots, \text{A}_m : \pi_m)$ be a DP in which $\mathbf{t} = (\text{A}_1^{\mathbf{v}_1} \ldots \text{A}_m^{\mathbf{v}_m} \rightarrow u)$, $\mathbf{u} \preceq \mathbf{a}$, and π_i are attributor DP, $1 \leq i \leq m$. Then:
$\|\pi\|^\sigma = \|\Gamma(\mathbf{x})\|^\sigma = W^\star$,
$\text{A}_i(\Gamma(\mathbf{x})) = \Gamma(\mathbf{x}_i)$, if $\pi_i = \Downarrow_{x_i} \pi_i'$, and $\text{A}_i(\Gamma(\mathbf{x})) = \|\pi_i\|^\sigma$ otherwise, $1 \leq i \leq m$,
$\Gamma(\mathbf{x}).\text{A}_i = \|\pi_i\|^\sigma$, $1 \leq i \leq m$.
Ex: See the DP in Fig. 4 and Example 12.

II.8. Relative attributor plans. Let $\pi = \iota(\Downarrow_x 0^\mathbf{t} \mid \pi_1)$ be a relative attributor plan in which $\mathbf{u} \preceq \mathbf{a}$ is an attributor type, $\mathbf{x}^\mathbf{u}$ is a global reference and $\pi_1 = \Downarrow_y \pi_1'$ is a sentential type DP. Then:
$\|\pi\|^\sigma = \|\Gamma(\mathbf{x})\|^\sigma = \perp$,
$\|\pi_1\|^\sigma = \|\Gamma(\mathbf{y})\|^\sigma$,
$\|\mathbf{rel}\|^\sigma(\Gamma(\mathbf{x}), \Gamma(\mathbf{y}))$ for a special relation **rel**.
Ex: *He was so$_{\Downarrow_x(0^{\mathbf{ndegr}})}$ glad, that ...*

5 On Dynamic DP Semantics

Dynamic semantics is defined through translation $\lceil .. \rceil$: for a discourse $\delta = (\pi_1, \ldots, \pi_n)$, $\lceil \delta \rceil = \lceil \pi_1 \rceil \ldots \lceil \pi_n \rceil$ is a process which, when applied to a starting context Σ_0, computes *dynamic extensions* of DP: $|\delta|_{\Sigma_0} = (|\pi_1|_{\Sigma_0}^{\Sigma_1}, |\pi_2|_{\Sigma_1}^{\Sigma_2}, \ldots, |\pi_n|_{\Sigma_{n-1}}^{\Sigma_n})$ ($|\pi_i|_{\Sigma_{i-1}}^{\Sigma_i}$ is the *d-extension* of π_i in Σ_i computed starting from Σ_{i-1}). The translation and the transitions are rather technical and will be published elsewhere. Here we only illustrate it by the process corresponding to the discourse $\delta = (\pi_1, \pi_2)$ where π_1 and π_2 are DP in Fig. 2,3.

Its intermediate data are shown in Tables 1,2 with columns: Context (current d-context), GRef (global object reference identifying a subplan), OId (identity of the created object), d-Extension elements (elements added to the d-extension of the object), LRef (local object reference), LVal (current value of the local object reference), Attributes (attribute value extension) and Semanteme (the root semanteme of the subplan). This computation executes two processes: $\lceil \pi_1 \rceil = p_1$ (see Table 1) and $\lceil \pi_2 \rceil = p_2$ (see Table 2). The computation of

Context	GRef	OId	d-Extension elements	LRef	LVal	Attributes	Semanteme
Σ_0	x_v	$o_v \in \mathbf{O^n}$	$(v)^*_{sh}$				VILLAGE
Σ_1	x_f	$o_f \in \mathbf{O^{n_a}}$	$\perp$	u_f	$\{\perp\}$		FARMER
Σ_2	s_1	$o_{loc} \in \mathbf{O^s}$	$\langle \text{DFM} : \perp, \text{DFS} : o_v \rangle$				loc
Σ_3	x_t	$o_t \in \mathbf{O^n}$	$\perp$	u_t	$\{\perp\}$		TRACTOR
Σ_4	s_2	$o_u \in \mathbf{O^s}$	$\langle S|\mathbf{A} : \perp, O|\mathbf{P} : \perp \rangle$			$o_u.\text{PSTATUS} = \text{DCL}^*$, etc.	USE
Σ_5	x_n	$o_n \in \mathbf{O^{n_a}}$	$\perp$	u_n	$\{\perp\}$		NEIGHBOR
Σ_6	s_3	$o_{sh} \in \mathbf{O^s}$	$\langle S|\mathbf{A} : \perp, O|\mathbf{P} : \perp, \text{CO-S}|\mathbf{A} : \perp \rangle$			$o_u.\text{PSTATUS} = \text{DCL}^*$, etc.	SHARE
Σ_7	s_4	$o_h \in \mathbf{O^s}$	$\langle S|\mathbf{A} : \perp, O|\mathbf{P} : \perp \rangle$			$o_h.\text{PSTATUS} = \text{DCL}^*$, etc.	HAVE

Table 1. Computation for the first DP π_1.

Context	GRef	OId	d-Extension elements	LRef	LVal	Attributes	Semanteme
Σ_8	y_f	$o_B \in \mathbf{O^{n_a}}$	$(\text{Bob})^*_{sh}$	u'_f	$\{o_B\}$		$(\text{Bob})_{sh}$
	x_f	o_f	$\perp, o_B$	u_f	$\{o_B\}$		
	s_2	o_u	$\langle S|\mathbf{A} : o_B, O|\mathbf{P} : \perp \rangle$				USE
	s_4	o_h	$\langle S|\mathbf{A} : o_B, O|\mathbf{P} : o_u \rangle$				HAVE
Σ_9	y_t	$o_{hv} \in \mathbf{O^n}$	o_{hv}	u'_t	$\{o_{hv}\}$	$o_{hv}.\text{STATE} = \text{OLD}^*$	HARVESTER
	x_t	o_t	$\perp, o_{hv}$	u_t	$\{o_{hv}\}$		TRACTOR
	s_1	o_u	$\langle S|\mathbf{A} : o_B, O|\mathbf{P} : o_{hv} \rangle$				USE
Σ_{10}	s_5	$o_{appurt} \in \mathbf{O^s}$	$\langle \text{DFM} : o_{hv}, \text{DFS} : o_B \rangle$				appurt
Σ_{11}	y_n	$o_T \in \mathbf{O^{n_a}}$	$(\text{Tom})^*_{sh}$	u'_n	$\{o_T\}$		$(\text{Tom})_{sh}$
	x_n	o_n	$\perp, o_T$	u_n	$\{o_T\}$		NEIGHBOR
	s_3	o_{sh}	$\langle S|\mathbf{A} : o_B, O|\mathbf{P} : o_{hv}, \text{CO-S}|\mathbf{A} : o_T \rangle$				SHARE
Σ_{12}	s_6	o'_{sh}	$\langle S|\mathbf{A} : o_B, O|\mathbf{P} : o_{hv}, \text{CO-S}|\mathbf{A} : o_T \rangle$			$o'_{sh}.\text{PSTATUS} = \text{DCL}^*$, etc.	SHARE

Table 2. Computation corresponding to the second DP π_2.

$p_1 = \lceil \pi_1 \rceil$ (see Table 1) is started in context Σ_0 in which there is an object $o_v = \gamma_{\Sigma_0}(x_v)$ referenced by $\Uparrow_{x_v}$ (*the village*). Then, in the course of seven consecutive transitions, it creates a new object o for each subplan identified by its determiner $\mathbf{D_x}$, binds the global reference x with o and adds the object to the accessible subset $\hbar_{\Sigma_1}(L_W)$ of the lexical class L_W corresponding to the head semanteme W of the subplan. In the case where W is a verbal, the process adds new facts to the shifted product related with L_W (the element of the column "Semanteme" identifies W and the corresponding subplan). For instance, in the transition to Σ_3 the process creates an object $o_t \in \mathbf{O^n}$ for TRACTOR with the *uncertain* extension $\{\perp\}$ (cf. with the transition to Σ_9, where it creates the *certain* extension $\{o_{hv}\}$). This is explained by the difference of access modes in the two determiners: *individual* **I** for TRACTOR and *holistic* **H** for HARVESTER. This difference manifests itself in the computation of p_2 (see Table 2). Viz., due to the determiner $(\mathbf{H} \Downarrow^1_{y_f \in x_f}, u'_f)$ applied to $(\text{Bob}^{n_a})_{sh}$, this computation changes Σ_7 to Σ_8, creates $o_B = \gamma_{\Sigma_8}(y_f) \in \mathbf{O^{n_a}}$ with extension $\theta_{\Sigma_8}(o_B) = \{(\text{Bob})^*_{sh}\}$, and, due to the relativized reference $y_f \dot\in x_f$, adds o_B to the extension $|o_f|^{\Sigma_8}$ and to $\hbar_{\Sigma_8}(L_{FARMER})$, and binds the local reference u_f with $\{o_B\}$ ($\lambda_{\Sigma_8}(u_f) = \{o_B\}$), whereby the shifted products for USE and HAVE are recomputed: $\langle S|\mathbf{A} : o_B, O|\mathbf{P} : \perp \rangle$ is added to the former and $\langle S|\mathbf{A} : o_B, O|\mathbf{P} : o_n \rangle$ is added to the latter. A similar effect is seen in the transitions to Σ_9 and to Σ_{11}.

This example shows that the individual determiners express a specific plurality-through-evidence: an object o_2 gets to the set-extension of a nominal object o_1 only due to a relativized determiner with relation $y \dot\in x$, where o_1 binds x and o_2 binds y, i.e. only due to a witness of this inclusion in the discourse.

Properties of DP semantics. The dynamic and the static DP semantics coincide in the corresponding contexts.

THEOREM 14. 1. Let $\delta = (\pi_1, \ldots, \pi_n)$ be a discourse, Σ_0 be an initial d-context, $|\delta|_{\Sigma_0} = (|\pi_1|_{\Sigma_0}^{\Sigma_1}, |\pi_2|_{\Sigma_1}^{\Sigma_2}, \ldots, |\pi_n|_{\Sigma_{n-1}}^{\Sigma_n})$ be the d-semantics of δ relative to Σ_0 and $\sigma_i = \langle \gamma_{\Sigma_i}, \lambda_{\Sigma_i}, \theta_{\Sigma_i}, \hbar_{\Sigma_i}\rangle$ be the s-contexts corresponding to d-contexts Σ_i. Then $|\pi_i|_{\Sigma_{i-1}}^{\Sigma_i} = \|\pi_i\|^{\sigma_i}$ for all $i, 0 < i \leq n$.

The dynamic DP semantics is monotonic:

THEOREM 15. Let $\delta^\infty = (\pi_1, \pi_2, \ldots)$ be an infinite discourse, $|\delta|_{\Sigma_0}^\infty = (|\pi_1|_{\Sigma_0}^{\Sigma_1}, |\pi_2|_{\Sigma_1}^{\Sigma_2}, \ldots)$ and $\Sigma' \leq^\circ \Sigma''$ iff $|o|^{\Sigma'} \subseteq |o|^{\Sigma''}$ and $\hbar_{\Sigma'}(L) \subseteq \hbar_{\Sigma''}(L)$ for all objects and classes. Then $\Sigma_0 \leq^\circ \Sigma_1 \leq^\circ \Sigma_2 \leq^\circ \ldots$ is an increasing chain.

6 Conclusion

One can see that the Speaker's stance discourse semantics outlined in this paper is different from all logical DRT-like semantics of discourse (cf. [7, 8, 9, 14]). The anaphora resolution is not included into the DP semantics because the referential relations are explicitly marked in DP using its determiners. Some of the referential relations established by these determiners, such as co-reference $\sim$, and attribute value comparison relations $<, >, =$, as well as the signs $+, -$ of verbal objects may introduce conflicts in the contexts. Checking of probable inconsistencies in the contexts is not required in DP semantics. This makes possible to apply it to correctly constructed DP with contradictory meaning, which is impossible in all kinds of logical theories of discourse.

Due to object-orientation, the DP semantics goes without quantifiers. Even so, there are certain similarities between the conventional quantifiers and the DP determiners. Creation of an object ($\gamma(x) = o$) is an analog of the existential quantifier. It is closer than the logical quantifier to the natural language "existence": *every entity mentioned in the discourse exists.* The access modes **H** and **I** correspond to two different concepts of universal quantification. The former, holistic, was always a problem to express in the traditional logical semantics. It allows one to adequately express the meaning of noun phrases as in *John likes (books)*$_{(\mathbf{H}\Downarrow_x^\omega u)BOOK}$ and provides a holistic interpretation for mass nominals as in *He needs more (water)*$_{\mathbf{H}\Downarrow_x^\omega WATER}$. The latter, individual, is rather close to $\forall$ from the point of view of extension constraining. It is not difficult to extend the DP-determiners by allowing complex relations constraining objects in the extension of nominals. For instance, the determiner $\mathbf{D}_c = (\mathbf{H} \Downarrow_x^\omega (\mathbf{H} \Downarrow_y^\omega \text{ROSE}, \mathbf{H} \Downarrow_z^\omega \text{LILIES}) (\mathbf{card}(y) > \mathbf{card}(z)))$ will be used in the O|P-subplan $\mathbf{D}_c \mathcal{A}_\cup \{\Uparrow_y, \Uparrow_z\}$ of *At least three girls gave (more roses than lilies)*$_{\mathbf{D}_c}$ *to John.* Such determiners are in practice comparable with so called cumulative quantifiers generally treated using generalized quantifiers (cf. [10, 11]). In our example, $\mathcal{A}_\cup\{\Uparrow_y, \Uparrow_z\}$ is a nominal aggregate with union extension: $\left|\mathcal{A}_\cup\{o_1, o_2\}\right|^\Sigma = |o_1|^\Sigma \cup |o_2|^\Sigma$. In the end, due to the "quantifier-freeness", the static DP semantics is fully compositional (DP-determiners are interpreted *in situ*, i.e. in verbals' argument positions).

BIBLIOGRAPHY

[1] N. Asher and J. Pustejovsky. *The Metaphysics of Words*. ms. Brandeis University and University of Texas, 1999.

[2] Johan Bos. Predicate logic unplugged. In P. Dekker and M. Stokhof, editors, *Proc. of the 10th Amsterdam Colloquium*, pages 133–142, 1995.

[3] Ann Copestake, Dan Flickinger, Rob Malouf, Susanne Riehemann, and Ivan Sag. Translation using minimal recursion semantics. In *Proc. of the 6th Int. Conf. on Theoretical and Methodological Issues in Machine Translation (TMI95)*, Leuven, Belgium, 1995.

[4] A. Dikovsky. A finite-state functional grammar architecture. In D. Leivant and Ruy de Queiroz, editors, *Logic, Language, Information and Computation. Proc. of the 14th Intern. Workshop WoLLIC 2007*, LNCS 4576, pages 131–146, Rio de Janeiro, Bresil, 2007. Springer Verlag.

[5] Alexander Dikovsky. Linguistic meaning from the language acquisition perspective. In *Proc. of the 8th Intern. Conf. "Formal Grammar 2003" (FG 2003)*, pages 63–76, Vienna, Austria, August 2003.

[6] Gottlob Frege. *Begriffsschrift, eine der arithmetischen nachgebildete Formelsprache des reinen Denkens*. Louis Nebert. Halle a. S., 1879.

[7] I. Heim. File change semantics and the familiarity theory of definiteness. In R. Bäuerle, C. Schwarze, and von A. Stechow, editors, *Meaning. Use and Interpretation of Language*. De Gruyter, Berlin, 1983.

[8] Hans Kamp and Uwe Reyle. *From Discourse to Logic: An introduction to modeltheoretic semantics, formal logic and Discourse Representation Theory*. Kluwer Academic Publishers, Dordrecht, Germany, 1993.

[9] Hans Kamp, Josef van Genabith, and Uwe Reyle. Discourse representation theory. In D.M. Gabbay and F. Guenthner, editors, *Handbook of Philosophical Logic*. Forthcoming, Second edition.

[10] E. Keenan. The semantics of determiners. In S. Lappin, editor, *The Handbook of Contemporary Semantic Theory*, pages 41–63. Blackwell, 1996.

[11] E. Keenan and D. Westerståhl. Generalized quantifiers in linguistics and logic. In van Benthem and ter Meulen [17], chapter 15, pages 837–893.

[12] Igor Mel'čuk. *Cours de morphologie générale. Volume II.* Presses de l'Université de Montréal/CNRS, Montréal, 1994.

[13] Igor A. Mel'čuk. Actants in semantics and syntax I-II. *Linguistics*, 42(1,2):1–66,247–291, 2004.

[14] R. Muskens, J. van Benthem, and A. Visser. Dynamics. In van Benthem and ter Meulen [17], chapter 10, pages 587–648.

[15] Reinhard Muskens. Order-independence and underspecification. In J. Groenendijk, editor, *Ellipsis, Underspecification, Events and More in Dynamic Semantics*. 1995. DYANA Deliverable R.2.2.C.

[16] James Pustejovsky. *The Generative Lexicon*. Cambridge:MIT Press, 1995.

[17] J. van Benthem and A. ter Meulen, editors. *Handbook of Logic and Language*. North-Holland Elsevier, The MIT Press, Amsterdam, Cambridge, 1997.

[18] Robert D. Van Valin, Jr. Generalized semantic roles and the syntax-semantics interface. In F. Corblin, C. Dobrovie-Sorin, and J.-M. Marandin, editors, *Empirical Issues in Formal Syntax and Semantics 2: Selected papers from the Colloque de Syntaxe et Semantique à Paris*, pages 373–388. 1999.

On the strength of some semi-constructive theories

SOLOMON FEFERMAN

ABSTRACT. Most axiomatizations of set theory that have been treated metamathematically have been based either entirely on classical logic or entirely on intuitionistic logic. But a natural conception of the set-theoretic universe is as an indefinite (or "potential") totality, to which intuitionistic logic is more appropriately applied, while each set is taken to be a definite (or "completed") totality, for which classical logic is appropriate; so on that view, set theory should be axiomatized on some correspondingly mixed basis. Similarly, in the case of predicative analysis, the natural numbers are considered to form a definite totality, while the universe of sets (or functions) of natural numbers are viewed as an indefinite totality, so that, again, a mixed semi-constructive logic should be the appropriate one to treat the two together. Various such semi-constructive systems of analysis and set theory are formulated here and their proof-theoretic strength is characterized. Interestingly, though the logic is weakened, one can in compensation strengthen certain principles in a way that could be advantageous for mathematical applications.[1]

1 Introduction

There are various foundational frameworks in which the full universe or domain of its objects is considered to be indefinite but for which certain predicates and logical operations on restricted parts of the universe are considered to be definite. For example, on one view in the case of set theory, each set is considered to be a definite (or "completed") totality, so that the membership relation and bounded quantification are definite, while the universe of sets at large is an indefinite (or "potential") totality. The idea carries over to frameworks with more than one universe, some of which may be regarded as definite while others are indefinite. For example, in the case of predicativity, the natural numbers are considered to form a definite totality, while the universe of sets (or functions) of natural numbers forms an indefinite totality. Thus quantification over the natural numbers is taken to be definite, but not quantification applied to variables for sets or functions of natural numbers. Most axiomatizations of set theory that have been treated metamathematically have been based either entirely on classical logic or entirely on intuitionistic logic, while almost all axiomatizations of predicative systems have been in classical logic. But it has been suggested on philosophical grounds that it is more appropriate to restrict

---

[1] For my friend and colleague Grisha Mints, on the occasion of his 70th birthday, June 7, 2009, with special appreciation for his steadfast support of logic at Stanford.

the application of classical logic to definite predicates and quantifiers and to take the basic logic otherwise to be intuitionistic. We shall show here, for various examples — including the ones that have been mentioned — that while this may provide a more philosophically satisfying formal model of the given foundational framework — there is no difference in terms of proof-theoretical strength from the associated system based on full classical logic; in that respect they are equally justified. On the other hand, the semi-constructive systems in general have a further advantage that they admit more formally powerful principles — such as the unrestricted axiom of choice — without increase of strength, and this can be advantageous when considering what mathematics can be accounted for in the given systems.

The initial stimulus for my work here was the paper of Coquand and Palmgren (2000) in which they give a constructive sheaf model for a theory of finite types over the natural numbers together with a domain of countable tree ordinals, formulated in intuitionistic logic plus the so-called *numerical omniscience scheme* for ϕ an arbitrary formula:

$$\text{(NOS)} \quad \forall n[\phi(n) \vee \neg \phi(n)] \to \forall n \phi(n) \vee \exists n \neg \phi(n)$$

A special case of this non-constructive principle is what Errett Bishop (1967) called the *limited principle of omniscience*:

$$\text{(LPO)} \quad \forall n f(n) = 0 \vee \exists n f(n) \neq 0$$

Bishop pointed out that all the results in classical analysis for which he found constructive substitutes follow from those substitutes plus LPO. In Feferman (2001) I reported determination of a bound on the proof-theoretical strength of the Coquand-Palmgren system by means of an extension of Gödel's *Dialectica* (functional) interpretation interpretation using non-constructive operators. That method goes back to Feferman (1971), with further applications in Feferman (1977), (1979); semi-constructive systems there played an essential intermediate role in the use of the method to determine the proof-theoretical strength of various classical systems. Here, by contrast, they are at the center of our attention.

What's needed in the following about Gödel's (D-)interpretation is reviewed in section 2; the reader is referred to the exposition of the basic interpretation and various of its extensions in Avigad and Feferman (1998) for more details. In section 3, we take up the particular extension by means of the non-constructive minimum operator, where it is also shown by a simple argument that (NOS) follows from (LPO) under the assumption of the Axiom of Choice (AC), as expressed in our finite type systems and verified by the D-intepretation. This is followed in section 4 by determination of the strength of some semi-constructive theories of finite type over the natural numbers, and then in section 5 the same augmented by the type of countable tree ordinals. Both sections 4 and 5 make direct use of previously established results. The main new results are in section 6, which deals with the strength of some semi-constructive systems of admissible set theory with strong choice principles plus a generalization of NOS called the Bounded Omniscience Scheme (BOS) and where the method of non-constructive D-interpretation is adapted in a new way. I conclude in section

7 with comparison with work on a various related systems and some open questions, in particular with strong systems of semi-intuitionistic set theory due to Poszgay (1971, 1972), Tharp (1971), Friedman (1973) and Wolf (1974) and, by contrast, a system conservative over PA, due to Friedman (1980).

The reader familiar with the material of Avigad and Feferman (1998) through its section 8 is encouraged to skip directly to the new results here in section 6, after taking note of Theorem 2 in section 3.2 below.

2 Review of Gödel's *Dialectica* interpretation

The main basic notions, notation and results concerning this interpretation are recalled here from Avigad and Feferman (1998). For each formula ϕ, the *double-negation* (or *negative*) *translation* of ϕ is denoted by ϕ^N, and the *Dialectica* (or *D-*) *interpretation* of ϕ is denoted by ϕ^D. The N-translation in general takes one from classical theories to formally intuitionistic theories, and the D-interpretation is then applied to take one to a quantifier free theory of functionals of finite type. The basic example is given by the N-translation of Peano Arithmetic PA into Heyting Arithmetic HA (Gödel, Gentzen) followed by the D-interpretation of HA into a quantifier-free theory T of primitive recursive functionals of finite type (Gödel). The latter extends directly to a D-interpretation of a finite type extension HA^ω of HA into T, and then by making a similar extension PA^ω of PA, we obtain a composite ND-interpretation of that system into T. Finally, T has a model in HEO, the Hereditarily Extensional (Recursive) Operations of finite type, and that can be formalized in PA. Thus all these systems are of the same proof-theoretical strength.

For the finite type theories involved, the finite type symbols (t.s.) $\sigma, \tau, \ldots$ are generated as follows: (1) 0 is a t.s. and (ii) if σ, τ are t.s., then so also is $\sigma \to \tau$. These theories have infinitely many variables $x^\tau, y^\tau, z^\tau, \ldots$ of each type τ; type superscripts are suppressed when there is no ambiguity. We occasionally use other kinds of letters like $f, g, \ldots n, m, \ldots$ as well as capital letters $X, Y, \ldots$ for variables of appropriate types. Terms $s, t, \ldots$ are generated from the variables and constants (to be described) by closure under application ts (or $t(s)$) when t is of a type $\sigma \to \tau$ and s is of type σ, the result being of type τ.

Application in terms such as rst is read by association to the left, while the t.s. $\rho \to \sigma \to \tau$ is read by association to the right. Each higher type symbol σ can be written in the form $\sigma = (\sigma_1 \to \ldots \to \sigma_k \to 0)$; then equality $s = t$ at type σ is informally regarded as an abbreviation for $sx_1 \ldots x_k = tx_1 \ldots x_k$ where the x_i are fresh variables of type σ_i for $i = 1, \ldots, k$. The formal axioms and rules that we take to govern equality at higher types are given by the so-called weakly extensional approach due to Spector and described in Avigad and Feferman (1998), p. 350.

The type $0 \to 0$ is also denoted 1, and the type $1 \to 0$ is denoted 2. The constant symbols include 0 of type 0 and Sc of type 1. In addition we have symbols K, S for the usual combinators in all appropriate types, satisfying equations of the form $Kst = s$ and $Srst = rt(st)$; the typed λ-calculus is then introduced by definition as usual. Finally we have constant symbols R for the recursors in all appropriate types, satisfying equations of the form

$$Rxy0 = x \text{ and } Rxyn' = yn(Rxyn)$$

where n is a variable of type 0 and $n' = \text{Sc}(n)$. The formulas $\phi, \psi, \ldots$ of T are generated from the equations $s = t$ and the falsity $\bot$ by closure under $\phi \wedge \psi$, $\phi \vee \psi$, and $\phi \to \psi$; then $\neg \phi$ is defined as $\phi \to \bot$. The formulas of PA$^\omega$ and HA$^\omega$ are in addition closed under universal and existential quantification, $\forall x \phi$ and $\exists x \phi$, w.r.t. variables x of all finite types. The underlying logic of PA$^\omega$ is classical while that of HA$^\omega$ and T is intuitionistic. The axioms in all three systems are as usual for 0 and Sc, and as indicated above for the symbols K, S and R. The induction axiom scheme is given as usual in these systems, while it is formulated as a rule in T. One may show that quantifier-free (QF) formulas (with all equations restricted to terms of type 0) are decided, i.e. satisfy the law of excluded middle (LEM) in all of these systems.

For a formula ϕ of the quantified finite type language, the D-interpretation of ϕ is of the form

$$\phi^D = \exists x \forall y \phi_D(x, y)$$

where x, y are sequences (possibly empty) of finite type variables and ϕ_D is a QF formula whose free variables are those of ϕ in addition to those of x and y. The inductive definition of the D-interpretation for formulas ϕ, ψ with ϕ^D as above, and $\psi^D = \exists u \forall v \psi_D(u, v)$ is as follows:

(i) For ϕ an atomic formula, x and y are both empty and $\phi^D = \phi_D = \phi$.

(ii) $(\phi \wedge \psi)^D = \exists x, u \forall y, v [\phi_D(x, y) \wedge \psi_D(u, v)]$

(iii) $(\phi \vee \psi)^D = \exists z, x, u \forall y, v [(z = 0 \wedge \phi_D(x, y)) \vee (z = 1 \wedge \psi_D(u, v))]$

(iv) $(\forall z \phi(z))^D = \exists X \forall y, z \phi_D(Xz, y, z)$

(v) $(\exists z \phi(z))^D = \exists x, z \forall y \phi_D(x, y, z)$

(vi) $(\phi \to \psi)^D = \exists U, Y \forall x, v [\phi_D(x, Yxv), \to \psi_D(Ux, v)]$

With $\neg \phi$ defined as $\phi \to \bot$, we have

(vii) $(\neg \phi)^D = \exists Y \forall x \neg \phi_D(x, Yx)$

The reasoning behind (iv) lies in the constructive acceptance of the Axiom of Choice, here taken as the following scheme in all finite types:

(AC) $\quad \forall x \exists y \phi(x, y) \to \exists Y \forall x \phi(x, Yx)$

The reasoning behind (vi) lies in a chain of steps[2], an intermediate one of which lies in transforming $[\exists x \forall y \phi_D(x, y) \to \exists u \forall v \psi_D(u, v)]$ into

(vi)* $\forall x \exists u \forall v \exists y [\phi_D(x, y) \to \psi_D(u, v)]$

Implicitly, that makes use of a principle called Independence of Premises (IP) that is not intuitionistically justified. Moreover, another one of the steps to (vi) implicitly makes use of the finite type forms of Markov's Principle,

(MP) $\quad \forall x (\neg \neg \exists y \phi \to \exists y \phi)$ for QF ϕ

---

[2]See Avigad and Feferman (1998) pp. 346-347 for all the steps involved.

which is also not justified intuitionistically. Nevertheless, Gödel's interpretation gives a constructive reduction of both AC and MP.[3] In the case of AC this is immediate, and in the case of MP, it is quite easy given that all QF formulas in the systems we're dealing with are decided. The additional power of the D-interpretation below comes from the fact that $(AC)^D$ and $(MP)^D$ are verified quite generally.

The following is the direct extension from HA to its finite type version of Gödel's main result (1958):

Theorem 1. If $HA^\omega + AC + MP$ proves ϕ and $\phi^D = \exists x \forall y \phi_D(x, y)$ then for some sequence of terms t of the same type as the sequence x of variables, T proves $\phi_D(t, y)$.

The main use of the recursor constants R is in verifying the D-interpretation of the induction axiom scheme in HA^ω.

By QF-AC we mean the scheme AC restricted to QF formulas ϕ. Since for such ϕ we have ϕ^N equivalent to ϕ, we see that $(QF\text{-}AC)^{ND}$ is also verified in T.

Corollary 1. $PA^\omega + QF\text{-}AC$ is N interpreted in HA^ω and so it is ND interpreted in T.

By an analysis of the reduction of the terms of T to normal form, one sees that its closed terms denote recursive functions defined by ordinal recursions on proper initial segments of the natural well-ordering of order type ϵ_0. Hence all the systems PA, PA^ω, HA, HA^ω and T have this same class of provably recursive functions.

3 The non-constructive minimum operator and interpretation of NOS

3.1 The non-constructive minimum operator

One way to arrange for arithmetical formulas to satisfy the Law of Excluded Middle, LEM, in a system based as a whole on intuitionistic logic is to make them equivalent to QF formulas by adjunction of a numerical quantification operator E_0 of type 2 satisfying the axiom

$$(E_0) \quad E_0 f = 0 \leftrightarrow \exists x (fx = 0)$$

for f, x variables of type 1 and 0, resp. In order for this to satisfy the ND-interpretation we need to verify the following two implications:

$$fx = 0 \rightarrow E_0 f = 0 \text{ and } E_0 f = 0 \rightarrow \exists x (fx = 0)$$

The first of these is automatically taken care of, and the N-interpretation of the second is taken care of by the verification of MP, but in order to get its further D-interpretation we need to have a functional X which satisfies $E_0 f = 0 \rightarrow f(Xf) = 0$, and hence $fx = 0 \rightarrow f(Xf) = 0$. To take care of this we adjoin a new constant symbol μ with axiom:

$$(\mu) \quad fx = 0 \rightarrow f(\mu f) = 0$$

---

[3] It also verifies the interpretation of IP, but we shall not make use of that fact.

We call μ the *non-constructive minimum operator*, though properly speaking that would need an additional axiom specifying that μf is the least x such that $fx = 0$ if $\exists x(fx = 0)$ and (say) is 0 otherwise; in fact, that is definable from μ using the primitive recursive bounded minimum operator.

3.2 The LPO axiom and the Numerical Omniscience Scheme

As stated, under the axiom (μ) every arithmetical formula is equivalent to a QF formula in intuitionistic logic; various consequences of this for semi-constructive systems incorporating that axiom will be dealt with in the next section. In particular, we can derive the NOS scheme for arithmetical formulas from that assumption. But in the presence of AC we can do even more. We here understand by the NOS, the scheme described in sec. 1 where we allow ϕ to be any formula of the language of HA^ω, and by LPO the statement given in sec. 1, where 'f' is a variable of type 1.

Theorem 2.

(i) $HA^\omega + (\mu)$ proves LPO.

(ii) $HA^\omega + (LPO) + AC$ proves NOS.

Proof. (i) is immediate, and for (ii) we note that if $\forall n[\phi(n) \vee \neg \phi(n)]$ holds, then so also does $\forall n \exists k[k = 0 \wedge \phi(n) \vee k = 1 \wedge \neg \phi(n)]$. Hence by AC there exists f of type 1 such that $\forall n[(f(n) = 0 \leftrightarrow \phi(n)) \wedge (f(n) = 1 \leftrightarrow \neg \phi(n))]$. □

4 Semi-constructive systems of finite type over the natural numbers

4.1 Primitive recursion in a type 2 functional and Kleene's variant

The system $HA^\omega + (\mu) + (AC)$ offers itself immediately for consideration as a semi-constructive system of interest; this is a predicative system that is somewhat stronger than PA. But we shall also consider systems using operators F of type 2 stronger than μ. Given any such F, Shoenfield defined a hierarchy H_α^F of functions for α less than the first ordinal not recursive in F, such that the 1-section of F (i.e., the totality of type 1 functions recursive in F) consists of all those functions that are primitive recursive in the usual sense in some such H_α^F. Now the normalization of terms of the system T augmented by such an F shows that its 1-section consists of all those functions primitive recursive in some H_α^F for $\alpha <\epsilon_0$. In particular, the 1-section of the functionals defined by closed terms of T augmented by μ consists of the functions in the HYP hierarchy up to (but not including) ϵ_0.

We shall also consider an interesting subsystem of $\hat{T}$ augmented by such type 2 functionals F, obtained by restricting the induction and recursion principles. The motivation for that restriction lies in the fact that the recursors R with values of higher type have a kind of impredicative character. For example, for values of $Rfgn$ of type 2, thought of as $\lambda h.Rfgnh$, we have $Rfgn' = gn(\lambda h.Rfgnh))$ and *that* evaluated at a given function h_1 makes prima-facie reference to *the* values of $Rfgn$ at all functions h. It is easily shown that non-primitive recursive functions such as the Ackermann function may be generated in this way. Kleene (1959) introduced restricted recursors $\hat{R}$ satisfying the recursion equations

$$\hat{R}xy0z \text{ and } \hat{R}xyn'z = y(\hat{R}xynz)$$

where z is a sequence of variables such that xz is of type 0. He showed that the 1-section of the functionals generated from 0, Sc, the K, S combinators and the $\hat{R}$ recursors by closure under application are exactly the primitive recursive functions. Thus taking $\hat{T}$ to be the subsystem of T with constants $\hat{R}$ in place of the constants R, and corresponding change of axioms, we have that the 1-section of $\hat{T}$ consists exactly of the primitive recursive functions in the usual sense. Now all this may be relativized to a type 2 functional F to show that the 1-section of $\hat{T}$ augmented by F consists exactly of the functions primitive recursive in H_n^F for some $n < \omega$. In particular, the 1-section of $\hat{T} + (\mu)$ consists of all the arithmetically definable functions.

By Res-PA$^\omega$ and Res-HA$^\omega$ we mean the systems using the $\hat{R}$ recursors in place of the R recursors and with the axiom of induction restricted to QF-formulas.

4.2 The strength of some semi-constructive systems based on the non-constructive minimum operator

We begin with semi-constructive variants of predicative systems, i.e. systems whose strength is at most that of ramified analysis up to the Feferman-Schütte ordinal Γ_0, or equivalently, the union of the (Π_1^0-CA$_\alpha$) systems for $\alpha < \Gamma_0$.

Theorem 3.

(i) The systems Res-HA$^\omega$ + (AC) + (MP) + (μ) and Res-PA$^\omega$ + (QF-AC) + (μ) are proof-theoretically equivalent to and conservative extensions of PA; furthermore they are conservative extensions of the 2nd order system ACA$_0$ for Π_2^1-sentences.

(ii) The systems HA$^\omega$ + (AC) + (MP) + (μ) and PA$^\omega$ + (QF-AC) + (μ) are proof-theoretically equivalent to — and conservative extensions for Π_2^1-sentences of — the 2nd-order systems (in decreasing order) Σ_1^1-DC, Σ_1^1-AC, and the union of the (Π_1^0-CA$_\alpha$) systems for $\alpha < \epsilon_0$.

(iii) The systems HA$^\omega$ + (AC) + (MP) + (μ) + (Bar-Rule) and PA$^\omega$ + (QF-AC) + (μ) + (Bar-Rule) are proof-theoretically equivalent to — and conservative extensions for Π_2^1 sentences of — the 2nd order systems (in decreasing order) Σ_1^1-DC + (Bar-Rule), Σ_1^1-AC + (Bar-Rule), and the union of the (Π_1^0-CA$_\alpha$) systems for $\alpha < \Gamma_0$.

(iv) There is no increase in strength when the NOS scheme is added to the semi-constructive systems in (i)-(iii).

Proofs. The result (i) is from Feferman (1977), (ii) is from Feferman (1971) and (iii) is from Feferman (1979). The ideas for their proofs are exposited in Avigad and Feferman (1998), sec. 8. Briefly, the proof of (i) uses the fact that the D-interpretation of the semi-constructive system Res-HA$^\omega$ + (AC) + (μ)

and the ND-interpretation of the classical system Res-PA$^\omega$ + (QF-AC) + (μ) both take us into $\hat{T} + \mu$, which is interpreted in PA preserving arithmetical sentences (as translated using μ). For the conservation statement, one notes that under the μ axiom, every Π_2^1 sentence is equivalent to one of the form $\forall f \exists g \phi(f, g)$, where ϕ is quantifier-free, hence if provable, it is preserved under the N-translation using (MP) and then under the D-interpretation one obtains a type 2 term t such that $\hat{T} + (\mu)$ proves $\phi(f, tf)$. That term defines $g = tf$ arithmetically from f. The main steps of the proof of (ii) follow the same lines, concluding with the interpretation in T + (μ), whose 1-section consists of the functions in the HYP hierarchy up to (but not including) ϵ_0, as described in 4.1 above. For (iii) the main new work goes first into the D-interpretation of HA$^\omega$ + (AC) + (MP) + (μ) + (Bar-Rule) in the extension of T + (μ) by two new rules, (BR) and (TR). These rules involve expressing in QF form, well-foundedness of any specific segment $\preceq_a$ of a given arithmetical well-ordering as the open formula $\forall x[\forall y(y \prec_a x \to y \in X) \to x \in X]$, denoted $I(\preceq_a, X)$, where X is a set-variable (i.e., a characteristic function at type 1). Then, for the natural well-ordering $\prec$ of order type Γ_0, the version BR of the Bar-Rule used in this context allows one to pass from $I(\preceq_a, X)$ for any specific a to the result $I(\preceq_a, \phi)$ of substituting in it any formula $\phi(x)$ of the system for the formula $x \in X$, while the rule (TR) allows one to introduce a transfinite recursor on the given segment under the same hypothesis. One gets up to each ordinal less than Γ_0 by a boot-strapping argument, and the proof that one doesn't go beyond is via a normalization argument. See Feferman (1979) pp. 87-89 for more details. Finally, (iv) is immediate by Theorem 2.

4.3 The strength of some semi-constructive systems based on μ plus the Suslin-Kleene operator

For a given f, let Tree(f) be the tree consisting of all finite sequence numbers s such that $f(s) = 0$. This tree is not well-founded if and only if $\exists g \forall x f(g \mid x) = 0$, where for any type 1 function g, $g \mid x$ is the number s of the finite sequence $g_0, \ldots, g(x-1)$. The Suslin-Kleene operator is the associated type 2 choice functional μ_1, obtained by taking the left-most descending branch in Tree(f) if that tree is not well-founded. It satisfies the axiom

$$(\mu_1) \quad \forall x f(g \mid x) = 0 \to \forall x f((\mu_1 f) \mid x) = 0$$

which may be re-expressed in QF form using the μ operator. From the work of Feferman (1977) and Feferman and Jäger (1983) one then obtains characterizations of the proof-theoretical strength of the semi-constructive systems HA$^\omega$ + (AC) + (MP) + (μ) + (μ_1), its restricted version, and its extension under the Bar-Rule, in a form analogous to Theorem 3. For example, in analogy to part (ii) of that theorem, the strength of HA$^\omega$ + (AC) + (MP) + (μ) + (μ_1) is characterized as that of the iterated Π_1^1-CA systems up to ϵ_0, which is the same as that of (Σ_2^1-DC). See Avigad and Feferman (1998) pp. 384-385 for full statement of results and indication of proofs. An alternative characterization may be given in terms of the iterated ID systems up to ϵ_0. And, finally, addition of the NOS comes for free by Theorem 2.

5 The strength of a semi-constructive theory of finite type over the natural numbers and countable tree ordinals.

Here we can draw directly on Avigad and Feferman (1998), sec. 9, which reports the work of an unpublished MS, Feferman (1968). The type structure is expanded by an additional ground type for abstract constructive countable tree ordinals, denoted Ω, and lower case Greek letters $\alpha, \beta, \gamma, \ldots$ are used to range over Ω. But now we use 'N' to denote the type symbol 0. The constants are augmented by 0_Ω of type Ω, Sup of type $(N \to \Omega) \to \Omega$, Sup^{-1} of type $\Omega \to (\Omega \to N)$, and for each σ, $R_{\Omega\sigma}$ of type $(\Omega \to (N \to \sigma) \to \sigma) \to \sigma \to \Omega \to \sigma$. The subscript '$\sigma$' is omitted from the ordinal recursor $R_{\Omega\sigma}$ when there is no ambiguity. The constant 0_Ω represents the one-point tree, and for f of type $(N \to \Omega)$, Sup(f) represents the tree obtained by joining together the subgrees fn for each natural number n. For $\alpha = \text{Sup}(f)$, $\text{Sup}^{-1}(\alpha) = f$ is the constructor of α; in that case we write α_n for $(\text{Sup}^{-1}(\alpha))n$. For each type σ the ordinal recursor R_Ω works to take an element a of type σ, a functional f of type $(\Omega \to (N \to \sigma) \to \sigma)$, and a tree ordinal α to an element $R_\Omega f a \alpha$ satisfying the recursion equations

$$(R_\Omega) \quad R_\Omega f a 0_\Omega = a, \text{ and for } \alpha \neq 0_\Omega, R_\Omega f a \alpha = f\alpha(\lambda n.R_\Omega f a \alpha_n)$$

We also take the language to include the constant μ. In it, we form three theories of countable tree ordinals of finite type, first a classical theory $\text{CO}^\omega_\Omega + (\mu)$, then a semi-intuitionistic theory $\text{SO}^\omega_\Omega + (\mu)$, both with full quantification at all finite types, and finally a quantifier free theory T_Ω.[4] The basic axioms of $\text{CO}^\omega_\Omega + (\mu)$ and $\text{SO}^\omega_\Omega + (\mu)$ are the same, consisting of the following:

(1) The axioms of $\text{HA}^\omega + (\mu)$, with the induction scheme extended to all formulas of the language;

(2) $\text{Sup}(f) \neq 0_\Omega$ and $\text{Sup}^{-1}(\text{Sup}(f)) = f$, for f of type $N \to \Omega$

(3) $\text{Sup}(\text{Sup}^{-1}(\alpha)) = \alpha$ for $\alpha \neq 0_\Omega$

(4) $(0_\Omega)_x = 0_\Omega$

(5) the (R_Ω) equations

(6) $\phi(0_\Omega) \wedge \forall \alpha[\alpha \neq 0_\Omega \wedge \forall x \phi(\alpha_x) \to \phi(\alpha)] \to \forall \alpha \phi(\alpha)$ for each formula $\phi(\alpha)$

The theory $T_\Omega + (\mu)$ has as axioms:

(1)* The axioms of $T + (\mu)$

(2)*-(5)* The same as (2)-(5)

(6)* The rule of induction on ordinals for QF formulas ϕ

---

[4] In Avigad and Feferman (1998), p. 387, we wrote OR^ω_1 for the system $\text{CO}^\omega_\Omega + (\text{QF-AC})$.

Note that this last is to be expressed in quantifier free form using the μ operator. In the next statement, ID_1 and $ID_1^{(i)}$ are respectively the classical and intuitionistic theory of non-iterated positive inductive definitions given by arithmetical $\phi(x, P^+)$.

Theorem 4. The following theories are all of the same proof-theoretical strength:

(i) ID_1

(ii) $CO_\Omega^\omega + (\mu) + $ (QF-AC)

(iii) $SO_\Omega^\omega + (\mu) + $ (AC) + (NOS)

(iv) $T_\Omega + (\mu)$

(v) $ID_1^{(i)}$

Proof. It is shown in Avigad and Feferman (1998) pp. 388-389 how to translate ID_1 into $CO_\Omega^\omega + (\mu)$. That system is then carried into $SO_\Omega^\omega + (\mu)$ by the N-translation. By a direct extension of the work described in secs. 2-4 above, we see that $SO_\Omega^\omega + (\mu) + $ (AC) + (NOS) is D-interpreted in $T_\Omega + (\mu)$; this also verifies the classical (QF-AC) under the ND-interpretation. Next, as in op. cit. pp. 390-391, $T_\Omega + (\mu)$ has a model in HRO(^2E), the indices of operations hereditarily recursive in ^2E in the sense of Kleene (1959), interpreting the type Ω objects as the members of a version O of the Church-Kleene ordinal notations. That model can be formalized in ID_1 so as to reduce $T_\Omega + (\mu)$ to ID_1. Finally, the reduction of ID_1 to $ID_1^{(i)}$ is due to Buchholz (1980), in fact to the theory of an accessibility inductive definition.[5]

The language of the theory W of Coquand and Palmgren (2000) is close to that of SO_Ω^ω, but does not contain the Sup^{-1} operator or the (μ) operator. Its axioms are essentially the same as those of SO_Ω^ω without the axioms for those two operators. In addition, it has three special choice axiom schemata, unique choice (AC!), countable choice (AC_0) and dependent choice (DC) — all of which follow from (AC) — as well as the Numerical Omniscience Scheme (NOS). Thus W is a subtheory of $SO_\Omega^\omega + $ (AC) + (NOS), and so the proof-theoretic strength of W is no greater than that of $ID_1^{(i)}$. Presumably, the latter (at least for accessibility inductive definitions) can be interpreted in W, but I have not checked that. The main part of Coquand and Palmgren (2000) is devoted to producing a constructive sheaf-theoretic model of W in Martin-Löf type theory with generalized inductive definitions; an obvious question is whether their argument provides an alternative reduction of W to $ID_1^{(i)}$. Finally, as noted in Theorem 2, NOS already follows in their system from LPO from countable choice.

6 Semi-constructive systems of set theory.

The basic idea for semi-constructive systems of set theory was stated in the introduction: each set is considered to be a definite totality, so that the membership relation and bounded quantification are definite, i.e. classical logic apply

---

[5] Avigad and Towsner (2009) have obtained an interesting alternative proof of the reduction of ID_1 to an accessibility $ID_1^{(i)}$, using a variant of the functional interpretation method.

to both, while the universe as a whole is considered to be indefinite, so that only intuitionistic logic applies to that. This suggests considering axiomatic systems of set theory based on intuitionistic logic for which it is assumed that classical logic applies to all Δ_0 formulas. The latter is accomplished by assuming the following restricted scheme for the Law of Excluded Middle,

$$(\Delta_0\text{-LEM}) \quad \phi \vee \neg\phi, \text{ for all } \Delta_0 \text{ formulas } \phi$$

In this context, we also take Markov's principle in the form:

$$(\text{MP}) \quad \neg\neg\exists x\phi \to \exists x\phi, \text{ for all } \Delta_0 \text{ formulas } \phi$$

Let IKPω be the system KP with logic restricted to be intuitionistic. To be more precise, IKPω takes the following as its non-logical axioms:

1. Extensionality

2. Unordered pair

3. Union

4. Infinity, in the specific form that there is a smallest set containing the empty set 0 and closed under the successor operation, $x' = x \cup \{x\}$.

5. Δ_0-Separation

6. Δ_0-Collection

7. The $\in$-Induction Rule

By 7, we mean the rule which allows us to infer $\forall x \psi(x)$ from $\forall x[(\forall y \in x)\psi(y) \to \psi(x)]$ for any formula $\psi(x)$. This is easily seen to imply the $\in$-Induction Scheme

$$\forall x[(\forall y \in x)\phi(y) \to \phi(x)] \to \forall x \phi(x)$$

by taking $\psi(x) = \{\forall z[(\forall y \in z)\phi(y) \to \phi(z)] \to \phi(x)\}$.

Some further schemata in the language of set theory shall be added to IKPω, first of all the Bounded Omniscience Scheme:

$$(\text{BOS}) \quad \forall x \in a[\phi(x) \vee \neg\phi(x)] \to \forall x \in a(\phi(x)) \vee \exists x \in a(\neg\phi(x))$$

for *all* formulas $\phi(x)$. The set-theoretical form of NOS is the special case of this in which $a = \omega$ the unique set specified by Axiom 4. We shall strengthen IKPω by (Δ_0-LEM) and BOS; but we can make a further considerable strengthening by adding the following form of the Axiom of Choice,

$$(\text{AC}_{\text{Set}}) \quad \forall x \in a \exists y \phi(x,y) \to \exists r[\text{Fun}(r) \wedge \text{Dom}(r) = a \wedge (\forall x \in a)\phi(x, r(x))]$$

for *all* ϕ, where Fun(r) expresses in usual set theoretic form that the binary relation r is a function, and Dom(r) = a expresses that a is the domain of r; both of these may be given as Δ_0 formulas. Note that in the presence of (AC$_{\text{Set}}$)

with axioms 1-3 and 5 we can infer Full Collection and Full Replacement i.e. these schemes for arbitrary formulas.

Theorem 5. The semi-constructive theory of sets, SCS = IKPω + (Δ_0-LEM) + (MP) + (BOS) + (AC$_{\text{Set}}$), is of the same strength as KPω and thence of ID$_1^{(i)}$. The same holds for a natural finite type extension SCS$^\omega$ over the universe of sets.

Proof. The proof is in three parts.

I. First, we show that KPω is interpretable in SCS via the N-translation. Since ϕ^N is provably equivalent to ϕ for every Δ_0 formula f in IKPω + (Δ_0-LEM), one readily checks that the N-translation of each of the axioms 1-5 holds in that subsystem of SCS. In the case of Δ_0-Collection, the N-translation is of the form

$$(\forall x \in a)\neg\neg\exists y \phi(x,y) \to \neg\neg\exists b (\forall x \in a)(\exists y \in b)\phi(x,y)$$

where ϕ is a Δ_0 formula. But then by (MP) this follows from Δ_0-Collection in SCS. Finally, the N-translation of an instance of the $\in$-induction rule is an instance of the same.

II. Next we introduce a new system T$_V$ and define a D-interpretation of SCS into T$_V$; by following through the interpretation, one may see what natural finite type extension SCS$^\omega$ of SCS is also verified in the process. The language of T$_V$ is typed, with a ground type V for sets, and function types $\sigma \to \tau$ for each types σ and τ. Variables for sets will be at the beginning or end of the alphabet, while variables for functions will generally be $f, g, h, \ldots$. Capital letters will be used for constants, except for 0 and ω; the constants are 0 (empty set), ω (natural numbers), D (disjunction operator), N (negation operator), E (characteristic function of equality), M (characteristic function of membership), C (bounded choice operator), P (unordered pair function), U (union function), S (separation operator), R^* (range operator) and R_σ (recursion operators). Terms are generated from variables and constants by closure under well-typed .application, ts. Atomic formulas are equations between terms, $s = t$, and membership of terms, $s \in t$. Formulas $\phi, \psi, \ldots$ are generated by closing the atomic formulas under the propositional operations and bounded quantification, $(\forall y \in t)\phi$ and $(\exists y \in t)\phi$. Truth values are represented in V by using 0 for True and any other value for False. The axioms of T$_V$ fall into three groups (A, B and C), as follows; these also implicitly determine the types of the various constants.

A. Equality and logical operation axioms.

1. (Decidability) $x = y \lor x \neq y$
2. (Equality) $Exy = 0 \leftrightarrow x = y$
3. (Membership) $Mxy = 0 \leftrightarrow x \in y$
4. (Disjunction) $Dxy = 0 \leftrightarrow x = 0 \lor y = 0$
5. (Negation) $Nx = 0 \leftrightarrow x \neq 0$
6. (Bounded choice) $x \in a \land fx = 0 \to Caf \in a \land f(Caf) = 0$

Note by 6 that $(\exists x \in a)fx = 0 \leftrightarrow Caf \in a \wedge f(Caf) = 0$. The following is then a direct consequence of the group A axioms.

Lemma 1. For each Δ_0 formula ϕ of set theory, all of whose free variables are among the list $\underline{x} = x_1, \ldots x_n$, we have a closed term t_ϕ such that the following is provable in T_V:
$$t_\phi(\underline{x}) = 0 \leftrightarrow \phi(\underline{x})$$

For the next group of axioms we write $a \subseteq b$ for $(\forall x \in a)(x \in b)$.

B. Set theoretic axioms.

  7. (Extensionality) $a \subseteq b \wedge b \subseteq a \to a = b$
  8. (Empty set) $\neg(x \in 0)$
  9. (Unordered pair) $x \in Pab \leftrightarrow x = a \vee x = b$
  10. (Union) $x \in Ua \leftrightarrow (\exists y \in a)(x \in y)$
  11. (Infinity) (i) $0 \in \omega \wedge (\forall x \in \omega)(x' \in \omega)$
      (ii) $0 \in a \wedge (\forall x \in a)(x' \in a) \to \omega \subseteq a$
  12. (Separation) $x \in Saf \leftrightarrow x \in a \wedge fx = 0$
  13. (Range) $y \in R^*af \leftrightarrow (\exists x \in a)(fx = y)$

As usual, for Axiom 11 in the preceding, we write $\{x, y\}$ for Pxy, $\{x\} = \{x, x\}$, $x \cup y = U\{x, y\}$, and $x' = x \cup \{x\}$. We also define $\langle x, y \rangle = \{\{x\}, \{x, y\}\}$ as usual in set theory, and use it to prove the following:

Lemma 2. There is a closed term Grph such that T_V proves
$$z \in (\text{Grph})af \leftrightarrow (\exists x \in a)(\exists y \in R^*af)[z = \langle x, y \rangle \wedge fx = y]$$

Proof. $(\text{Grph})af$ is the graph of f restricted to a, considered as a set; it is formed by separation from the Cartesian product $a \times (R^*af)$. This depends on the proof in general of the existence of Cartesian products $a \times b$, as follows. First let g be such that for each x, y, $gxy = \langle x, y \rangle$, so that gx is $\lambda y.\langle x, y \rangle$. Then for $x \in a$, $gx : b \to \{x\} \times b$ and $R^*(b, gx) = \{x\} \times b$. Finally, take $h = \lambda x.R^*(b, gx)$ so that $a \times b = U(R^*(a, h))$. $\square$

In the following I shall write $f|a$ for $(\text{Grph})af$.

The final group of axioms is for recursion and induction. The latter is formulated as a rule in a way specifically to enable the D-interpretation of the $\in$-Induction scheme in KPω.

C. Recursion Axiom and Induction Rule.

  14. (Recursion) For each type σ, R_σ is of type $(V \to V \to \sigma) \to (V \to \sigma)$. Then for f a variable of type $(V \to V \to \sigma)$ and x of type V and for $g = R_\sigma f$ we have the equation
  $$gx = f(g|x)x$$

15. (Induction) Suppose that $\theta(x, g, u)$ is a formula and that G and Z are closed terms for which the following has been inferred:

$$(\forall y \in x)\theta(y, Gy, Zxu) \to \theta(x, Gx, u)$$

Then we may infer $\theta(x, Gx, u)$.

NB. In the preceding, g and u may be sequences of variables (possibly empty) of arbitrary type, while x is of type V.

This completes our description of the system T_V.

Lemma 3. SCS is D-interpreted in T_V.

Proof. The D-interpretation of each of the axioms 1-6 of IKPω in T_V is straightforward. Furthermore, by the general facts about the D-interpretation established in section 2 above, we obtain without further work the D-interpretations of (Δ_0-LEM), (MP), and (AC), this last in the functional form $\forall x \exists y \phi(x, y) \to \exists f \forall x \phi(x, fx)$, where ϕ is an arbitrary formula. To obtain the D-interpretation of (AC$_{\text{Set}}$) from this, suppose $(\forall x \in a)\exists y \phi(x, y)$. Then under the D-interpretation, we also have $\forall x \exists y (x \in a \to \phi(x, y))$, so there exists an f such that $(\forall x \in a)\phi(x, fx)$. Let $r = f|a$; then by Lemma 2, Fun(r) and Dom$(r) = a$ and $(\forall x \in a)\phi(x, r(x))$, as required by (AC$_{\text{Set}}$). To prove the D-interpretation of BOS, we argue just as for Theorem 2 in the proof of NOS, but now combining AC with the bounded choice operator C instead of the operator μ.

So the only thing left to deal with is the D-interpretation of the $\in$-Induction Rule 7 of IKPω. For that, let $\psi(x)^D = \exists g \forall u \psi_D(x, g, u)$, where g, u are sequences of variables (possibly empty) of various types. We write θ for ψ_D. Then to form the D-interpretion of the hypothesis of the $\in$-Induction Rule, we pass through the following sequence of formulas

$$\forall x\{(\forall y \in x)\exists h \forall w \theta(y, h, w) \to \exists f \forall u \theta(x, f, u)\},$$

$$\forall x\{\exists g \forall w \forall y[y \in x \to \theta(y, gy, w)] \to \exists f \forall u \theta(x, f, u)\},$$

$$\forall g, x \exists f \forall u \exists w, y\{[y \in x \to \theta(y, gy, w)] \to \theta(x, f, u)\},$$

$$\exists f', y', w' \forall g, x, u\{[y'gxu \in x \to \theta(y'gxu, g(y'gxu), w'gxu)] \to \theta(x, f'gx, u)\}.$$

So finally, by induction hypothesis we have closed terms F, Y, W, such that the following is provable in T_V:

$$[Ygxu \in x \to \theta(Ygxu, g(Ygxu), Wgxu)] \to \theta(x, Fgx, u).$$

Then the following is also provable in T_V:

$$(\forall y \in x)\theta(y, gy, Wgxu) \to \theta(x, Fgx, u).$$

Now apply the Recursion Axiom of T_V to obtain G satisfying the equation $Gx = F(G|x)x$. Substituting $G|x$ for g throughout the preceding, and taking $Zxu = W(G|x)u$, it follows that $(\forall y \in x)\theta(y, Gy, Zxu)$ has been inferred. Hence by the Induction Rule 15 of T_V we may infer $\theta(x, Gx, u)$, which is the D-interpretation of $\forall x \psi(x)$.

III. To complete the proof of Theorem 5, we need to interpret T_V in a system of strength KPω. This is provided by the system of Operational Set Theory, OST, for a type-free applicative structure over set theory introduced in Feferman (2001a); see also Feferman (2006) and Jäger (2007). The language of OST extends the language L of set theory by a binary operation symbol A for application, a unary relation symbol $\downarrow$ for definedness and various constants. The terms $r, s, t, \ldots$ of the extended language are generated from the variables $a, b, c, \ldots, f, g, h, \ldots, x, y, z$ and constants by closing under application, $A(s, t)$. We write st for $A(s, t)$, and think of s as a partial function (coded as a set) whose value at t exists if $(st)\downarrow$ holds; this allows interpretation of a partial combinatory type-free calculus in OST. The logic of OST is the classical logic of partial terms due to Beeson.[6] The axioms of OST come in four groups:

(1) Axioms for the applicative structure given by the (partial) combinators k, s.

(2) Axioms for logical operations for negation, disjunction and bounded quantification, along with the characteristic function for membership, as in T_V.

(3) Basic set-theoretic axioms for extensionality, empty set, unordered pair, union, infinity and the ∈-Induction Scheme, as in KPω.

(4) Operational set-theoretic axioms for Separation, Range (or Replacement) as in T_V; in addition there is a Universal Choice operator C satisfying $\exists x(fx = 0) \to (Cf)\downarrow \wedge f(Cf) = 0$.

For each type symbol σ of T_V, we define $M_\sigma(x)$ inductively as follows to express in the language of OST that x is an object of type σ:

(i) $M_V(x)$ is $(x = x)$

(ii) $M_{\sigma \to \tau}(x)$ is $\forall y[M_\sigma(y) \to xy\downarrow \wedge M_\tau(xy)]$

We may treat the predicates M_σ as classes and write $f : M_\sigma \to M_\tau$ for $M_{\sigma \to \tau}(f)$. The translation of the constants of T_V into those of OST except for the recursors is immediate; for each of these we may check that if the constant is of type σ then its translation is a closed term of OST that is provably in M_σ.

So now consider any recursor R_σ; this is of type $(V \to V \to \sigma) \to (V \to \sigma)$. As its interpretation we make use of the type-free form of the recursion theorem that is a consequence of the applicative axioms of OST; this provides a closed term **rec** such that for any f, $\mathbf{rec}f\downarrow$ and for $g = \mathbf{rec}f$ and any x, we have $gx \simeq fgx$, i.e. either both sides are defined and equal or both are undefined. We also make use of Lemma 5 of Feferman (2006), according to which there is a closed term **fun** such that for any f, x such that $(\forall y \in x)fy\downarrow$ we have $\mathbf{fun}fx\downarrow$, and $\mathbf{fun}fx$ is the graph of f restricted to x considered as a set; in other words **fun** may be taken as the interpretation of Grph and we also write

---

[6]See Troelstra and van Dalen (1988) pp. 50-51, where Et is written for $t\downarrow$ and the logic of partial terms is called E-logic.

$f|x$ for **fun**fx. Finally, using the recursor **rec**, we obtain a closed term **r** such that $\mathbf{r}f\downarrow$ for all f, and for $g = \mathbf{r}f$, the following is provable:

$$gx \simeq f(g|x)x$$

We claim each R_σ can be translated by this same term **r**. That is, no matter what σ we take, we have

$$\mathbf{r} : (V \to V \to M_\sigma) \to (V \to M_\sigma)$$

For, suppose given any $f : (V \to V \to M_\sigma)$; and let $g = \mathbf{r}f$. It is to be shown that $g : (V \to M_\sigma)$, i.e. that for each x, $gx\downarrow$ and gx is in M_σ. This is proved by $\in$-induction on x; if it holds for all $y \in x$, then **fun**$gx\downarrow$, i.e. $g|x$ is in V, so by assumption, $f(g|x)x$ is in M_σ, and hence the same holds for gx. QED

Lemma 4. Under this translation, T_V is interpreted in OST.

Proof. The verification of all the axioms of T_V by the corresponding axioms of OST up to those for Recursion and Induction are immediate. The Recursion axiom is taken care of in the way just described, so it is only left to check the Induction Rule. So suppose that $\theta(x, v, u)$ is a formula which is a translation of a formula of T_V for which v is a sequence of variables of type $\sigma = \sigma_1, \ldots, \sigma_n$ and u is a sequence of variables of type $\tau = \tau_1, \ldots, \tau_m$; we write $M_\sigma(v)$ for the conjunction of statements $M_{\sigma_i}(v_i)$ and similarly for $M_\tau(u)$. Suppose further that G and Z are closed terms of OST for which the following has been inferred:

$$\forall x \forall u \{M_\sigma(Gx) \wedge [M_\tau(u) \to M_\tau(Zxu)] \wedge [(\forall y \in x)\theta(y, Gy, Zxu) \to \theta(x, Gx, u)]\}$$

Then we conclude

$$\forall x\{(\forall y \in x)\forall u[M_\tau(u) \to \theta(y, Gy, u)] \to \forall u[M_\tau(u) \to \theta(x, Gx, u)\}$$

Thus by the Induction Scheme in OST we conclude $\forall x \forall u[M_\tau(u) \to \theta(x, Gx, u)]$, which verifies the translation of the conclusion of the Induction Rule in T_V. □

We may now complete the proof of Theorem 5 by means of the fact established in Feferman (2006) (and in another way in Jäger (2007)) that OST is of the same proof-theoretical strength as KPω. Finally, the fact that KPω is of the same proof-theoretical strength of ID$_1$ is due to Jäger (1982); it is then of the same strength as ID$_1^{(i)}$ by Buchholz (1980). □

If the power set operation is considered as a definite operation, which is suggested by one philosophical view of set theory which still regards the universe of all sets as an indefinite totality, we are led to a semi-constructive system for which we can prove the following theorem in the same way as was done for Theorem 5.

Theorem 6. The system IKPω + (Pow) + (Δ_0-LEM) + (MP) + (BOS) + (AC) has proof-theoretical strength between the classical systems KPω + (Pow) and KPω + (Pow) + (V=L).

This makes use of the result proved in Jäger (2007) that the proof-theoretical strength of OST + (Pow) is bounded by that of KPω + (Pow) + (V=L). It is conjectured but it is not known whether the strength of the latter is the same as that of KPω + (Pow); the standard argument to eliminate the Axiom of Constructibility does not apply in any obvious way.

7 A miscellany of related work and questions

7.1 Kohlenbach's "Lesser" NOS

Kohlenbach (2001) considers the following weakening of NOS that he calls the Lesser Numerical Omniscience Scheme:

(LNOS) $\quad \forall n[(\phi(n) \vee \neg\phi(n)) \wedge (\psi(n) \vee \neg\psi(n))] \wedge \neg\exists n\phi(n) \wedge \neg\exists n\psi(n)$

$$\to \forall n \neg\phi(n) \vee \forall n \neg\psi(n)$$

His main result is that the semi-constructive system Res-HA$^\omega$ + (AC) + (MP) + (LNOS) is conservative over PRA for Π_2^0 sentences. The proof is by means of functional interpretation combined with the method of majorization. Kohlenbach also shows that the system in question proves WKL, i.e. König's Lemma for binary trees ("weak König's Lemma"). Ferreira and Oliva (2005) have introduced another method, called *bounded functional interpretation*, which they show may be used to obtain the same results in a simpler way. It would be interesting to see if their majorization and/or bounding techniques can be used to amplify the results of the present paper.

Kohlenbach (2008), p. 154, has also observed that WKL implies KL over Res-HA$^\omega$ + AC$_{0,0}$, so in such contexts, the difference between "weak" and "usual" König's Lemma disappears; this is in accord with the advantage of beefing up constructive and semi-constructive systems stressed here.

Since PRA is considered by many to be the limit of finitism, it would also be interesting to produce a natural semi-constructive system of finite type over the natural numbers for which all bounded formulas are decidable and whose proof-theoretical strength is equal to that of PRA. Finally, one may speculate that there are suitable such systems equivalent in strength to feasible arithmetic.

7.2 Friedman's system ALPO

Friedman (1980) introduced a semi-constructive system ALPO (for "Analysis with the Limited Principle of Omniscience") in the language of set theory with the natural numbers as a set of urelements, for which the main result is conservation of ALPO over PA for all arithmetic sentences.[7] For comparison with the system SCS = IKP$_\omega$ + (Δ_0-LEM) + (MP) + (BOS) + (AC) of Theorem 5 above, here are the axioms of ALPO: A. Ontological (urelements and sets), B. Urelement extensionality, C. Successor axioms, D. Infinity, E. Sequential induction, F. Sequential recursion, G. Pairing (unordered), H. Union, I. Exponentiation, J. Countable choice, K. Δ_0-separation, L. Strong collection, M. Limited principle of omniscience. By E is meant that any sequence (i.e. function) a of natural numbers which is such that $a(0) = 0 \wedge \forall n(a(n) = 0 \to a(n') = 0)$ then $\forall n(a(n) = 0)$. Axiom E guarantees definition by primitive recursion. The axiom J is of course a consequence of AC$_{\text{Set}}$ in our system, as is the strong collection axiom L (i.e. collection applied to arbitrary formulas). Other than Axiom I, all of these are thus derivable in SCS. That axiom asserts the existence for any sets a, b, of the set of all functions from a to b, which is not a consequence of SCS or even of its finite type extension SCS$^\omega$ (at least not

---

[7] I was reminded of Friedman's work by Jeremy Avigad.

in any obvious way). In his paper, Friedman makes use of a special model-theoretic argument in order to eliminate Axiom I before completing the proof that ALPO is conservative over PA. It would be of great interest to see whether the methods of functional interpretation employed here can be adapted to prove the same. Note that Axiom I does follow from the power set axiom used in the extension of SCS for Theorem 6.

7.3 Burr's interpretation of KPω

A useful variant functional interpretation due to Shoenfield (1967) sec. 8.3 in $\forall\exists$ form that is sometimes used applies directly to a classical system without requiring initial passage through the N-translation. The straightforward attempt to give such an interpretation of KPω meets an immediate obstacle if the constant 0 is to be part of the language; namely, it follows from provability of $\forall x \exists y (x \neq 0 \to y \in x)$ that one must have a term $t(x)$ such that $x \neq 0 \to t(x) \in x$ is provable in the target QF system. In other words one must have a non-constructive ("choice") operator for bounded quantification (of the sort provided in the system T_V by the bounded choice operator C). In order to avoid this, Burr (1998, 2000) gives a further Diller-Nahm (1974) $\forall\exists$-variant interpretation of KPω in a QF theory of primitive recursive set functionals of finite type. It is quite different from the interpretation given here in sec. 6, but there may be interesting relationships that are worth pursuing.

7.4 Some systems of semi-intuitionistic set theory with the power set axiom

The study of such subsystems of ZF formulated in intuitionistic logic with LEM for bounded formulas was apparently initiated by Poszgay (1971, 1972) and then studied more systematically by Tharp (1971), Friedman (1973) and Wolf (1974).[8] Poszgay had conjectured that his system is as strong as ZF, but Tharp and Friedman proved its consistency in ZF using a modification of Kleenes method of realizability. Wolf established the equivalence in strength of several related systems. The first is K_1, a system with axioms of Extensionality, Pairing, Union, Infinity and Power Set, the full Induction Scheme, and with Replacement restricted to formulas in which all quantifiers are bounded or subset bounded. K_2 is K_1 + LEM, and K_3 is K_1 plus a certain strong axiom scheme of Transfinite Recursive Definitions which implies the Full Replacement and Collection axiom schemes; finally K_3^* is K_3 + MP. (In this notation, what Tharp and Friedman proved is consistency in ZF of an extension of K_1 plus Full Replacement and the usual Axiom of Choice.) Wolf's main results include equiconsistency of K_1, K_2 + V=L, and K_3^*. The system K_3^* is close in many respects to the system IKPω + (Pow) + (Δ_0-LEM) + (MP) + (BOS) + (AC$_{\text{Set}}$) dealt with here in Theorem 6, except for BOS and AC$_{\text{Set}}$ (Full Axiom of Choice scheme), and which also implies Full Replacement and Collection. It should be of interest to make a detailed comparison between these systems and of the methods involved.

---

[8] I am indebted to Harvey Friedman and Robert Wolf for bringing this work to my attention, after the body of this paper was completed.

7.5 Mathematics in semi-constructive systems

Coquand and Palmgren (2000) give a couple of examples of mathematical theorems in their semi-constructive system for countable tree ordinals (described in sec. 5 above) that can be provided with a constructive foundation via their constructive sheaf-theoretic model of the system. The first is König's Lemma for binary trees; but in fact, as shown by Kohlenbach in the work described in 7.1 above, a much, much weaker system (conservative over PRA) suffices to do the same. The second is Dickson's Lemma, according to which if $u : \mathbb{N} \to \mathbb{N}$ and $v : \mathbb{N} \to \mathbb{N}$ are two sequences of natural numbers then there exist $p < q$ such that $u(p) \leq u(q)$ and $v(p) \leq v(q)$. That follows from a prior lemma, that for any sequence $u : \mathbb{N} \to \mathbb{N}$ of natural numbers, there exists a sequence $n_0 < n_1 < n_2 < \ldots$ such that $u(n_0) \leq u(n_1) \leq u(n_2) \leq \ldots$. To obtain Dickson's Lemma from this, one first finds a strictly increasing sequence of natural numbers on which u is increasing, and then a strictly increasing subsequence of that on which v is increasing, to get a sequence $n_0 < n_1 < n_2 < \ldots$ on which both u and v are increasing. We may then take $p = n_0$ and $q = n_1$. Again, what is needed can be proved in a much, much weaker system, namely that of Theorem 3(i) conservative over PA. The truth of Dickson's Lemma implies that we can obtain p, q as recursive functionals of u and v; simply search for the first p, q which make it true. The constructive model of Coquand and Palmgren can hardly be expected to provide more useful information about the complexity of that functional.

More generally, all of the semi-constructive systems treated in Theorem 3 are candidates of potential interest in which to carry out predicative mathematics. The actual pursuit of that part of mathematics in various classical systems of explicit mathematics, as described, e.g., in (Feferman 1975 and Feferman and Jäger 1993, 1996) as well as in theories of finite type over the natural numbers (Feferman 1977, 1979) make systematic use of explicit witnessing data. For example, a uniformly continuous function on a closed interval of real numbers is treated as a pair consisting of a function of real numbers on that interval and a uniform modulus of continuity functions. As pointed out by Friedman at the outset of his (1980) article, such padding is unnecessary in semi-constructive systems in which the Axiom of Choice holds in sufficiently strong form, as it does in ALPO and in the various systems considered here. How far this freedom takes us is another matter, but the actual development of predicative mathematics in these systems should certainly be revisited in that light. In addition, one should see how much mathematics can be conveniently carried out in the impredicative semi-constructive systems of secs. 5 and 6. Finally, it would be worth pursuing the formulation and determination of the proof-theoretical strength of semi-constructive systems of explicit mathematics and operational set theory, neither of which has been directly handled here, and in both of which mathematics can in general be carried out in a more flexible manner than in typed systems or even in set theoretical systems.

Acknowledgements: I wish to thank the two referees for their useful comments on a draft of this paper, and Shivaram Lingamneni for his help in transforming it into a LaTeX file.

References

J. Avigad and S. Feferman (1998), Gödels functional ("Dialectica") Interpretation, in (S. Buss, ed.) Handbook of Proof Theory, Elsevier, Amsterdam, 337-405.

J. Avigad and H. Towsner (2009), Functional interpretation and inductive definitions, J. Symbolic Logic (to appear).

E. Bishop (1967), Foundations of Constructive Analysis, McGraw Hill, New York.

W. Buchholz (1980), The $\Omega_{\mu+1}$-rule, in (W. Buchholz, et al., eds.) Iterated inductive definitions and subsystems of analysis. Recent proof theoretical studies, Lecture Notes in Mathematics 897, 188-233.

W. Burr (1998), Functionals in Set Theory and Arithmetic, Doctoral Dissertation, Mnster.

W. Burr (2000), A Diller-Nahm-style functional interpretation of $KP\omega$, Archive for Mathematical Logic 39, 599-604.

T. Coquand and E. Palmgren (2000), Intuitionistic choice and classical logic, Archive for Mathematical Logic 39, 53-74.

J. Diller and W. Nahm (1974), Eine Variante zur Dialectica Interpretation der Heyting Arithmetik endlicher Typen, Arch. Math. Logik u. Grundlagenforschung 16, 49-66.

S. Feferman (1968), Ordinals associated with theories for one inductively defined set. (Unpublished notes.)

S. Feferman (1971), Ordinals and functionals in proof theory, in Proc. Int'l Cong. Mathematicians, Nice 1970, vol. 1, Gauthier-Villars, Paris, 229-233.

S. Feferman (1975), A language and axioms for explicit mathematics, in (J. N. Crossley, ed.), Algebra and Logic, Lecture Notes in Mathematics 450, 87-139.

S. Feferman (1977), Theories of finite type related to mathematical practice, in (J. Barwise, ed.) Handbook of Mathematical Logic, North-Holland, Amsterdam, 913-971.

S. Feferman (1979), A more perspicuous system for predicativity, in (K. Lorenz, et al. eds.) Konstruktionen vs. Positionen I, de Gruyter, Berlin , 87-139.

S. Feferman (2001), On the strength of some systems with the numerical omniscience scheme, (abstract) Bull. Symbolic Logic 7, 111.

S. Feferman (2001a), Notes on operational set theory, I. Generalization of small large cardinals in classical and admissible set theory. http://math.stanford.edu/feferman/papers/OperationalST-I.pdf.

S. Feferman (2006), Operational set theory and small large cardinals, Proceedings of WoLLIC 06, to appear in Information and Computation; see http://dx.doi.org/10.1016/j.ic.2008.04.007.

S. Feferman and G. Jäger (1993), Systems of explicit mathematics with nonconstructive -operator, I, Annals of Pure and Applied Logic 65, 243-263.

S. Feferman and G. Jäger (1996), Systems of explicit mathematics with nonconstructive -operator, II, Annals of Pure and Applied Logic 79, 37-52.

F. Ferreira and P. Oliva (2005), Bounded functional interpretation, Annals of Pure and Applied Logic 135, 73-112.

H. Friedman (1973), Some applications of Kleene's methods for intuitionistic systems, Lecture Notes in Mathematics 337.

H. Friedman (1980), A strong conservative extension of Peano Arithmetic, in (J. Barwise, et al., eds.) The Kleene Symposium, North-Holland, Amsterdam, 113-122.

G. Jäger (1982), Zur Beweistheorie der Kripke-Platek-Mengenlehre ber den natürlichen Zahlen, Archiv f. Math. Logik u. Grundlagenforschung 22.

G. Jäger (2007), On Feferman's operational set theory OST, Annals of Pure and Applied Logic 150, 19-39.

U. Kohlenbach (2001), Intuitionistic choice and restricted classical logic, Mathematical Logic Quarterly 47, 455-460.

U. Kohlenbach (2008), Applied Proof Theory. Proof interpretations and their use in mathematics, Springer-Verlag, Berlin.

L. Pozsgay (1971), Liberal intuitionism as a basis for set theory, in Axiomatic Set Theory, Proc. Symp. Pure Math. XIII, Part 1, 1971, 321-330.

L. Pozsgay (1972) Semi-intuitionistic set theory, Notre Dame J. of Formal Logic 13, 546-550.

J. R. Shoenfield (1967), Mathematical Logic, Addison-Wesley, Reading, MA.

L. H. Tharp (1971), A quasi-intuitionistic set theory, J. Symbolic Logic 36, 456-460.

A. S. Troelstra and D. van Dalen (1988), Constructivism in Mathematics. An introduction, vol. 1, North-Holland, Amsterdam.

R. S. Wolf (1974), Formally Intuitionistic Set Theories with Bounded Predicates Decidable, PhD Thesis, Stanford University.

On the logical analysis of proofs based on nonseparable Hilbert space theory

ULRICH KOHLENBACH[1]

1 Introduction

Starting in [15] and then continued in [9, 17, 24] and [18], general logical metatheorems were developed that guarantee the extractability of highly uniform effective bounds from proofs of theorems that hold for general classes of structures such as metric, hyperbolic, CAT(0), normed or Hilbert spaces. To obtain uniformity e.g. w.r.t. parameters that range over metrically bounded (but not compact) sets it is crucial to exploit the fact that the proof to be analyzed does not use any separability assumption on the underlying spaces (as e.g. the existence of uniform bounds for the very statement of separability would yield the total boundedness of bounded metric spaces, see [18] for a detailed discussion of this issue). In order to do so we developed in [15] formal systems $\mathcal{T}^\omega[X,\ldots]$ that treat such abstract spaces X as atoms added to fragments $\mathcal{T}^\omega$ of full (though only weakly extensional) analysis $\mathcal{A}^\omega$ by adding a new base type X for variables ranging over X and all the finite types built upon $\mathbb{N}$ and X. The metatheorems developed in the aforementioned papers and the book [18] were based on novel extensions of (monotone [12]) functional interpretation in the sense of Gödel and Spector. These theorems made it possible to explain some concrete proof unwindings that had been carried out in fixed point theory resulting in unexpectedly uniform bounds (see e.g. [13, 14, 21]) and paved the way for many new applications in fixed point theory, geodesic geometry, ergodic theory and topological dynamics that were guaranteed to be possible by these proof theoretic results (see e.g. [1, 3, 4, 7, 8, 16, 20, 23, 22]). For applications of related forms of 'proof mining' in proof theory itself see Mints [25, 26].

$\mathcal{A}^\omega$ is a very strong formal system for analysis as it contains the full axiom schemas of dependent and countable choice in all types and so full arithmetical comprehension over numbers. While the latter feature makes it possible to formalize virtually all proofs in ordinary analysis for **separable** spaces it is not clear a priori how much e.g. of Hilbert space theory for $(X,\langle\cdot,\cdot\rangle)$ can be carried out in the absence of any separability condition on X in $\mathcal{A}^\omega[X,\langle\cdot,\cdot\rangle,\mathcal{C}]$ (see [18] for the definition of this theory) or even fragments $\mathcal{T}^\omega[X,\langle\cdot,\cdot\rangle,\mathcal{C}]$.

In this paper, we show that general orthogonal projection arguments, the Riesz representation theorem and the weak sequential compactness of the unit ball in X can be proved in such systems. As an application we show that a theorem

---
[1]Dedicated to Grigori Mints for his 70th Birthday. The author has been supported by the German Science Foundation (DFG Project KO 1737/5-1).

of Browder [5] (stating the convergence of a certain explicit iteration sequence) can be proved and use this to obtain (by applying a metatheorem from [18]) a highly uniform effective rate of metastability in the sense of Tao [29, 30] on that convergence.

All undefined notions and notations in this paper (including the representation of real numbers by number theoretic functions $f : \mathbb{N} \to \mathbb{N}$ and the corresponding operations $+_\mathbb{R}, -_\mathbb{R}, \ldots$ and relations $<_\mathbb{R}, \leq_\mathbb{R}, \ldots$ on these representatives) are understood as in [18] on which this paper relies. We denote the type for the natural numbers $\mathbb{N} := \{0, 1, 2, \ldots\}$ (resp. for number theoretic functions $f : \mathbb{N} \to \mathbb{N}$) by 0 (resp. 1). Since all theories used in this paper are weakly extensional in the sense of [18] we drop the prefix 'WE-' in their name and e.g. write PA^ω instead of WE-PA^ω.

2 Main results

DEFINITION 1. A formula $A(x^X) \in \mathcal{L}(\mathcal{T}^\omega[X, \|\cdot\|, \ldots])$ (for some theory $\mathcal{T}^\omega$ in the language of functionals of all finite types such as PA^ω) defines

1. a nonempty subset of X if

   (a) $\forall x^X, y^X \, (A(x) \land x =_X y \to A(y))$,

   (b) $\exists x^X \, A(x)$;

2. a nonempty convex subset of X if in addition to (a) and (b)

   $$\forall \lambda^1, x^X, y^X \, (A(x) \land A(y) \to A((1 -_\mathbb{R} \tilde{\lambda}) \cdot_X x +_X \tilde{\lambda} \cdot_X y)),$$

   where $\lambda \mapsto \tilde{\lambda}$ is the construction for the representation of $[0, 1]$ from [18](Definition 4.24);

3. a linear subspace of X if (a) and

   $$A(0_X) \land \forall \alpha^1, \beta^1, x^X, y^X \, (A(x) \land A(y) \to A(\alpha \cdot_X x +_X \beta \cdot_X y));$$

4. a closed nonempty convex subset (resp. closed linear subspace) of X if in addition to the conditions for nonempty convex subsets (resp. linear subspaces) we have that

   $$\forall x_{(\cdot)}^{0 \to X} \, (\forall n \, A(x_n) \land \\ \forall m^0, n^0, k^0 \, (m, n \geq k \to \|x_m - x_n\|_X \leq_\mathbb{R} 2^{-k}) \to A(C(x_{(\cdot)}))).$$

   Here C is the operator from [18] that maps each fast converging Cauchy sequence in X to its limit thereby expressing the completeness of X.

We say that A defines provably in some theory $\mathcal{T}$ whose language is contained in $\mathcal{L}(\text{PA}^\omega[X, \|\cdot\|])$ resp. of $\mathcal{L}(\text{PA}^\omega[X, \|\cdot\|, C])$ any of the concepts above if the corresponding formulas are provable in $\mathcal{T}$. In fact, instead of PA^ω this definition can also be applied to any fragment (or extension) as long as the ingredients to formulate the respective formulas above are contained in the language.

REMARK 2. Definition 1.1 resp. 1.2 can also be applied to formulas in $\mathcal{L}(\mathrm{PA}^\omega[X,d])$ resp. $\mathcal{L}(\mathrm{PA}^\omega[X,d,W])$, where $\mathrm{PA}^\omega[X,d]$ (resp. $\mathrm{PA}^\omega[X,d,W]$) is the extension of PA^ω by an abstract metric (resp. hyperbolic) space X (see [18] for details).

Convention on notation: From now on we will adopt a more informal notation and e.g. write $\|\cdot\|, k \in \mathbb{N}, \lambda \in \mathbb{R}, x \in X$ and $\alpha x + \beta y$ instead of $\|\cdot\|_X, k^0, \lambda^1, x^X$ and $\alpha \cdot_X x +_X \beta \cdot_X y$ and only use (partially) the latter when the precise logical form matters.

PROPOSITION 3.

1. $\mathrm{PA}^\omega[X, \|\cdot\|\,]$ proves that for every nonempty subset $S \subseteq X$ given by a formula A as in the definition above (using $\exists y \in S \ldots$, resp. $\forall y \in S \ldots$, as shorthand for $\exists y^X (A(y) \wedge \ldots)$ resp. $\forall y^X (A(y) \to \ldots)$) the following holds:

    (1) $\forall k \in \mathbb{N} \forall x \in X \exists y \in S \forall z \in S\, (\|x - y\| < \|x - z\| + 2^{-k})$.

    $(\mathrm{PA}^\omega + \mathrm{AC}^{0,0})[X, \|\cdot\|\,]$ proves that $d := \inf\limits_{z \in S}\{\|x - z\|\}$ exists.

2. $\mathcal{A}^\omega[X, \langle \cdot, \cdot \rangle, \mathcal{C}]$ proves that for every closed nonempty convex subset C given by a formula A the following holds:

    (2) $\forall x \in X \exists ! y \in C \forall z \in C\, (\|x - y\| \leq \|x - z\|)$.

    In particular, $d := \inf\limits_{z \in C}\{\|x - z\|\}$ not only exists but is attained by some $y \in C$.

    The result holds a-fortiori for closed linear subspaces of X that are given by a formula A.

Proof: 1. Suppose that (1) fails for some $k \in \mathbb{N}, x \in X$ i.e.

(3) $\forall y \in S \exists z \in S\, (\|x - y\| \geq \|x - z\| + 2^{-k})$.

Let $\widehat{y} \in S$. Then by induction on $n \in \mathbb{N}$ one shows that for every n there exists a sequence, i.e. a function $f^{0 \to X}$, such that $f(0) =_X \widehat{y}$ and

(4) $\forall i \leq n (f(i) \in S) \wedge \forall i < n\, (\|x - f(i)\| \geq \|x - f(i+1)\| + 2^{-k})$.

Now let $n > \|x - \widehat{y}\| \cdot 2^k$. Then there exists a function $f^{0 \to X}$ such that

$$\|x - \widehat{y}\| = \|x - f(0)\| \geq \|x - f(n)\| + n \cdot 2^{-k} > \|x - f(n)\| + \|x - \widehat{y}\|,$$

which is a contradiction.
The last claim of '1.' follows from the fact that for every $k \in \mathbb{N}$ there exists a rational number r encoded by, say, m such that r is a 2^{-k-1} rational approximation to $\|x - y\|$, where y is from (1) for $k + 1$. Now $\mathrm{AC}^{0,0}$ gives a fast converging Cauchy sequence (r_k) (encoded by (m_k)) that represents d.

2. Let $x \in X$ and $\widehat{y} \in C$. Applying to (1) countable choice $\mathrm{AC}^{0,X}$ yields a sequence (y_k) in C with

(5) $\forall k \in \mathbb{N} \forall z \in C\, (\|x - y_k\| < \|x - z\| + 2^{-k})$.

It is not hard to verify that for uniformly convex normed spaces with modulus of convexity $\eta : (0,2] \to (0,1]$ the function

$$\Phi(\varepsilon) := \frac{\varepsilon}{4} \cdot \eta(\varepsilon/(D+1)),$$

where $\|x - \widehat{y}\| \leq D \in \mathbb{N}$, is a so-called modulus of uniqueness for the projection, i.e.

(6)
$$\forall \varepsilon \in (0,1), z_1, z_2 \in C$$
$$(\bigwedge_{i=1}^{2} (\forall z \in C (\|x - z_i\| \leq \|x - z\| + \Phi(\varepsilon)) \to \|z_1 - z_2\| \leq \varepsilon)$$

(see e.g. proposition 17.4 in [18] for – a stronger version of – this). In the case of a Hilbert space it is well-known that (for the best modulus η)

$$\eta(\varepsilon) = 1 - \sqrt{1 - \varepsilon^2/4} \geq \varepsilon^2/8 \quad (\varepsilon \in (0,2]).$$

Now define

$$\Psi(k) := \min l \left[2^{-l} \leq \frac{2^{-3k}}{32(D+1)^2} \right].$$

Then – using (5) and (6) – $(y_{\Psi(k)})_k$ is a Cauchy sequence with Cauchy rate 2^{-k}. Hence by the completeness axiom $(\mathcal{C})$ of $\mathcal{A}^\omega[X, \langle \cdot, \cdot \rangle, \mathcal{C}]$ the limit y exists in X and – by the condition that A defines a closed convex subset C – also in C. It is now easy to verify that

$$\|x - y\| \leq \|x - z\| \quad \text{for all } z \in C.$$

The uniqueness of y immediately follows from (6). $\square$

REMARK 4. Proposition 3.2 also holds for the theory $\mathcal{A}^\omega[X, \|\cdot\|, \eta, \mathcal{C}]$ of uniformly convex Banach spaces with a modulus of uniform convexity η.

The standard proof that, in the case where X is a Hilbert space and $\mathcal{L}$ a closed linear subspace, $x - y$ for the point y as in proposition 3.2 is orthogonal to $\mathcal{L}$ goes through in our formal context without problems. In fact, we have the following quantitative version:

PROPOSITION 5. $\widehat{\mathrm{PA}}^\omega_{|}[X, \langle \cdot, \cdot \rangle]$ proves the following: if $\mathcal{L}$ is a linear subspace given by a formula A, then

$$\begin{cases} \forall \varepsilon > 0 \, \forall K \geq 1 \, \forall x \in X \, \forall y, z \in \mathcal{L} \\ (\|z\| \leq K \wedge \|x - y\|^2 \leq \|x - (y + \alpha z)\|^2 + \frac{\varepsilon^2}{K^2} \to |\langle x - y, z \rangle| \leq \varepsilon), \end{cases}$$

where

$$\alpha := \frac{\langle x - y, z \rangle}{\max \left(\left(\varepsilon/(2 \max(\|x\|, \|y\|, 1)) \right)^2, \|z\|^2 \right)}.$$

In particular:

$$\forall x \in X \forall y \in \mathcal{L} \, (\forall z \in \mathcal{L} \, (\|x - y\| \leq \|x - z\|) \to \forall z \in \mathcal{L}(\langle x - y, z \rangle = 0)).$$

Proof: Case 1:
$$\|z\|^2 \leq \bigl(\varepsilon/(2\max(\|x\|,\|y\|,1))\bigr)^2.$$
Then
$$\|z\| \leq \frac{\varepsilon}{2\max(\|x\|,\|y\|,1)}$$
and so
$$|\langle x-y,z\rangle| \leq |\langle x,z\rangle| + |\langle y,z\rangle| \leq \|x\|\cdot\|z\| + \|y\|\cdot\|z\| \leq \varepsilon.$$
Case 2:
$$\|z\|^2 > \bigl(\varepsilon/(2\max(\|x\|,\|y\|,1))\bigr)^2.$$
Then
$$\alpha = \frac{\langle x-y,z\rangle}{\|z\|^2}.$$
Hence
$$\begin{aligned}\|x-y\|^2 &\leq \|x-(y+\alpha z)\|^2 + \tfrac{\varepsilon^2}{K^2} = \\ \langle (x-y)-\alpha z, (x-y)-\alpha z\rangle &+ \tfrac{\varepsilon^2}{K^2} = \\ \|x-y\|^2 - 2\alpha\langle x-y,z\rangle + \alpha^2\|z\|^2 &+ \tfrac{\varepsilon^2}{K^2} = \\ \|x-y\|^2 - 2\tfrac{\langle x-y,z\rangle^2}{\|z\|^2} + \tfrac{\langle x-y,z\rangle^2}{\|z\|^2} &+ \tfrac{\varepsilon^2}{K^2}.\end{aligned}$$
Thus
$$\frac{\langle x-y,z\rangle^2}{\|z\|^2} \leq \frac{\varepsilon^2}{K^2}$$
and so (using that $\|z\| \leq K$)
$$\langle x-y,z\rangle^2 \leq \varepsilon^2.$$
Hence $|\langle x-y,z\rangle| \leq \varepsilon$. □

A particularly relevant instance of proposition 3 (needed below to reduce the general case of weak compactness to the separable case) is the following one where we project to a closed linear subspace
$$\mathcal{L} := \overline{\mathrm{Lin}_\mathbb{R}\{x_n : n \in \mathbb{N}\}}$$
generated from a sequence (x_n) in X. Clearly, $\mathcal{L}$ can (provably in $\mathcal{A}^\omega[X,\langle\cdot,\cdot\rangle,\mathcal{C}]$) be represented by a formula A_L, namely

$$A_L(x) :\equiv \begin{cases} \exists f^1\,((\varphi(k,f,(x_n)))_k \text{ is a Cauchy sequence in } \mathcal{L}_\mathbb{Q} \text{ with rate } 2^{-k} \\ \wedge\, x =_X \lim_{k\to\infty} \varphi(k,f,(x_n))), \end{cases}$$

where
$$\mathcal{L}_\mathbb{Q} := \bigcup_{n\in\mathbb{N}} \{r_0 x_0 + \ldots + r_n x_n : (r_0,\ldots,r_n) \in \mathbb{Q}^{n+1}\},$$
$$\varphi(k,f,(x_n)) := \sum_{i=0}^{lth(f(k))-1} q_{(f(k))_i} \cdot_X x_i.$$

and $(q_n)_{n \in \mathbb{N}}$ is some primitive recursive standard enumeration of $\mathbb{Q}$ (we identify q_n with its canonical embedding into $\mathbb{R}$).
In this case it is, however, easier to work directly with the countable dense subset $\mathcal{L}_\mathbb{Q} \subset \mathcal{L}$:

PROPOSITION 6. $\mathcal{A}^\omega[X, \langle \cdot, \cdot \rangle, \mathcal{C}]$ *proves the following: for every sequence* (x_n) *in* X *and every* $x \in X$ *there exists the projection of* x *to the closed linear subspace*

$$\mathcal{L} := \overline{\text{Lin}_\mathbb{R}\{x_n : n \in \mathbb{N}\}}$$

generated from $\{x_n : n \in \mathbb{N}\}$, *i.e.*

$$\forall x \in X \exists! y \in \mathcal{L} \forall z \in \mathcal{L} (\|x - y\| \leq \|x - z\|).$$

Moreover, instead of the full axiom schema DC *of dependent choice (or full countable choice combined with higher induction) the proof only needs (for given* x *and* (x_n)) *a fixed instance of* Π_1^0-AC *as well as* Σ_1^0-*induction (that alone suffices to prove the approximate version 3.1 in this case) and hence can be carried out, in particular, in* $\mathcal{T}^\omega[X, \langle \cdot, \cdot \rangle, \mathcal{C}]$, *where* $\mathcal{T}^\omega := \widehat{\text{PA}}^\omega \upharpoonright + \text{AC}_{ar}^{0,0}$.

Proof: Since $\mathcal{L}_\mathbb{Q}$ is dense in $\mathcal{L}$ we can represent elements of $\mathcal{L}$ as Cauchy sequences w.r.t. $\|\cdot\|$ of elements in $\mathcal{L}_\mathbb{Q}$ with Cauchy rate 2^{-k}. Let $(y_k)_{k \in \mathbb{N}}$ some primitive recursive (in (x_n)) standard enumeration of $\mathcal{L}_\mathbb{Q}$.
The existence of an ε-projection of x to $\mathcal{L}$ (in the sense of proposition 3.1) can then equivalently be formulated as

$$\forall x^X, k^0 \exists n^0 \forall m^0 (\|x - y_n\| \leq \|x - y_m\| + 2^{-k})$$

which can be proved by Σ_1^0-IA.
We then use Π_1^0-AC to form a sequence $(y_{f(k)})_{k \in \mathbb{N}}$ of 2^{-k}-projections and finish the proof as in the case of proposition 3.2. $\square$

DEFINITION 7. A formula $A_L(x^X, y^1)$ in $\mathcal{L}(\mathcal{T}^\omega[X, \ldots])$ is said to represent a bounded linear functional $L : X \to \mathbb{R}$ of a normed space X if

(7) $\quad \exists C^0 \forall x^X \exists y^1 (A_L(x, y) \land |y|_\mathbb{R} \leq_\mathbb{R} (C)_\mathbb{R} \cdot_\mathbb{R} \|x\|_X)$,

(8) $\quad \forall x^X \forall y^1, y_2^1 (A_L(x, y_1) \to (A(x, y_2) \leftrightarrow y_1 =_\mathbb{R} y_2)))$,

(9) $\quad \begin{aligned} &\forall \alpha_1^1, \alpha_2^1, y_1^1, y_2^1 \forall x_1^X, x_2^X \\ &\begin{cases} (A_L(x_1, y_1) \land A_L(x_2, y_2) \to \\ A_L(\alpha_1 \cdot_X x_1 +_X \alpha_2 \cdot_X x_2, \alpha_1 \cdot_\mathbb{R} y_1 +_\mathbb{R} \alpha_2 \cdot_\mathbb{R} y_2)). \end{cases} \end{aligned}$

In the following we write, more informally, '$C \cdot \|x\|$' instead of '$(C)_\mathbb{R} \cdot_\mathbb{R} \|x\|_X$' etc.

PROPOSITION 8. *Provably in* $\widehat{\text{PA}}^\omega \upharpoonright [X, \|\cdot\|]$, *every formula* A_L *that represents a bounded linear functional* $L : X \to \mathbb{R}$ *satisfies*

1. $x_1 =_X x_2 \land y_1 =_\mathbb{R} y_2 \land A_L(x_1, y_1) \to A_L(x_2, y_2)$,

2. $A_L(x_1, y_1) \land A_L(x_2, y_2) \to |y_1 -_\mathbb{R} y_2| \leq_\mathbb{R} C \cdot \|x_1 - x_2\|$, *where* C *is as in* (7).

Proof: 1. Let C by as in (7) and assume $x_1 =_X x_2$, $y_1 =_\mathbb{R} y_2$ and $A_L(x_1, y_1)$. By (7) we get $\exists \hat{y}^1 A_L(x_2, \hat{y})$ and so by (9) $A_L(x_1 -_X x_2, y_1 -_\mathbb{R} \hat{y})$. Again by (7) we get the existence of a $\tilde{y}^1$ such that

$$A_L(x_1 -_X x_2, \tilde{y}) \wedge |\tilde{y}| \leq_\mathbb{R} C \cdot \|x_1 -_X x_2\| =_\mathbb{R} 0.$$

Hence $\tilde{y} =_\mathbb{R} 0$ and so, by (8), $y_1 - \hat{y} =_\mathbb{R} 0$, i.e. $\hat{y} =_\mathbb{R} y_1 =_\mathbb{R} y_2$. Again by (8) this gives $A_L(x_2, y_2)$.

2. Let C be as in (7) and assume $A_L(x_1, y_1)$ and $A_L(x_2, y_2)$. Then by (9) $A_L(x_1 -_X x_2, y_1 -_\mathbb{R} y_2)$. From (7) we get

$$\exists y^1 \left(A_L(x_1 -_X x_2, y) \wedge |y| \leq_\mathbb{R} C \cdot \|x_1 -_X x_2\| \right).$$

(8) now yields that $y_1 -_\mathbb{R} y_2 =_\mathbb{R} y$ and so $|y_1 -_\mathbb{R} y_2| \leq_\mathbb{R} C \cdot \|x_1 -_X x_2\|$. □

PROPOSITION 9. $\mathcal{A}^\omega[X, \langle \cdot, \cdot \rangle, C]$ *proves the Riesz representation theorem in the following schematic form: if a formula $A(x^X, y^1)$ in $\mathcal{L}(\mathcal{A}^\omega[X, \langle \cdot, \cdot \rangle, C])$ represents a bounded linear functional $L : X \to \mathbb{R}$, then there exists a point v^X such that*

$$L(x) =_\mathbb{R} \langle v, x \rangle \text{ for all } x^X.$$

Proof: We can follow essentially the standard textbook proof:
Case 1: $L \equiv 0$, i.e. $\forall x^X, y^1 (A_L(x,y) \to y =_\mathbb{R} 0)$ (or – equivalently – $\forall x^X \exists y^1 (A_L(x,y) \wedge y =_\mathbb{R} 0)$). Then take $v := 0_X$.
Case 2: $\exists x_0^X (|L(x_0)| >_\mathbb{R} 0)$, i.e. $\exists x_0^X \exists y^1 (A_L(x_0, y) \wedge |y| >_\mathbb{R} 0)$. Reasoning in $\mathcal{A}^\omega[X, \langle \cdot, \cdot \rangle, C]$ one easily shows that $Kern(L) := \{x \in X : L(x) =_\mathbb{R} 0\}$ is a closed linear subspace of X that is given by the formula

$$A_{Kern(L)}(x) :\equiv \forall y^1 (A_L(x,y) \to y =_\mathbb{R} 0).$$

By proposition 3.2 there exists a (unique) point $w \in Kern(L)$ such that

$$\|x_0 - w\| = \text{dist}(x_0, Kern(L)).$$

Now consider $v_0 := x_0 - w$. Then $|L(v_0)| =_\mathbb{R} |L(x_0)| >_\mathbb{R} 0$ and so $\|v_0\| >_\mathbb{R} 0$ by (7). By proposition 5 we have

$$\forall z \in Kern(L) (\langle v_0, z \rangle =_\mathbb{R} 0).$$

Define $v_1 := \frac{v_0}{\|v_0\|}$. Then again $\langle v_1, z \rangle =_\mathbb{R} 0$ for all $z \in Kern(L)$. Now put $a := L(v_1)$ and, finally, $v := a \cdot_X v_1$. Using that $L(L(x) \cdot_X v_1 -_X L(v_1) \cdot_X x) =_\mathbb{R} 0$ we obtain

$$0 =_\mathbb{R} \langle v_1, L(x) \cdot_X v_1 -_X L(v_1) \cdot_X x \rangle =_\mathbb{R} L(x) -_\mathbb{R} a \cdot \langle v_1, x \rangle,$$

i.e.
$$L(x) =_\mathbb{R} \langle a \cdot_X v_1, x \rangle =_\mathbb{R} \langle v, x \rangle$$

for all $x \in X$. □

REMARK 10.

1. Definition 7 can be relativized to x^X being taken from a closed linear subspace $\mathcal{L}$ of X that is given via a formula $A_{\mathcal{L}}$ in the sense of definition 1.4. In this sense the previous proposition also applies to linear functionals $L : \mathcal{L} \to \mathbb{R}$.

2. In the following we will need the Riesz representation theorem for bounded linear functionals $L : \mathcal{L} \to \mathbb{R}$ on **separable** closed linear subspaces of X of the form
$$\mathcal{L} := \overline{\text{Lin}_{\mathbb{R}}\{x_n : n \in \mathbb{N}\}},$$
where (x_n) is a sequence in X.

In the case of separable Banach spaces it has been shown in [2] that the Riesz representation theorem can be proved from arithmetical comprehension over the weak base system RCA_0 used in reverse mathematics (see [28]). With the same proof we can establish the Riesz representation theorem for spaces of the form $\mathcal{L}$ above and bounded linear operators $L : \mathcal{L} \to \mathbb{R}$ that are given directly as a functional (rather than a representing formula) in $(\widehat{\text{PA}}^\omega\restriction + \text{AC}_{ar}^{0,0})[X, \langle \cdot, \cdot \rangle, \mathcal{C}]$. Here the crucial observation is that if (y_k) is dense in $\mathcal{L}$ (e.g. some standard enumeration of $\mathcal{L}_\mathbb{Q}$) and $y \in Kern(L)$ with $L(y) = 1$ (for $Kern(L) = \{0\}$ things are trivial), then the sequence defined by $w_k := y_k - L(y_k) \cdot y$ is dense in $Kern(L)$ (see [2], p.168) so that one can reason as in the proof of proposition 6 with w_n instead of y_n.

THEOREM 11. Let $\mathcal{T}^\omega := \widehat{\text{PA}}^\omega\restriction + \text{AC}_{ar}^{0,0}$. Then $\mathcal{T}^\omega[X, \langle \cdot, \cdot \rangle, \mathcal{C}]$ proves that the closed unit ball $B_1(0)$ in X is weakly sequentially compact, i.e. that for every sequence (x_n) in $B_1(0)$ there exists a point $v \in B_1(0)$ and a subsequence $(x_{n_k})_k$ of (x_n) such $\langle x_{n_k}, w \rangle$ converges to $\langle v, w \rangle$ for a every $w \in X$.
Instead of $B_1(0)$ and with $\mathcal{T}^\omega := \text{PA}^\omega + \text{AC}_{ar}^{0,0}$ one may have any bounded closed convex subset $C \subset X$ that is given by a formula of $\mathcal{L}(\mathcal{T}^\omega[X, \langle \cdot, \cdot \rangle, \mathcal{C}])$.

Proof: Let (x_n) be a sequence in $B_1(0)$. Consider again the separable closed linear subspace of X
$$\mathcal{L} := \overline{\text{Lin}_{\mathbb{R}}\{x_n : n \in \mathbb{N}\}}$$
and the countable dense subset
$$\mathcal{L}_\mathbb{Q} := \bigcup_{n \in \mathbb{N}} \{r_0 x_0 + \ldots + r_n x_n : (r_0, \ldots, r_n) \in \mathbb{Q}^{n+1}\}$$
given by some primitive recursive (in (x_n)) standard enumeration (y_k). A bounded linear functional $L : \mathcal{L} \to \mathbb{R}$ can be recovered from its restriction to $\mathcal{L}_\mathbb{Q}$, i.e. from $L^- : \mathbb{N} \to \mathbb{R}$ defined by $L^-(k) := L(y_k)$. In this way we can represent such functionals that are bounded by 1 as points $z = (a_n)_n$ in $\prod_{n \in \mathbb{N}}[-\|y_n\|, \|y_n\|]$ that satisfy

$$(*) \quad y_k =_X r_1 \cdot_X y_i +_X r_2 \cdot_X y_j \to a_k =_\mathbb{R} r_1 a_i +_\mathbb{R} r_2 a_j$$

for all $i, j, k \in \mathbb{N}$ and $r_1, r_2 \in \mathbb{Q}$ (see [6]).
Using arithmetical comprehension we can show in ACA_0 (see [6] or [28]) and

hence in $\mathcal{T}^\omega$ that $\prod_{n\in\mathbb{N}}[-\|y_n\|,\|y_n\|]$ as well as its subset of points satisfying $(*)$ is sequentially compact, i.e. that every sequence in this space has a convergent subsequence w.r.t. the product metric on $\prod_{n\in\mathbb{N}}[-\|y_n\|,\|y_n\|]$ (the functional interpretation of this fact can be realized by functionals involving only primitive recursion at type 0 and bar recursion of lowest type $B_{0,1}$; see [27] for an explicit construction of the solution functionals).

We now consider the following sequence (L_n) linear functionals $\mathcal{L} \to \mathbb{R}$ that are all bounded by 1 :
$$L_n(x) := \langle x_n, x\rangle.$$

Let (L_n^-) be the corresponding sequence of points in $\prod_{n\in\mathbb{N}}[-\|y_n\|,\|y_n\|]$ that satisfy $(*)$ and $(L_{n_k}^-)$ a convergent subsequence with limit L^-. Clearly, L^- also represents a linear functional $L : \mathcal{L} \to \mathbb{R}$ that is bounded by 1.

By the Riesz representation theorem applied to $\mathcal{L}$ (instead of X) L is represented by an element $v \in \mathcal{L}$ (provably in $(\widehat{\text{PA}}^\omega{\upharpoonright} + \text{AC}_{ar}^{0,0})[X, \langle\cdot,\cdot\rangle, \mathcal{C}]$ by remark 10.2) i.e.
$$\forall w \in \mathcal{L}\,(L(w) =_\mathbb{R} \langle v, w\rangle).$$

Hence (for all $w \in \mathcal{L}$)
$$\langle x_{n_k}, w\rangle =_\mathbb{R} L_{n_k}(w) \stackrel{k\to\infty}{\longrightarrow} L(w) =_\mathbb{R} \langle v, w\rangle.$$

Now let $x \in X$ and apply proposition 6 to get the (by proposition 5 orthogonal) projection $x_L \in \mathcal{L}$ of x onto $\mathcal{L}$. Then
$$\langle x_{n_k}, x\rangle =_\mathbb{R} \langle x_{n_k}, x_L\rangle \text{ and } \langle v, x\rangle =_\mathbb{R} \langle v, x_L\rangle.$$

Hence
$$\langle x_{n_k}, x\rangle \stackrel{k\to\infty}{\longrightarrow} \langle v, x\rangle.$$

It is easy to see in $\widehat{\text{PA}}^\omega{\upharpoonright}[X, \langle\cdot,\cdot\rangle]$ that $\|v\| \leq 1$: Suppose that $\|v\| = 1 + \varepsilon$ for some $\varepsilon > 0$. Then
$$\langle x_{n_k}, v\rangle \stackrel{k\to\infty}{\longrightarrow} \langle v, v\rangle = (1+\varepsilon)^2 \geq 1 + 2\varepsilon,$$

whereas $\langle x_{n_k}, v\rangle \leq \|x_{n_k}\| \cdot \|v\| \leq 1 + \varepsilon$ for all $k \in \mathbb{N}$ which is a contradiction. If instead of $B_1(0)$ one has a bounded closed convex subset $C \subset X$ given by a formula of the language, then one can argue essentially as above with some K satisfying $K \geq \|x\|$ for all $x \in C$ replacing 1 and $\prod_{n\in\mathbb{N}}[-\|y_n\|,\|y_n\|]$ replaced by $\prod_{n\in\mathbb{N}}[-K\cdot\|y_n\|, K\cdot\|y_n\|]$ However, to show that $v \in C$ is somewhat more involved. Usually one applies a Hahn-Banach separation theorem here to show that closed convex sets are weakly closed but in our situation there is a more elementary proof of this fact via Mazur's theorem which implies that there is a sequence (z_n) of finite convex combinations of $\{x_{n_k} : k \in \mathbb{N}\}$ which strongly converges to v. Since, by the convexity of C, we have that $z_n \in C$ for all $n \in \mathbb{N}$, the closedness of C yields that $v \in C$. The proof of Mazur's theorem for Hilbert spaces, as given e.g. in [11], can easily be formalized in $(\mathrm{PA}^\omega + \mathrm{AC}_{ar}^{0,0})[X, \langle \cdot, \cdot \rangle, \mathcal{C}]$. □

THEOREM 12 (F. Browder [5]). *Let X be a Hilbert space and $U : X \to X$ be a nonexpansive mapping. Assume that there exists a nonempty bounded closed convex subset $C \subset X$ such that U maps C into itself. For $v_0 \in C$ and $t \in (0,1)$ let $U_t(x) := tU(x) + (1-t)v_0$ and u_t be the unique fixed point of this strict contraction. Then (u_t) converges strongly to a fixed point $p \in C$ of U as $t \to 1$.*

REMARK 13. The theorem by Browder states, furthermore, that p is the unique fixed point of U in C that is closest to v_0.

PROPOSITION 14. *Browder's proof of his theorem above can, for closed bounded convex subsets C that are given by a formula A_C in the sense of definition 1.4, be proved in $\mathcal{A}^\omega[X, \langle \cdot, \cdot \rangle, \mathcal{C}]$ (under the assumption that U has a fixed point).*

Proof: We just sketch how to formalize the proof here (a detailed explicit logical analysis of the proof – together with an extraction of χ in proposition 16 below – has to be devoted to another paper): we first note that the fact that we may use only the weak form of extensionality is no problem as nonexpansive mappings U trivially are extensional. The proof of Browder's theorem first proceeds ('lemma 1') by forming the projection u_0 of v_0 onto the (nonempty) set F of all fixed points of U in C. It is an easy consequence of the uniform convexity of X that F is convex. Moreover, F obviously is closed and given by the formula $A_F(u^X) :\equiv (A_C(u^X) \wedge U(u) =_X u)$. Hence we can apply proposition 3.2 to establish the existence of $u_0 \in F$.

A second – perfectly elementary – 'lemma 2' shows that every weak limit of an approximate fixed point sequence is a fixed point of U. The proof is then concluded by showing that for any sequence $(k_j)_{j \in \mathbb{N}}$ with $k_j \to 1$ one can find a subsequence of (u_{k_j}) converging strongly to u_0 (in this proof it is used that (u_{k_j}) always possesses a weakly convergent subsequence whose weak limit $v \in C$ by 'lemma 2' even is in F). With theorem 11 in place, it is easily verified that the (nontrivial but – modulo the weak compactness – elementary) proof of this formalizes in our formal framework which we skip here. Thus it only remains to formalize the proof that the latter fact implies that u_t tends to u_0 for $t \to 1$. Let (k_j) be a sequence with $\lim_{j \to \infty} k_j = 1$ and assume that for some $\varepsilon > 0$

$$\forall n \exists j > n \, (\|u_{k_j} - u_0\| >_\mathbb{R} \varepsilon).$$

Then Σ_1^0-$\mathrm{AC}^{0,0}$ (and hence QF-$\mathrm{AC}^{0,0}$) yields a function g such that

$$\forall n \, (g(n) > n \wedge \|u_{k_{g(n)}} - u_0\| >_\mathbb{R} \varepsilon).$$

Then $k_{g(n)} \stackrel{n \to \infty}{\to} 1$, but $(u_{k_{g(n)}})_n$ does not contain a subsequence that converges to u_0. □

REMARK 15. From a well-known theorem of Browder-Göhde-Kirk it follows that U always has fixed points. However, for our application of the logical metatheorem (corollary 6.8 in [9]) below we do not have to consider the proof of this theorem but only the proof relative to the assumption of a fixed point.

In fact, the logical metatheorem we will use below allows one to convert this proof into a new one which only uses the (trivial) existence of approximate fixed points so that the need to use the Browder-Göhde-Kirk theorem disappears altogether.

THEOREM 16. *Under the assumptions of theorem 12 with $C := B_1(0)$ there exists a computable functional $\chi : \mathbb{N} \times \mathbb{N}^{\mathbb{N}} \to \mathbb{N}$ (that is independent from X, U and $v_0 \in B_1(0)$) such that*

$$\forall k \in \mathbb{N} \forall g : \mathbb{N} \to \mathbb{N} \exists n \leq \chi(k, g) \forall i, j \in [n; n + g(n)] \, (\|x_i - x_j\| < 2^{-k}),$$

where $x_i := u_{t_i}$ with $t_i := 1 - \frac{1}{i+1}$ and $[n; n + m] := \{n, n+1, \ldots, n+m\}$. Similarly, for any sequence (t_n) in $(0, 1)$ that converges towards 1 where then the bound depends also on a (majorant of) rate of metastability of that convergence

$$\forall n \in \mathbb{N} \forall g \in \mathbb{N}^{\mathbb{N}} \forall i \in [\chi(g, n); \chi(g, n) + g(\chi(g, n))] \, (|1 - t_i| \leq \frac{1}{n+1})$$

and a function $h : \mathbb{N} \to \mathbb{N}$ such that $\forall n \in \mathbb{N} \, (t_n \leq 1 - \frac{1}{h(n)+1})$.

Proof: We first note that x_i can be explicitly defined as a functional in v_0, U and i in the language of $\mathcal{A}^{\omega}[X, \langle \cdot, \cdot \rangle, \mathcal{C}]$ using the Picard iteration of U and the completion operator $\mathcal{C}$.
By proposition 14 we have

$$\mathcal{A}^{\omega}[X, \langle \cdot, \cdot \rangle, \mathcal{C}] \vdash$$
$$\begin{cases} \forall v_0 \in B_1(0) \, \forall U : X \to X \, \forall k \in \mathbb{N} \, (U \text{ n.e. } \wedge \, U(B_1(0)) \subseteq B_1(0) \\ \wedge \, Fix(U) \neq \emptyset \to \exists n \in \mathbb{N} \forall m \in \mathbb{N} \, (\|x_n - x_m\| <_{\mathbb{R}} 2^{-k})) \end{cases}$$

and hence

$$\mathcal{A}^{\omega}[X, \langle \cdot, \cdot \rangle, \mathcal{C}] \vdash$$
$$\begin{cases} \forall v_0 \in B_1(0) \, \forall U : X \to X \, \forall k \in \mathbb{N} \, \forall g : \mathbb{N} \to \mathbb{N} \\ (U \text{ n.e. } \wedge \, U(B_1(0)) \subseteq B_1(0) \wedge \, Fix(U) \neq \emptyset \to \\ \exists n \in \mathbb{N} \forall i, j \in [n; n + g(n)] \, (\|x_i - x_j\| <_{\mathbb{R}} 2^{-k})), \end{cases}$$

where

$$\forall i, j \in [n; n + g(n)] \, (\|x_i - x_j\| < 2^{-k})$$

is equivalent to a Σ_1^0-formula over already $\mathcal{T}^{\omega}[X, \langle \cdot, \cdot \rangle]$ with

$$\mathcal{T}^{\omega} := \widehat{\mathrm{PA}}^{\omega}\!\!\upharpoonright + \text{ QF-AC}^{0,0}$$

since only the bounded collection principle for Σ_1^0-formulas is needed.
We note that the condition '$U(B_1(0)) \subseteq B_1(0)$' can be written as a $\forall$-formula

$$\forall x^X \, (\|U(\check{x})\| \leq_{\mathbb{R}} 1), \text{ where } \check{x} := \frac{x}{\max(1, \|x\|)}.$$

The claim of the theorem now follows immediately from corollary 6.8 in [9] (together with the treatment of completeness conditions from [18], pages 433-434). Here one takes v_0 in place of z and C is being treated trivially as the whole space X (i.e. by adding a universal axiom $\forall x^X \, (\chi_C(x^X) =_0 0)$), where $c_X := 0_X$, so that we can take $b := 2$. □

REMARK 17. Since (as discussed already in remark 15) the above proof does not use anymore the existence of a fixed point of U, it can be used in itself as an alternative proof of the existence of a fixed point: since (x_n) satisfies the no-counterexample version of the Cauchy property it is a Cauchy sequence and its limit clearly must be a fixed point of U.

REMARK 18. An issue to be devoted to further research is whether the closed convex set C can be treated completely abstract (via a characteristic function as in corollary 6.8 from [9]) which would mean that (in contrast to closed convex sets given by a formula in the sense of definition 1.4) one is allowed to use only a quantifier-free rule of extensionality

$$\frac{A_0 \to s =_X t \wedge s \in C}{A_0 \to t \in C} \quad (A_0 \text{ quantifier-free})$$

rather than the full extensionality condition from definition 1.1.(a). If the rule suffices one could extract an effective bound χ that would not depend on C except for some norm upper bound on $M \geq \|v\|$ for all $v \in C$.

Comments added in proof: Subsequent work ([19]) that we have carried out after the present paper was finished sheds new light on the logical status of Browder's proof as well as the theorem as such:

1. As a final **result** of the actual logical analysis of Browder's proof, the use of weak compactness in the end disappears and the extraction process yields even a primitive recursive (in the sense of Kleene) bound on the metastable version of Browder's theorem (see [19] for a discussion of this phenomenon). In fact, the extraction works for arbitrary bounded closed convex subsets $C \subset X$ (instead of $B_1(0)$) and the bound depends on C only via an upper bound on the diameter of C.

2. A different proof of Browder's theorem that avoids already weak compactness from the beginning can be obtained from Halpern [10]. The – much simpler – logical analysis of this proof again yields a primitive recursive rate of metastability (see again [19]).

BIBLIOGRAPHY

[1] Avigad, J., Gerhardy, P., Towsner, H., Local stability of ergodic averages. Trans. Amer. Math. Soc. **362**, pp. 261-288 (2010).
[2] Avigad, J., Simic, K., Fundamental notions of analysis in subsystems of second-order arithmetic. Ann. Pure Appl. Logic **139**, pp. 138-184 (2006).
[3] Briseid, E.M., A rate of convergence for asymptotic contractions. J. Math. Anal. Appl. **330**, pp. 364-376 (2007).
[4] Briseid, E.M., Fixed points of generalized contractive mappings. Journal of Nonlinear and Convex Analysis **9**, pp. 181-204 (2008).
[5] Browder, F.E., Convergence of approximants to fixed points of nonexpansive nonlinear mappings in Banach spaces. Arch. Rational Mech. Anal. **24**, pp. 82-90 (1967).
[6] Brown, D.K., Functional analysis in weak subsystems of second order arithmetic. PhD Thesis. The Pennsylvania State University. University Park, PA (1987).
[7] Gerhardy, P., A quantitative version of Kirk's fixed point theorem for asymptotic contractions. J. Math. Anal. Appl. **316**, pp. 339-345 (2006).

[8] Gerhardy, P., Proof mining in topological dynamics. Notre Dame Journal of Formal Logic **49**, pp. 431-446 (2008).
[9] Gerhardy, P., Kohlenbach, U., General logical metatheorems for functional analysis. Trans. Amer. Math. Soc. **360**, pp. 2615-2660 (2008).
[10] Halpern, B., Fixed points of nonexpanding maps. Bull. Amer. Math. Soc. **73**, pp., 957-961 (1967).
[11] Hunter, J.K., Applied Analysis. World Scientific,439 pp. (2001).
[12] Kohlenbach, U., Analysing proofs in analysis. In: W. Hodges, M. Hyland, C. Steinhorn, J. Truss, editors, *Logic: from Foundations to Applications. European Logic Colloquium* (Keele, 1993), pp. 225–260, Oxford University Press (1996).
[13] Kohlenbach, U., A quantitative version of a theorem due to Borwein-Reich-Shafrir. Numer. Funct. Anal. and Optimiz. **22**, pp. 641-656 (2001).
[14] Kohlenbach, U., Uniform asymptotic regularity for Mann iterates. J. Math. Anal. Appl. **279**, pp. 531-544 (2003).
[15] Kohlenbach, U., Some logical metatheorems with applications in functional analysis. Trans. Amer. Math. Soc. vol. 357, no. 1, pp. 89-128 (2005).
[16] Kohlenbach, U., Some computational aspects of metric fixed point theory. Nonlinear Analysis **61**, pp. 823-837 (2005).
[17] Kohlenbach, U., A logical uniform boundedness principle for abstract metric and hyperbolic spaces. Electronic Notes in Theoretical Computer Science (Proceedings of WoLLIC 2006) **165**, pp. 81-93 (2006).
[18] Kohlenbach, U., Applied Proof Theory: Proof Interpretations and their Use in Mathematics. Springer Monographs in Mathematics. xx+536pp., Springer Heidelberg-Berlin, 2008.
[19] Kohlenbach, U., On quantitative versions of a theorem due to F.E. Browder. Preprint 2010, submitted.
[20] Kohlenbach, U., Lambov, B., Bounds on iterations of asymptotically quasi-nonexpansive mappings. In: G-Falset, J., L-Fuster, E., Sims, B. (eds.), Proc. International Conference on Fixed Point Theory, Valencia 2003, pp. 143-172, Yokohama Press 2004.
[21] Kohlenbach, U., Leuştean, L., Mann iterates of directionally nonexpansive mappings in hyperbolic spaces. Abstract and Applied Analysis, vol. 2003, no.8, pp. 449-477 (2003).
[22] Kohlenbach, U., Leuştean, L., A quantitative mean ergodic theorem for uniformly convex Banach spaces. Ergodic Theory and Dynamical Systems **29**, pp. 1907-1915 (2009).
[23] Kohlenbach, U., Leuştean, L., Asymptotically nonexpansive mappings in uniformly convex hyperbolic spaces. J. European Math. Soc. **12**, pp. 71-92 (2010).
[24] Leuştean, L., Proof mining in $\mathbb{R}$-trees and hyperbolic spaces. Electronic Notes in Theoretical Computer Science (Proceedings of WoLLIC 2006) **165**, pp. 95-106 (2006).
[25] Mints, G.E., Unwinding a non-effective cut elimination proof. In: Grigoriev, D., Harrison, J., Hirsch, E.A. (Eds.): Computer Science - Theory and Applications, First International Computer Science Symposium in Russia, CSR 2006, St. Petersburg, Russia, June 8-12, 2006, Proceedings. Springer LNCS **3967**, pp. 259-269 (2006).
[26] Mints, G.E., Proof search tree and cut elimination. In: Avron, A., Dershowitz, N., Rabinovich, A. (eds.), Pillars of Computer Science 2008, Springer LNCS 4800, pp. 521-536 (2008).
[27] Safarik, P., Kohlenbach, U., The Interpretation of the Bolzano-Weierstraß Principle Using Bar Recursion. To appear in: Math. Log. Quart.
[28] Simpson, S.G., Subsystems of Second Order Arithmetic. Perspectives in Mathematical Logic. Springer-Verlag. xiv+445 pp. 1999.
[29] Tao, T., Soft analysis, hard analysis, and the finite convergence principle. Essay posted May 23, 2007. Appeared in: 'T. Tao, Structure and Randomness: Pages from Year One of a Mathematical Blog. AMS, 298pp., 2008'.
[30] Tao, T., Norm convergence of multiple ergodic averages for commuting transformations. Ergodic Theory and Dynamical Systems **28**, pp. 657-688 (2008).

Polynomial-Time Decidability of a Bounded Universal Theory of Congruences Modulo a Prime Number

Nikolai K. Kosovsky

1 Introduction: It Is Important to Find Examples of Similar Practically Useful Problems with Drastically Different Computational Complexity

For many problems from the class NP, we know their computational complexity. In particular:

- for some of these problems, we can prove that they are NP-complete, while

- for some other problems, there is a polynomial-time algorithm for solving them – i.e., we can prove that these problems belong to the class P.

There are many problems, however, for which we do not know whether they are NP-complete or in P (or neither). For these problems, most research effort currently concentrates on analyzing these problems one by one. In addition to this analysis of *individual* problems, it is also desirable to get a good feel for the classes of NP-complete and P problems *as a whole*.

In geometric terms, a good way to understand the area is to find its boundary. From this viewpoint, to get a better general understanding of which problems are NP-complete and which are solvable in polynomial time, it is desirable to "map" the "boundary" between NP-complete and polynomial-time problems.

What do we mean by a boundary? A point from a set belongs to the boundary of this set if there are arbitrarily close points which belong to the complement of this set. Thus, to find a point on a boundary of a set means, in effect, that we find two close points, one of which belongs to the set and the other one belongs to its complement.

Thus, a good way to "map" the class of NP-complete problems – and thus, to get a better understanding of which problems are NP-complete – is to find examples of very similar problems with drastically different computational complexity. For example, it is desirable to find NP-complete problems for which very similar problems are solvable in polynomial time. It is natural to call two problems similar when they are instances of a parameterized family of problems, corresponding to very close values of parameters.

Examples of such pairs of problems are well known; for example (see, e.g., [3]):

- 3-SAT, propositional satisfiability problems for 3-CNF formulas (CNF formulas in which every clause has ≤ 3 literals) is NP-complete, while

- a similar problem 2-SAT, propositional satisfiability problems for 2-CNF formulas (CNF formulas in which every clause has ≤ 2 literals) can be solved in polynomial time.

Here, the integer values 2 and 3 of the corresponding parameter are maximally close.

A similar example was produced in our previous paper [6]:

- the problem of checking,
  - given an integer-valued matrix A with n columns and an integer-valued m-dimensional vector $B \geq -1$,
  - whether an inequality $Ax \geq B$ has an integer-value solution $x = (x_1, \ldots, x_n)$,

  is NP-complete, while

- for $B \geq 0$, this problem can be solved in polynomial time.

From the practical viewpoint, our objective is to make as much progress as possible in solving practical problems. From this viewpoint, when providing examples of close problems that would help us map the class of NP-complete problems, we should pay special attention to problems closely related to practice. As emphasized in [1], congruences modulo a given number appear in many practical applications.

In this paper, following our previous work [4, 5], we provide examples of problems – related to congruences –

- which are very similar but

- which have drastically different computational complexity.

While searching for such pairs of practice-related problems, we also have found new polynomial-time algorithms. Since these algorithms solve practice-related problems, we hope that these new algorithms will eventually lead to practical applications.

2 Relevant Results About Congruences Modulo m: A Brief Overview

In this paper, we will deal:

- with integer-valued polynomial *congruences* of the form

$$P(x) \equiv P'(x) \pmod{m},$$

  where $P(x)$ and $P'(x)$ are polynomials, and

- with *incongruences*, i.e., statements of the form $P(x) \not\equiv P'(x) \pmod{m}$.

To avoid trivial NP-completeness results (see details below), we assume that a polynomial is explicitly represented as a linear combination of monomials. In other words, we allow a polynomial $2x_1^2 \cdot x_2 + x_1 \cdot x_2^2 + x_1^2 + x_2$, but we do not allow expressions like $x_1^2 \cdot (x_2 + x_1 \cdot x_2^2) + x_1^2 + x_2$.

REMARK 1. For convenience, we also assume:

- that all the coefficients at the monomials are non-zeros,
- that the monomials are ordered in the decreasing order of their total degree, and
- that within each value of a total degree, the monomials $c \cdot x^d$, where $x^d \stackrel{\text{def}}{=} x_1^{d_1} \cdot \ldots \cdot x_n^{d_n}$, are ordered in the decreasing lexicographic order of the degree tuples $d = (d_1, \ldots, d_n)$.

From this viewpoint, $x_1^2 \cdot x_2 + x_1 \cdot x_2^2 + x_1^2 + x_2$ is a correct representation since:

- all the coefficients are non-zeros,
- the 3-rd degree terms $x_1^2 \cdot x_2$ and $x_1 \cdot x_2^2$ come first, followed by the 2-nd degree term x_1^2 and by a 1-st degree term x_2, and
- between the two 3-rd degree monomials, $x_1^2 \cdot x_2$ comes first since the corresponding degree tuple $(2, 1)$ precedes $(1, 2)$ in the decreasing lexicographic order.

Note that every linear combination of monomials can be re-ordered in this form in polynomial time, so this additional assumption has no effect on whether the corresponding problem is NP-complete or solvable in polynomial time.

The reason why we restricted ourselves to polynomials which are explicit linear combinations of monomials is that, as we mentioned, without this restriction the problem of checking the solvability of an incongruence is NP-complete. To be more precise, for every integer m, the following problem is NP-complete: given a polynomial $A(x)$ (with parentheses allowed), check whether the incongruence $A(x) \not\equiv 0 \pmod{m}$ has a solution $x \in \{0, 1\}^*$ (i.e., a solution in which each component x_i of the integer-valued tuple $x = (x_1, \ldots, x_n)$ is either 0 or 1); this result is listed as Corollary to Theorem 2 in [5].

In contrast, for restricted polynomials (linear combinations of monomials), the problem can already be solved in polynomial time ([5], Theorem 1). Specifically, for every prime number p, the following problem can be solved in polynomial time:

- given a (restricted) polynomial $P(x)$,
- check whether the incongruence $P(x) \not\equiv 0 \pmod{p}$ has a solution

$$x \in \{0, 1\}^*,$$

i.e., the problem of checking whether the following formula is true:

$$\exists x_{\in \{0,1\}^*} \; P(x) \not\equiv 0 \pmod{p}.$$

Note that if, instead of a single incongruence, we consider multiple incongruences, the problem again becomes NP-complete. Specifically, for every integer m, the following problem is NP-complete ([5], Theorem 2):

- given (restricted) polynomials $P_i(x)$ and $P'_i(x)$, $1 \leq i \leq k$,
- check whether the following formula is true:

$$\exists x \in \{0,1\}^* \, (P_1(x) \not\equiv P'_1(x) \pmod{m} \, \& \, \ldots \, \& \, P_k(x) \not\equiv P'_k(x) \pmod{m}).$$

We will use these results in our proofs.

3 First Result: Polynomial-Time Decidability of a Bounded Universal Theory of Congruences Modulo a Prime Number

In this section, instead of simply considering individual incongruences and their conjunctions, we consider arbitrary Boolean combinations of polynomial congruences $P_i(x) \not\equiv P'_i(x) \pmod{p}$. Specifically, we allow Boolean combinations using $\vee$, $\&$, and $\neg$.

For every prime number p and a natural number Q, by $B_{p,Q}$ we denote the class of all closed formulas of the type $\forall x \in \{0,1\}^* R(x)$, where $R(x)$ is a Boolean combination of polynomial congruences $P_i(x) \equiv P'_i(x) \pmod{p}$ in which the total number of $\vee$ and $\&$ connectives does not exceed Q.

For example, the formula

$$\forall x_1 \in \{0,1\} \forall x_2 \in \{0,1\} (\neg(x_1 \cdot x_2 \equiv 0 \pmod{p})) \vee x_1 \equiv 0 \pmod{p} \vee x_2 \equiv 0 \pmod{p})$$

belongs to $B_{p,Q}$ if $Q \geq 2$.

THEOREM 2. *For every prime number p and integer Q, there exists a polynomial-time algorithm that, given a formula from the class $B_{p,Q}$, checks whether this formula is true.*

Proof. Let us describe the desired algorithm. We start by simplifying the given formula.

1°. First, we take into account that we are only interested in the values $x_i \in \{0,1\}$ for which $x_i^{d_i} = x_i$ for all $d_i \geq 1$.

Thus, without changing the truth value of the formula, in each of the monomials of each of the polynomials, we can reduce each factor $x_i^{d_i}$ with $d_i \geq 1$ to x_i.

2°. Second, we replace each polynomial congruence $P_i(x) \equiv P'_i(x) \pmod{p}$ with an equivalent congruence $D_i(x) \equiv 0 \pmod{p}$, where $D_i(x) \stackrel{\text{def}}{=} P_i(x) - P'_i(x)$. After this replacement, all congruences have the form $D_i(x) \equiv 0 \pmod{p}$, with a zero right-hand side.

3°. Third, we use the De Morgan law to replace each conjunction $F \& F'$ with $\neg(\neg F \vee \neg F')$ (and delete whatever double negations we get). After this replacement, we have a Boolean formula that only contains $\vee$ and $\neg$. In this formula, the number of the resulting negations does not exceed $O(Q)$ – i.e., for fixed Q, is bounded by a constant.

4°. Next, we reduce the problem to the case when the Boolean combination $R(x)$ is simply a congruence. We will do it by eliminating the propositional connectives one by one: starting with the inside connectives, we replace each Boolean combination of congruences with an equivalent single congruence.

4.1°. Let us start with eliminating disjunctions ∨. Since p is a prime number, the product of two numbers is equal to 0 mod p if and only if one of the factors is equal to 0 mod p. Thus,

$$(P(x) \equiv 0 \pmod{p}) \vee (P'(x) \equiv 0 \pmod{p}) \Leftrightarrow P(x) \cdot P'(x) \equiv 0 \pmod{p}.$$

After we expand the product $P(x) \cdot P'(x)$ into a linear combination of monomials, we replace each term $x_i^{d_i}$ for which $d_i \geq 1$ with x_i.

Since we need to represent the product as a linear combination of monomials, the product of two polynomials with N terms each can result in a product consisting of N^2 terms.

4.2°. Let us now deal with negation. The negation $P(x) \not\equiv 0 \pmod{p}$ means that the remainder of dividing $P(x)$ by p is equal to 1, 2, ..., or $p-1$; so, one of the values $P(x) + 1$, $P(x) + 2$, ..., or $P(x) + (p-1)$ is equal to 0 mod p. Since p is a prime number, this is equivalent to the product of these expressions being equal to 0 mod p:

$$P(x) \not\equiv 0 \pmod{p} \Leftrightarrow (P(x) + 1) \cdot \ldots \cdot (P(x) + p - 1) \equiv 0 \pmod{p}.$$

After computing the product of the polynomials, we perform a similar reduction of $x_i^{d_i}$ with $d_i \geq 1$ to x_i.

4.3°. Finally, let us show that the above reduction leads to an at most polynomial-size increase in the length of the original formula.

Indeed, the number of the reduction steps is bounded by a constant times the number Q of connectives in the original formula – i.e., by a constant. At each step, we perform either 1 or $p-2$ multiplications, so the total number of multiplications is bounded by $O(p \cdot D)$, i.e., also by a constant.

After each multiplication, the original size L of the formula gets at most squared, to $\leq L^2$. After two steps, we get $\leq (L^2)^2 = L^{(2^2)}$; after s steps, we get a new formula of the size $\leq L^{(2^s)}$. Since s is bounded by a constant, we thus get a polynomial size increase.

5°. After all these reductions, we get a formula of the form

$$\forall x \in \{0,1\}^* \ P(x) \equiv 0 \pmod{p},$$

where $P(x)$ is a sum of monomials $P(x) = \sum a_{(k)} \cdot x^{v_k}$ with different tuples $v_k = (v_{k1}, \ldots, v_{kn})$ consisting of 0s and 1s.

REMARK 3. Here, as above, x^{v_k} stands for $x_1^{v_{k1}} \cdot \ldots \cdot x_n^{v_{kn}}$.

Let us show that $P(x)$ is always equal to 0 mod p if and only if all the coefficients a_k are equal to 0 mod p. This will enable us to check the above formula (which is equivalent to the original one) in polynomial time – by simply checking the values of the coefficients.

Of course, if all a_k are zeroes mod p, we get $P(x) = 0 \pmod{p}$. Let us prove the inverse by contradiction. Let us assume that $P(x) = 0 \pmod{p}$ for all $x \in \{0,1\}^*$, and some coefficients a_k are different from 0 mod p. Each 0-1 tuple v_k can be viewed as a characteristic function of a subset of the set $\{1,\ldots,n\}$; let us denote this set by S_k. There is always a subset S_{k_0} that is not contained in any other subset S_k. We can then take

- $x_i = 1$ for all i from this subset S_{k_0} – i.e., for all i for which $v_{k_0 i} = 1$, and

- $x_i = 0$ for all other values of i.

One can easily check that for this choice of values x_i, we get $P(x) = a_k$. Since we assumed that $P(x) \equiv 0 \pmod{p}$ and $a_k \not\equiv 0 \pmod{p}$, we thus get a contradiction – proving that all the coefficients a_k are actually equal to 0 mod p.

The theorem is proven. ∎

REMARK 4. In the above result, we assumed that the prime number p and the total number Q of propositional connectives are bounded. It turns out that without these assumptions the decision problem becomes NP-complete:

- when we keep Q bounded, but allow p to be part of the input, the problem is NP-complete even if we consider a single congruence $P(x) \equiv 0 \pmod{p}$; the proof of this NP-completeness is given in Section 5 of [5];

- when we keep p fixed, but allow Q to be part of the input, then NP-completeness follows from the above-mentioned result about the NP-completeness of checking solvability of systems of incongruences (Theorem 2 from [5]).

4 Additional Results: Case of Integer-Valued Variables

Let us show that the above results can be extended to the case when variables x_i can take arbitrary integer values – not only 0s and 1s. Indeed, since we are considering congruences mod p, it is sufficient to consider values $x_i \in \{0, 1, \ldots, p-1\}$. Each of these values x_i can be represented by its binary digits: $x_i = x_{i0} \cdot 2^0 + x_{i1} \cdot 2^1 + \ldots + x_{iN} \cdot 2^N$, where $x_{ij} \in \{0,1\}$ and N is the smallest integer for which $2^{N+1} \geq p$.

Substituting the corresponding expressions in each of the polynomials $P(x)$, we get a new polynomial in which all the variables x_{ij} take only values 0 and 1. The existence of the values x_i that make the value of the original polynomial equal to 0 mod p (or non-0 mod p) is equivalent to the existence of the values $x_{ij} \in \{0,1\}$ that make the new polynomial take the same values.

This is one of the ideas behind the following result. In its statement, for every prime number p and natural number Q, by $C_{p,Q}$ we denote the class of all closed formulas of the form $\forall x\, R(x)$, where $R(x)$ is a Boolean combination of polynomial congruences $P_i(x) \equiv P'_i(x) \pmod{p}$ in which the total number of ∨ and & connectives does not exceed Q, and where x runs over arbitrarily tuples of integers.

THEOREM 5. *For every prime number p and integer Q, there exists a polynomial-time algorithm that, given a formula from the class $C_{p,Q}$, checks whether this formula is true.*

REMARK 6. This result can be deduced from the Theorem (presented below) by using the equivalence

$$(P \equiv 0 \ (\mathrm{mod} \ p) \vee Q \equiv 0 \ (\mathrm{mod} \ p)) \Leftrightarrow P \cdot Q \equiv 0 \ (\mathrm{mod} \ p).$$

In particular, for the cases

- when no Boolean combinations are used and
- when we only consider one negation,

we get the following results:

COROLLARY 7. *For every p, there exists a polynomial-time algorithm that, given a polynomial $P(x)$, checks whether the incongruence $P(x) \not\equiv 0 \ (\mathrm{mod} \ p)$ has a solution x, i.e., checks whether the following formula is true:*

$$\exists x \, (P(x) \not\equiv 0 \ (\mathrm{mod} \ p)).$$

COROLLARY 8. *For every p, there exists a polynomial-time algorithm that, given a polynomial $P(x)$, checks whether the congruence $P(x) \equiv 0 \ (\mathrm{mod} \ p)$ has a solution x, i.e., checks whether the following formula is true:*

$$\exists x \, (P(x) \equiv 0 \ (\mathrm{mod} \ p)).$$

These results are corollaries of the following theorem:

THEOREM 9. *There exists a polynomial-time algorithm that, given a prime number p and a polynomial $P(x)$, checks whether the incongruence*

$$P(x) \not\equiv 0 \ (\mathrm{mod} \ p)$$

has a solution x, i.e., checks whether the following formula is true:

$$\exists x \, (P(x) \not\equiv 0 \ (\mathrm{mod} \ p)).$$

Proof.

$1°$. The proof of this theorem is based on the following auxiliary result: for $P(x) = \sum a_k \cdot x^{v_k}$ for which $v_{ki} < p$ for all i, we have

$$\forall x \, (P(x) \equiv 0 \ (\mathrm{mod} \ p)) \Leftrightarrow \forall k (a_k \equiv 0 \ (\mathrm{mod} \ p)).$$

Once this result is proven, to check whether $\forall x \, (P(x) \equiv 0 \ (\mathrm{mod} \ p))$, it is sufficient to check whether whether all the coefficients a_k are equal to 0 mod p – which, of course, requires polynomial (actually linear) time.

$2°$. To prove the auxiliary result, it is sufficient to consider the case of polynomials of a single variable x_1.

Indeed, with respect to x_1, the original polynomial has the form $P(x) = P_0 + P_1 \cdot x_1 + \ldots + P_k \cdot x_1^k$, where P_i are polynomials of $x_2, \ldots, x_n$. For every combination of values $x_2, \ldots, x_n$, the corresponding polynomial $P_0 + P_1 \cdot x_1 + \ldots + P_k \cdot x_1^k$ is equal to 0 mod p. Thus, from the result about the one-variable case, we conclude that all its coefficients P_i are equal to 0 mod p. We can then represent each P_i as a polynomial of x_2, and, by applying the same one-variable case, conclude that all the coefficients are equal to 0 mod p, etc.

$3°$. To prove the auxiliary result for the case of a single variable, we first use Fermat's Little Theorem $x^p \equiv x \pmod{p}$ to reduce all the powers of x to power x^d with $d \leq p-1$. Thus, we have a polynomial $P(x) = a_0 + a_1 \cdot x + \ldots + a_{p-1} \cdot x^{p-1}$ of degree $\leq p - 1$ which is equal to 0 mod p for all x. So, over the field $Z/pZ = \{0, 1, \ldots, p-1\}$ of all remainders mod p, this polynomial has p different roots: $x = 0$, $x = 1$, $\ldots$, and $x = p - 1$.

It is known that over every field – including the field Z/pZ of all remainders mod p – a polynomial of a degree d with non-zero coefficients can have no more than d roots. Since $d \leq p - 1 < p$, this means that all the coefficients a_k should indeed be zeroes mod p. The auxiliary statement is proven, and so is the theorem. ∎

BIBLIOGRAPHY

[1] V. Arvind and T. C. Vijayarughavan, "The complexity of Solving Linear Equations over a Finite Ring", In: V. Diekert and B. Durand (eds.), *Proc. of the 22nd Annual Symposium on Theoretical Aspects of Computer Science STACS'2005, Stuttgart, Germany, February 24–26, 2005*, Springer, Berlin-Heidelberg, 2005, pp. 472–484.

[2] D.-Z. Du and K.-I. Ko, *Theory of Computational Complexity*, Wiley, New York, 2000.

[3] M. R. Garey and D. S. Johnson, *Computers and Intractability: A Guide to the Theory of NP-Completeness*, Freeman, New York, 1979.

[4] T. M. Kosovskaya and N. K. Kosovsky, "Complexity of consistency checking of a system of linear integer comparisons," *Problems of Contemporary Mathematics: Proceedings of the International Conference in Honor of 200th Anniversary of Karl Gustav Jacob Jacobi and 750th Anniversary of Kaliningrad-Köningsberg*, Kaliningrad, April 4–8, 2005, Kaliningrad Univ. of Humanities Publ., 2005, pp. 157–158.

[5] N. K. Kosovskii and T. M. Kosovskaya, "The Number of Steps for Construction of a Boolean Solution to Polynomial Congruences and Systems of Polynomial Congruences," *Vestnik of St. Petersburg University: Mathematics*, 2007, Vol. 40. No. 3, pp. 218–223.

[6] N. K. Kosovsky and A. V. Tishkov, "Polynomial Algorithms for Determining Consistency of Systems of Strict and Non-Strict Inequalities in Rational Numbers," In: V. P. Chuvakov (ed.), *Topical Problems of Contemporary Mathematics*, Vol. 3, Novosibirsk, 1997, pp. 95–100.

Forty Five Years of Friendship
BORIS KUSHNER

Grigori Mints is 70! This cannot be was my first reaction, when colleagues approached me with the idea of writing some reminiscences about Grisha (his informal name in friends' circle). And still here it is. Everyone in our generation is approaching age numbers that sound rather biblically.

My reminiscences will not be about the scholarly achievements of Grisha. Those are first rate and deserve a special comprehensive article which I am sure will be written. I will try to look back through times past. And here I have to rely on my memory that confidently keeps emotional content of events, but is a rather fragile instrument when it comes to dates, and partly to names. I apologize beforehand if my memory fails me occasionally.

Both Grisha and I began our research careers as members of a school of constructive mathematics that had been established by A. A. Markov, Jr. not long before we arrived on the scene.

A few words about the background of Markovs School are due. A. A. Markov, Jr. (1903–1979) was the only son of the great Russian mathematician A. A. Markov, Sr. (1856–1922). In 1924 he received a degree in physics from the State University of Leningrad (formerly Saint Petersburg, then Petrograd, then Leningrad, then again Saint Petersburg such were the winds of History!). He developed into a Renaissance type of the mathematician; the corpus of his scholarly works covers various areas of mathematics, mechanics and theoretical physics. In all those endeavors he kept a strong interest in philosophical issues concerning mathematics. So, it is not surprising that A. A. Markov, Jr. began, after the WW2, to develop his own direction in mathematics that is known today under the name Markov's Constructive Mathematics (or Russian Constructive Mathematics). (For more detail see my papers [1, 2, 3, 4, 5, 6, 7]).

At the end of the 1950s, Markov moved from Leningrad to Moscow, where he became the Chair of the newly organized Department of Mathematical Logic at Moscow State University (the first such department in the Soviet Union). It was a bifurcation point for the Russian Constructive School. From now on it would have two main branches: in Moscow and in Leningrad (self-contained branches in Armenia and Czechoslovakia were to come into existence later). Nikolai Aleksandrovich Shanin headed the Leningrad branch. He was an excellent mathematician-topologist and made fundamental contributions to constructive logic, semantics and constructive mathematical analysis. He was (as he is today) a charismatic scholar and deeply influenced the research of his students. In all fairness one should speak about Shanin's constructive school, or, in a wider context, about Markov-Shanin Constructive Mathematics.

Nikolai Aleksandrovich was a born leader; he had both the talent and the temperament to lead. He always was open for discussion, and I cherish memories of my hot debates with him on various logical topics. I have to give credit to both our leaders Markov and Shanin for their democratic attitude. At that time I was just a young man, a "freshman" researcher, but still felt free to challenge their opinions. I never noticed a condescending attitude from their side.

I myself graduated from the School of Mathematics at Moscow State Lomonosov University in 1964. The same year my first mathematical paper was published in the journal Soviet Doklady (Doklady Akademii Nauk SSSR). Those were rather romantic times for Markov's Constructive Mathematics and for all of us involved. Our hopes were high and the future looked exciting and bright. Our leader A. A. Markov, Jr. was just in his early sixties; he was full of energy and his extraordinary talent and personality inspired his students. Almost every single day was marked by a discovery, and I somehow recall those times as a kind of uninterrupted spring.

The political climate was relatively mild in comparison with Stalin's era. It was the very end of the so called Khruschev's "thaw". Solzhenitsyn's "One day of Ivan Denisovitch" had just been published (1962). A group of highly gifted poets, who would later be called shestidesyatniki (a rough translation: "poets of the sixties"), such as Evtushenko, Voznesensky, Akhmadulina recited their poems in large halls, sometimes even in stadiums. Evgeny Evtushenko wrote and, by a miracle and the courage of his publisher, published his famous poem "Baby Yar" (1961) about the mass execution of Jews near Kiev by Nazis; Shostakovich used these verses in his 13th Symphony (1962) that was premiered in Moscow in spite of attempts of the authorities to derail the performance ...

Self-proclaimed grass-root poets gathered on Mayakovsky square (a prominent place in downtown of Moscow, named after the renowned Russian poet Vladimir Mayakovsky (1893–1930)) around the monument of the poet and read their works to each other and to everyone who was eager to listen. Militia (Russian name for police) could be seen around, but did not interfere. It was a time, when songs by Bulat Okudzhava, Vladimir Vysotsky, Aleksand Galich and other poets-singers ("bards") were widely distributed by home-made magnetic tapes (so called "Magnitizdat" (Audiotape Publishing)) ... Sure, the totalitarian State was still in place. Nikita Khruschev was an unpredictable politician. When it comes to art one can recall, e.g., the disgusting political campaign against Boris Pasternak after he had been awarded the Nobel Prize and other not so widely known actions of this kind.

Still, an optimist could believe that the tough totalitarian control over lives and thoughts of Soviet citizens was giving way ... Alas, the clouds were already gathering ...

It was in 1963 that I heard for the first time the name "Mints" from my teachers A. A. Markov, Jr. and N. M. Nagorny (1928–2007). They brought to my attention his publication (1962) in Soviet Doklady. The subject was differentiation in the constructive (in Markov's sense) setting. Grisha proved that the derivative of a differentiable constructive function could not be effectively obtained from a Gödel number of this function. The information that is sufficient to compute values of the function in question is not sufficient to effectively find its derivative. One had to add the effective module of differentiability to the initial data to fulfill that task. It was one of the first results of that kind.

Still some time was to elapse before I met Grisha in person. After all, I lived in Moscow and Grisha resided in Leningrad. The occasion came during following year (1964) with the First Symposium on Automated Deduction. The Symposium took place in a small town Trakai near Vilnius, Lithuania. The town was a rather scenic one, being situated at a magnificent lake that boasted an island with a medieval, well-preserved castle. It was still close to the beginning of the computer era, at least in the USSR. Today, when the word "computer" is part of the Russian language and a Russian-speaker uses it quite freely, it is hard to imagine the technology available in the 1960s. I recall my first experience with "EVM" (an abbreviation for "elektronnaya vychislitel'naya mashina", i.e. "electronic computing machine"). It was called "BESM-2M" (an abbreviation for "bystrodeystvuyuschaya elektronnaya schetnaya mashina" ("a speed action electronic computing machine-2M")). It was a vacuum-tube machine with 4 (!) kilobytes of RAM. The mechanical-electronic monster together with its air-conditioning equipment occupied a considerable part of the building of the Computing Center of the Academy of Sciences of USSR (Moscow). The initial data had to be presented on punch cards, and huge bunches of them were seen in the anteroom of the machine hall.

Those machines (both in the USA and in the USSR) were first built to perform enormous calculations for nuclear weapon programs. Later it turned out that the perhaps more valuable trait of those devices was their capacity to deal with all types of digitally represented information. The very term "computer" sounds today rather misleading. Perhaps the term "information machine" would be more accurate.

Shanin's group presented in Trakai first results of their pioneering research on automated deduction, i.e. in using computers to obtain formal proofs in axiomatic systems (they began with the propositional calculus). And a group of researchers from Moscow demonstrated their first practical programs for playing chess. All that was in line with developing non-calculating applications of computing machines.

Mints delivered an impressive talk. If my memory does not fail me, it was a review of research around Herbrand's theorem. Even today I clearly see his inspired face and hear his low-pitch calm and confident voice. I introduced myself in a break and we talked for 10-15 minutes. He presented me with an offprint of his paper on differentiation and inscribed it "To a reader from the author". Well, I was not just a reader, my first paper was due soon, but in fact it happened to be the first offprint I had ever received from an author. I liked Mints and our conversation, but I had no idea that it was the beginning of a lifelong friendship.

In those times there were very few EVM-s in the Soviet Union (and, by the way, machine time was expensive), so when the Leningrad group began its computer experiments with automated deduction they rented time on the (then advanced) EVM BESM-6 of the Computing Center of the Academy of Sciences (Moscow). Remote access to a computer was not known yet. (I remember the first experiments with teletypes inside of the building of the Computing

Center.) Therefore "Leningraders" had frequently to travel to Moscow. As I worked in the Computing Center in the division of Mathematical Logic (headed by A. A. Markov) I got to know closely leading Leningrad logicians, such as Nikolai A. Shanin, Sergey Yu. Maslov (1939–1982), Grigori E. Mints, Vladimir P. Orevkov, Anatol' O. Slisenko ...

Participation in various conferences, workshops etc strengthened the ties between the Moscow and Leningrad branches of Constructive Mathematics. Not like today, conferences and workshops lasted then sometimes up to 10-12 days. They offered plenty of time for non-formal discussions (often through the night), music making, sightseeing, hiking etc ...I should add that there always existed a kind of competition between the two Russian capitals, the old one – Moscow, and the young Nordic, more European one – Saint Petersburg. A touch of this rivalry could be felt also in Mathematics, including our small community in constructive mathematics. It added color to our relations, some jokes and friendly teasing. Grisha took active part in all those events.

Grisha always impressed us by his enormous mathematical erudition. It seemed that he was closely familiar with every paper related to proof theory. As his Moscow counterpart in such all-encompassing knowledge, I have to mention the late Albert G. Dragalin (1941–1998). It was a pleasure to witness their discussions!

In turn, we Moscovites traveled to Leningrad to give talks in seminars of the Leningrad branch of the Steklov Institute. It offered me opportunities for getting to know Grisha more closely.

In the meantime the political climate was changing. The Brezhnev regime replaced Khrushchev's. The antisemitism that had always been present in Russia intensified. It was supported, behind closed doors, by the Communist Party and by the government structures the Party held under complete control. The semi-official antisemitism was one of the manifestations of the "telephone law": a phone call from a district Party committee determined important personnel decisions with no regard to written laws.

In academia that meant difficulties in getting jobs that would match the scholarly achievements of a Jewish applicant, problems with promotions and so on. Each position for such an applicant had to be secured by special, usually informal efforts of influential academic supporters. From time to time information about such discussions behind the closed doors of party and administration offices would leak out. Some exchanges sounded Kafkaesque, and some were rather tragic-comical ...

To secure a good education for their children was a serious problem for Jewish parents. There was an unwritten, but efficient ban on admission of "non-Arian" children to the prestigious Schools.

Unfortunately, the Soviet mathematical community was not free of antisemitism and a number of excellent mathematicians (including some of my university contemporaries) disgraced themselves by torturing and flunking Jewish children on admission tests. They offered them special, so-called "Jewish problems", i.e. very hard problems practically unsolvable in the environment of an oral examination.

To that one could add the awakening of national feelings, a resurrection of interest in Jewish culture, Judaism and the Jewish State, Israel. No wonder that many Jewish families wanted to leave the Soviet Union for good and applied for exit visas. So did Mints in the early 1980s, and his request (as many others in those years) was denied. He joined the group known in the West as "refuseniks" (a strange English half-translation, half-imitation of the Russian term "otkaznik"). The situation of a refusenik was usually hard. Such an individual would find himself/herself in a rather suspended mode of life: neither here, nor there.

Under those circumstances Grisha showed an admirable strength of character. First of all, in a quite noble manner, he resigned his position in the Leningrad Steklov Institute. It was to prevent problems with the administration and Party committee for the department of the institute he belonged to, i.e., for his colleagues and friends. For several years he supported his family by private tutoring and other non-research jobs.

My family and I recall with special warmth Grisha's visits to Moscow in those years, when he stayed with us. It would usually begin with a morning phone call, advising us that Grisha was going to take a day train. Such trains arrived in Moscow between 10 and 11 pm. So I had time after work to buy a bottle of wine and – most importantly – to put a typewriter on my desk. Grisha was always doing Mathematics! It was so natural, that it seemed to be a part of his physiology. A well-known Russian writer Yuri Olesha entitled one of his books as "Not a single day without a line". Grisha definitely lived up to this "commandment" ("a line" should be replaced by a formula or a theorem) . . . With the same stamina he met many personal problems.

And still, there is some justice in this world! After several painful years of suspended existence, Grisha was invited to Tallinn, the capital of Estonia. The totalitarian clutch in the national republics was usually not as tight, and Estonian colleagues were able to secure in the mid-1980s a prominent academic position for Grisha in the Institute of Cybernetics of the Estonian Academy of Sciences. In this institution he obtained internationally recognized results in proof theory and automated deduction.

Our further meetings with Grisha took place already in the USA to where I had moved with my family in 1989. Grisha enjoyed the highest esteem of his colleagues all over the world, so I was not surprised to learn (in 1991) that he had been offered a tenured full-professor position at Stanford University, California. It was a joy to visit him several times in his house in Palo Alto, to eat oranges from trees in his garden, to take long walks and to feed donkeys in a local park. What a pleasure to see a life-long friend and excellent scholar approaching his 70th birthday – that is going to be celebrated by the worldwide logical community. There is a justice in this world, indeed. Many happy returns, dear Grisha!

Note: It is my pleasant obligation to thank Professor Wilfried Sieg for tremendous help with editing this essay. Needless to say, that I bear full responsibility for the errors that may remain.

BIBLIOGRAPHY

[1] B.A. Kushner, Lectures on Constructive Mathematical Analysis, American Mathematical Society, translations of mathematical monographs, vol. 60, Providence, Rhode Island, 1984 (translation from Russian).

[2] B.A. Kushner, Markov's Constructive Mathematical Analysis: the Expectations and Results. In "Mathematical Logic", P.P. Petkov, Ed., Plenum Press, New York-London (1990), 53-58.

[3] B.A. Kushner, Markov and Bishop, An essay. In "Golden Years of Moscow Mathematics". American Mathematical Society – London Mathematical Society, (1993), 179–197.

[4] B.A. Kushner, Kurt Goedel and the constructive mathematics of A.A. Markov. Goedel'96: Logical Foundations of Mathematics, Computer Science and Physics – Kurt Goedel's Legacy, Brno, Czech Republic, August 1996, Proceedings, Petr Hajek, Ed. Lectures Notes on Logic, 50–63, Springer 1996.

[5] B.A. Kushner, Markov's Constructive Analysis; a Participant's View. Theoretical Computer Science, 219, 267–285, 1999.

[6] B.A. Kushner, The Constructive Mathematics of A.A. Markov, American Mathematical Monthly, vol. 113, No 6, June-July 2006.

[7] B.A. Kushner, Uchitel', In: Zametki po Evreiskoi Istorii, http://berkovich-zametki.com/2007/Starina/Nomer5/Kushner1.htm, 2007. In Russian.

Weak interpolation in extensions of minimal logic

LARISA MAKSIMOVA

ABSTRACT. A weak version of interpolation in extensions of Johansson's minimal logics is defined, and its equivalence to a weak version of Robinson's joint consistency is proved. We show that, in contrast to superintuitionistic logics, the weak interpolation property WIP is non-trivial in propositional logics extending minimal logic. We find some criteria for validity of WIP in extensions of the minimal logic.[1]

Interpolation theorem proved by W. Craig [2] in 1957 for classical first order logic was a source of many research results devoted to interpolation problem in classical and non-classical logical theories [1, 13, 14, 4]. Now interpolation is considered as a standard property of logics and calculi like consistency, completeness and so on. For intuitionistic predicate logic and for the predicate version of Johansson's minimal logic the interpolation theorem was proved by K. Schütte [17]. Some variants of the interpolation theorem were proved for intuitionistic logic in [14].

In this paper we consider a variant of the interpolation property in minimal logic and its extensions. Minimal logic, introduced by I. Johansson [6], has the same positive fragment as intuitionistic logic but has no special axioms for negation. In contrast to classical and intuitionistic logics, minimal logic admits non-trivial theories containing some proposition together with its negation.

The original definition of interpolation admits different analogs which are equivalent in classical logic but are not equivalent in other logics. It is known that in classical theories the interpolation property is equivalent to the joint consistency RCP, which arises from the joint consistency theorem proved by A. Robinson [16] for classical predicate logic. It was shown by D. Gabbay [3] that in intuitionistic predicate logic the full version of RCP does not hold. But some weaker version of RCP is valid, and this weaker version is equivalent to CIP in all superintuitionistic predicate logics.

In this paper we concentrate on the weak interpolation property WIP introduced in [7]. We prove that WIP is equivalent to some weak version WRP of Robinson consistency property in all extensions of minimal logic. In [7] we noted that all propositional superintuitionistic logics have WIP, although it does not hold for superintuitionistic predicate logics. Since only finitely many propositional superintuitionistic logics possess CIP [8], WIP and WRP are not

---

[1] Supported by Russian Foundation for Basic Research, (project 09-01-0009) and by ADTP "Development of the Scientific Potential of Higher School" of the Russian Federal Agency for Education (Grant 2.1.1.419).

equivalent to CIP and RCP over intuitionistic logic. Here we show that WIP is non-trivial in propositional extensions of minimal logic.

We find a counter-example to WIP in J-logics. Also we prove that a large subclass of J-logics has the weak interpolation property. We define a J-logic Gl and state that the problem of weak interpolation in J-logics is reducible to the same problem over Gl. In section 6 an algebraic criterion for WIP in J-logics is given.

1 Interpolation and joint consistency

If $\mathbf{p}$ is a list of non-logical symbols, let $A(\mathbf{p})$ denote a formula whose non-logical symbols are in $\mathbf{p}$, and $\mathcal{F}(\mathbf{p})$ the set of all such formulas.

Let L be a logic, and $\vdash_L$ the deducibility relation in L. Suppose that $\mathbf{p}, \mathbf{q}, \mathbf{r}$ are disjoint lists of non-logical symbols, and $A(\mathbf{p},\mathbf{q},x)$, $B(\mathbf{p},\mathbf{r})$ are formulas. The Craig interpolation property CIP and the deductive interpolation property IPD are defined as follows:

CIP. If $\vdash_L A(\mathbf{p},\mathbf{q}) \to B(\mathbf{p},\mathbf{r})$, then there exists a formula $C(\mathbf{p})$ such that $\vdash_L A(\mathbf{p},\mathbf{q}) \to C(\mathbf{p})$ and $\vdash_L C(\mathbf{p}) \to B(\mathbf{p},\mathbf{r})$.

IPD. If $A(\mathbf{p},\mathbf{q}) \vdash_L B(\mathbf{p},\mathbf{r})$, then there exists a formula $C(\mathbf{p})$ such that $A(\mathbf{p},\mathbf{q}) \vdash_L C(\mathbf{p})$ and $C(\mathbf{p}) \vdash_L B(\mathbf{p},\mathbf{r})$.

In [7] *the weak interpolation property* was introduced:

WIP. If $A(\mathbf{p},\mathbf{q}), B(\mathbf{p},\mathbf{r}) \vdash_L \bot$, then there exists a formula $A'(\mathbf{p})$ such that $A(\mathbf{p},\mathbf{q}) \vdash_L A'(\mathbf{p})$ and $A'(\mathbf{p}), B(\mathbf{p},\mathbf{r}) \vdash_L \bot$.

In all extensions of minimal logic we have

$$\text{CIP} \Leftrightarrow \text{IPD} \Rightarrow \text{WIP}.$$

In classical predicate logic CIP is equivalent to the Robinson consistency property

RCP. Let T_1, T_2 be two consistent L-theories in the languages $\mathcal{L}_1, \mathcal{L}_2$ respectively. If $T_1 \cap T_2$ is a complete L-theory in the common language $\mathcal{L}_1 \cap \mathcal{L}_2$, then $T_1 \cup T_2$ is L-consistent.

The same equivalence holds in all classical modal logics [4]. We recall the definitions. By $\Gamma \to_L A$ we denote deducibility of A from Γ in L by rule R1. Then $\Gamma \to_L B$ holds if and only if there exist $n \geq 0$ and formulas $A_1, \ldots, A_n \in \Gamma$ such that

$$L \vdash (A_1 \& \ldots \& A_n) \to B.$$

We say that a set Γ is L-consistent if $\Gamma \not\to_L \bot$. A set T of formulas of the language $\mathcal{L}$ is said to be an L-theory of this language if it is closed under $\to_L$, that is, $T \to_L A$ for $A \in \mathcal{L}$ implies $A \in T$. An L-theory T of the language $\mathcal{L}$ is complete in $\mathcal{L}$ if $A \in T$ or $\neg A \in T$ for any formula $A \in \mathcal{L}$.

It was proved in [7] that in classical modal logics RCP is equivalent to

RCP'. Let T_1, T_2 be two L-theories in the languages $\mathcal{L}_1, \mathcal{L}_2$ respectively, $\mathcal{L}_0 = \mathcal{L}_1 \cap \mathcal{L}_2$, $T_{i0} = T_i \cap \mathcal{L}_0$. If the set $T_{10} \cup T_{20}$ in the common language $\mathcal{L}_0$ is L-consistent, then $T_1 \cup T_2$ is L-consistent.

2   J-logics and theories

In extensions of intuitionistic predicate logic the Craig interpolation property is equivalent to a weaker version RCP″ of Robinson's consistency property [3]. It was proved by Gabbay [3] that the general form RCP of Robinson's property fails in intuitionistic predicate logic. The notion of an intuitionistic theory was defined as a pair (T, F), where T was a set of "true" formulas and F a set of "false" formulas. And in RCP we wanted to keep all true and all false formulas of both theories (T_1, F_1) and (T_2, F_2), which was not always possible. The weaker property RCP″ required an additional condition $F_1 \subseteq F_2$, in particular, F_1 should be in the common language. By analogy with RCP′, in [7] we defined a version WRP of Robinson's consistency property, where a theory was identified with the set of its "true" formulas. It was proved that WRP is equivalent to WIP and is much weaker than CIP in the case of superintuitionistic logics. Moreover, all propositional superintuitionistic logics possess WIP.

In this paper we consider extensions of Johansson's minimal logic. Minimal logic JQ is axiomatized by negation-free axiom schemata of intuitionistic predicate logic.

Let L be any axiomatic extension of minimal logic. Due to the deduction theorem, $\to_L$ is the same as $\vdash_L$. We define an *L-theory* as a set T closed with respect to $\vdash_L$. An L-theory is *consistent* if it does not contain the constant $\bot$. It is clear that an L-theory T is the same as the theory $(T, \{\bot\})$ in the sense of Gabbay [3].

Thus we can define *the weak Robinson property WRP* as follows:

WRP. Let T_1 and T_2 be two L-theories in the languages $\mathcal{L}_1$ and $\mathcal{L}_2$ respectively, $\mathcal{L}_0 = \mathcal{L}_1 \cap \mathcal{L}_2$, $T_{i0} = T_i \cap \mathcal{L}_0$. If the set $T_{10} \cup T_{20}$ in the common language is L-consistent, then $T_1 \cup T_2$ is L-consistent.

The following theorem is an analog of [3, Theorem 8.32] proved for intermediate logics by Gabbay.

Theorem 2.1. *For any (predicate or propositional) extension L of the minimal logic, WIP is equivalent to WRP.*

Proof. Assume L has WIP and prove WRP. Let T_1, T_2 be two L-theories in the languages $\mathcal{L}_1, \mathcal{L}_2$ respectively, $\mathcal{L}_0 = \mathcal{L}_1 \cap \mathcal{L}_2$, $T_{i0} = T_i \cap \mathcal{L}_0$. Suppose that $T_1 \cup T_2$ is L-inconsistent. Then there exist formulas $A \in T_1$ and $B \in T_2$ such that
$$A, B \vdash_L \bot.$$

By WIP there is a formula C of the common language $\mathcal{L}_0$ such that
$$A \vdash_L C \text{ and } C, B \vdash_L \bot.$$

Then we get
$$T_1 \vdash_L C \text{ and } T_2 \vdash_L C \to \bot$$

and so $T_{10} \cup T_{20}$ is L-inconsistent.

Now assume that L has WRP. Let A, B be arbitrary formulas in the languages $\mathcal{L}_1, \mathcal{L}_2$ respectively and

$$A, B \vdash_L \bot.$$

Denote by T_1 the L-theory in the language $\mathcal{L}_1$ with the only axiom A and by T_2 the L-theory in the language $\mathcal{L}_2$ axiomatized by B. Then $T_1 \cup T_2$ is L-inconsistent. By WRP the theory $T_{i1} \cup T_{i2}$ in the common language $\mathcal{L}_0 = \mathcal{L}_1 \cap \mathcal{L}_2$, where $T_{i0} = T_i \cap \mathcal{L}_0$, is L-inconsistent, that is,

$$T_{10} \cup T_{20} \vdash_L \bot.$$

So there is a formula C in the common language $\mathcal{L}_0$ such that

$$C \in T_{10} \text{ and } C, T_{20} \vdash_L \bot.$$

By definition of T_i, T_{i0} we obtain

$$A \vdash_L C \text{ and } C, B \vdash_L \bot.$$

∎

Corollary 2.2. *If a (predicate or propositional) extension of minimal logic has CIP, then it has WRP.*

Proof. It is clear by the deduction theorem that CIP implies WIP. Then the statement immediately follows from Theorem 2.1. ∎

3 Propositional J-logics

In this section we study propositional J-logics.

In [8] a description of all propositional superintuitionistic logics with interpolation property was obtained. There are only finitely many superintuitionistic logics with this property. All positive logics with the interpolation property were described in [9], where a study of this property was initiated for extensions of Johansson's minimal logic too. Minimal logic and intuitionistic logic have the Craig interpolation property [17].

The language of the logic J contains $\&, \vee, \to, \bot, \top$ as primitives; negation is defined by $\neg A = A \to \bot$; $(A \leftrightarrow B) = (A \to B) \& (B \to A)$. A formula is said to be *positive* if contains no occurrences of $\bot$. The logic J can be axiomatized by the calculus that has the same axiom schemas as the positive intuitionistic calculus Int$^+$, and the only rule of inference is modus ponens [6]. By a J-*logic* we mean an arbitrary set of formulas containing all the axioms of J and closed under modus ponens and substitution rules. We denote

$$\text{Int} = \text{J} + (\bot \to p), \ \text{Cl} = \text{Int} + (p \vee \neg p), \ \text{Neg} = \text{J} + \bot.$$

A logic is *non-trivial* if it differs from the set of all formulas. A J-logic is *superintuitionistic* if it contains intuitionistic logic Int, and *negative* if it contains the logic Neg; L is *paraconsistent* if contains neither Int nor Neg. One can prove that a logic is negative if and only if it is not contained in Cl. For any J-logic L we denote by $E(L)$ the family of all J-logics containing L.

It was proved in [7] that all propositional superintuitionistic logics possess the weak interpolation property WIP. Evidently, all negative logics also have this property. We will prove the following fact:

Theorem 3.1. For any J-logic L the following are equivalent:

1. L has WIP,
2. $L \cap L_1$ has WIP for any negative logic L_1,
3. $L \cap \text{Neg}$ has WIP.

Proof. $1 \Rightarrow 2$. Let L possess WIP, $L_1 \vdash \bot$ and
$$L \cap L_1 \vdash A(\mathbf{p},\mathbf{q}) \to (B(\mathbf{p},\mathbf{r}) \to \bot).$$
Then
$$L \vdash A(\mathbf{p},\mathbf{q}) \to (B(\mathbf{p},\mathbf{r}) \to \bot).$$
Since L has WIP, there is a formula $C(\mathbf{p})$ such that
$$L \vdash A(\mathbf{p},\mathbf{q}) \to C(\mathbf{p}) \text{ and } L \vdash C(\mathbf{p}) \to (B(\mathbf{p},\mathbf{r}) \to \bot).$$
The former condition implies
$$L \cap L_1 \vdash A(\mathbf{p},\mathbf{q}) \to C(\mathbf{p}) \vee \bot.$$
It follows from the latter that
$$L \vdash C(\mathbf{p}) \vee \bot \to (B(\mathbf{p},\mathbf{r}) \to \bot).$$
In addition,
$$L_1 \vdash C(\mathbf{p}) \vee \bot \to (B(\mathbf{p},\mathbf{r}) \to \bot).$$
Therefore
$$L \cap L_1 \vdash C(\mathbf{p}) \vee \bot \to (B(\mathbf{p},\mathbf{r}) \to \bot).$$
Thus $C(\mathbf{p}) \vee \bot$ is an interpolant of the given formula in $L \cap L_1$.
$2 \Rightarrow 3$. Obvious.
$3 \Rightarrow 1$. Let $L \cap \text{Neg}$ have WIP and
$$L \vdash A(\mathbf{p},\mathbf{q}) \to (B(\mathbf{p},\mathbf{r}) \to \bot).$$
Then
$$L \cap \text{Neg} \vdash A(\mathbf{p},\mathbf{q}) \to (B(\mathbf{p},\mathbf{r}) \to \bot).$$
There exists an interpolant $C(\mathbf{p})$ for this formula in $L \cap \text{Neg}$. It is clear that $C(\mathbf{p})$ is also an interpolant in L. ∎

The well known theorem of Glivenko [5] says that a propositional formula of the form $\neg A$ is intuitionistically valid if and only if it is a classical two-valued tautology. In [18, 15] a J-logic
$$\text{Ljp}' = \text{J} + \neg\neg(\bot \to p)$$
was investigated. It was proved that for this logic the following analog of Glivenko's theorem holds:

Lemma 3.2. $\text{Cl} \vdash \neg A \iff \text{Ljp}' \vdash \neg A$.

Evidently, any superintuitionistic logic contains Ljp'. By analogy with [7] we can prove

Theorem 3.3. Any propositional J-logic containing $J + \neg\neg(\bot \to p)$ possesses WIP.

Proof. Let L contain $J + \neg\neg(\bot \to p)$ and
$$A(\mathbf{p},\mathbf{q}), B(\mathbf{p},\mathbf{r}) \vdash_L \bot.$$

If L contains the logic Neg, then $\bot$ is an interpolant for this formula.

Assume that $\bot$ does not belong to L. Then L is contained in Cl, and there is $C(\mathbf{p})$ such that
$$A(\mathbf{p},\mathbf{q}) \vdash_{\text{Cl}} C(\mathbf{p}) \text{ and } C(\mathbf{p}), B(\mathbf{p},\mathbf{r}) \vdash_{\text{Cl}} \bot.$$

It follows that
$$\text{Cl} \vdash \neg(A(\mathbf{p},\mathbf{q})\&\neg C(\mathbf{p})) \text{ and } \text{Cl} \vdash \neg(\neg\neg C(\mathbf{p})\& B(\mathbf{p},\mathbf{r})).$$

By Lemma 3.2 we obtain
$$\text{Ljp}' \vdash \neg(A(\mathbf{p},\mathbf{q})\&\neg C(\mathbf{p})) \text{ and } \text{Ljp}' \vdash \neg(\neg\neg C(\mathbf{p})\& B(\mathbf{p},\mathbf{r})).$$

It follows that the same formulas are provable in L, and so we have
$$A(\mathbf{p},\mathbf{q}) \vdash_L \neg\neg C(\mathbf{p}) \text{ and } \neg\neg C(\mathbf{p}), B(\mathbf{p},\mathbf{r}) \vdash_{\text{Cl}} \bot.$$
∎

We note that the logic Ljp' itself possesses CIP.

Proposition 3.4. The logic Ljp' has CIP.

Proof. Note that the formula $A(p) = \neg\neg(\bot \to p)$
is J-*conservative*, that is,

$A(p), A(q) \vdash A(p\&q);\ A(p), A(q) \vdash A(p \vee q);\ A(p), A(q) \vdash A(p \to q);\ \vdash A(\bot)$.

It follows that this formula preserves CIP [4], and so CIP holds in the logic Ljp' because J has CIP. ∎

Corollary 3.5. Any propositional J-logic containing $J + (\bot \vee (\bot \to p))$ possesses WIP.

Proof. Immediate from Theorem 3.3, because $\neg\neg(\bot \to p)$ is derivable from $(\bot \vee (\bot \to p))$ in J. ∎

For predicate logics neither Theorem 3.3 nor Corollary 3.5 holds. In [7] a predicate superintuitionistic logic without WIP was found. Of course, the formula $(\bot \vee (\bot \to p))$ is a theorem of that logic, which extends the minimal logic. To be precise, we consider a superintuitionistic predicate logic that is not H^*-complete. Such a logic was built by N.-Y.Suzuki [19]. It does not possess WIP due to the following fact:

Proposition 3.6. Let L be a (predicate or propositional) extension of the minimal logic. If L has the weak interpolation property then L is H^*-complete.

Proof. Recall that a logic L is said to be H^*-*complete* if for all formulas A and B having no predicate symbols or free object variables in common,

$$L \vdash A \vee B \Rightarrow (L \vdash \neg\neg A \text{ or } L \vdash \neg\neg B).$$

We show that WIP implies H^*-completeness. Assume that L has WIP and $L \vdash A \vee B$. Then $\neg A, \neg B \vdash_L \bot$, and by WIP there is a variable-free formula C such that $\neg A \vdash_L C$ and $C, \neg B \vdash_L \bot$. But any variable-free formula is equivalent to $\bot$ or $\top$ in the minimal logic. If C is equivalent to $\bot$, then $\vdash_L \neg\neg A$; if C is equivalent to $\top$, then $\vdash_L \neg\neg B$. ∎

4 Reduction of WIP

Theorem 3.3 cannot be extended to all J-logics. The picture changes when we turn to extensions of the logic

$$\text{Gl} = \text{J} + (p \vee (p \to \bot)) = \text{J} + (p \vee \neg p).$$

It was proved in [10] that this logic has CIP. In the following section we find an extension of Gl without WIP. Here we prove that the problem of weak interpolation is reducible to the same problem in extensions of Gl.

Consider extensions of Gl in more detail. We note that for any extension of Gl the following analog of Glivenko's theorem holds.

Lemma 4.1. *For any J-logic L and any formula A:*

$$L + (p \vee \neg p) \vdash \neg A \iff L \vdash \neg A.$$

Proof. Let a formula $\neg A$ be a theorem of $L+(p\vee\neg p)$. The proof of this formula can be viewed as a derivation in L from a finite set $\Gamma = \{(B_i \vee \neg B_i) \mid i \leq n\}$ for suitable formulas B_i. It follows that $\Gamma \vdash_L \neg A$, and by the deduction theorem we obtain

$$L \vdash (B_1 \vee \neg B_1) \to \ldots \to ((B_n \vee \neg B_n) \to (A \to \bot)\ldots),$$

which is equivalent to

$$L \vdash A \to ((B_1 \vee \neg B_1) \to \ldots \to ((B_n \vee \neg B_n) \to \bot)\ldots).$$

Since the formula $((q \vee \neg q) \to \bot) \leftrightarrow \bot$ is provable in J, it follows that $(A \to \bot)$ is provable in L, that is, $\neg A$ is a theorem of L.

The converse is obvious. ∎

Lemma 4.2. *Let L be a J-logic, $L' = L + (p \vee \neg p)$, Γ a set of formulas and A a formula. Then*

$$\Gamma \vdash_{L'} \neg A \iff \Gamma \vdash_L \neg A.$$

Proof. Let $\Gamma \vdash_{L'} \neg A$. Then by the deduction theorem $L' \vdash (B_1 \& \ldots \& B_k) \to (A \to \bot)$ for some $B_1, \ldots, B_k \in \Gamma$. It is equivalent to $L' \vdash \neg(B_1 \& \ldots \& B_k \& A)$. Then $L \vdash \neg(B_1 \& \ldots \& B_k \& A)$ by Lemma 4.1, and so $\Gamma \vdash_L \neg A$. ∎

Now we prove that the problem of weak interpolation in J-logics can be reduced to the same problem over Gl.

Theorem 4.3. For any J-logic L, the logic L has WIP if and only if $L+(p\vee\neg p)$ has WIP.

Proof. Let L have WIP, $L' = L + (p \vee \neg p)$ and

$$A(\mathbf{p},\mathbf{q}), B(\mathbf{p},\mathbf{r}) \vdash_{L'} \bot.$$

Then by the deduction theorem

$$A(\mathbf{p},\mathbf{q}) \vdash_{L'} \neg B(\mathbf{p},\mathbf{r}),$$

and by Lemma 4.2

$$A(\mathbf{p},\mathbf{q}) \vdash_{L} \neg B(\mathbf{p},\mathbf{r}).$$

Thus

$$A(\mathbf{p},\mathbf{q}), B(\mathbf{p},\mathbf{r}) \vdash_{L} \bot.$$

Then there exists an interpolant $C(\mathbf{p})$ in L, which is also an interpolant in L'. For the converse, let $L' = L + (p \vee \neg p)$ have WIP and

$$A(\mathbf{p},\mathbf{q}), B(\mathbf{p},\mathbf{r}) \vdash_{L} \bot.$$

Then

$$A(\mathbf{p},\mathbf{q}), B(\mathbf{p},\mathbf{r}) \vdash_{L'} \bot.$$

By WIP there is an interpolant $C(\mathbf{p})$ such that

$$A(\mathbf{p},\mathbf{q}) \vdash_{L'} C(\mathbf{p}) \text{ and } C(\mathbf{p}), B(\mathbf{p},\mathbf{r}) \vdash_{L'} \bot.$$

It follows that $A(\mathbf{p},\mathbf{q}) \vdash_{L'} \neg\neg C(\mathbf{p})$ and by Lemma 4.2

$$A(\mathbf{p},\mathbf{q}) \vdash_{L} \neg\neg C(\mathbf{p}).$$

Furthermore, by the deduction theorem we obtain $B(\mathbf{p},\mathbf{r}) \vdash_{L'} \neg C(\mathbf{p})$. Hence $B(\mathbf{p},\mathbf{r}) \vdash_{L} \neg C(\mathbf{p})$ by Lemma 4.2. Then we derive $B(\mathbf{p},\mathbf{r}) \vdash_{L} \neg\neg\neg C(\mathbf{p})$ and

$$\neg\neg C(\mathbf{p}), B(\mathbf{p},\mathbf{r}) \vdash_{L'} \bot.$$

Thus $\neg\neg C(\mathbf{p})$ is an interpolant in L. ∎

5 Counter-example to WIP in J-logics

To find a counter-example to WIP, we use an algebraic semantics. For extensions of minimal logic the algebraic semantics is built using the so-called J-*algebras*, that is, algebras $\mathbf{A} = \langle A; \&, \vee, \rightarrow, \bot, \top \rangle$ satisfying the conditions:

- $\langle A; \&, \vee, \rightarrow, \bot, \top \rangle$ is a lattice with respect to $\&, \vee$ having a greatest element $\top$, where
- $z \leq x \rightarrow y \iff z \& x \leq y$,

- ⊥ is an arbitrary element of A.

A formula A is said to be *valid* in a J-algebra **A** if the identity $A = \top$ is satisfied in **A**.

A J-algebra is called a *Heyting algebra* if ⊥ is the least element of A, and a *negative algebra* if ⊥ is the greatest element of A. A one-element J-algebra **E** is said to be *degenerate*; it is the only J-algebra that is both a negative algebra and a Heyting algebra. A J-algebra **A** is *non-degenerate* if it contains at least two elements; **A** is said to be *well connected* (or *strongly compact*) if for all $x, y \in \mathbf{A}$ the condition $x \vee y = \top \Leftrightarrow (x = \top$ or $y = \top)$ is satisfied. An element a of **A** is called an *opremum of* **A** if it is the greatest among the elements of **A** different from $\top$. By B_0 we denote the two-element Boolean algebra $\{\bot, \top\}$.

We build a counter-example to WIP in J-logics. The definition of H^*-completeness is given in Proposition 3.6.

Theorem 5.1. *There exists a J-logic, which contains* $\mathrm{Gl} = \mathrm{J} + (p \vee (p \to \bot))$ *and does not possess the weak interpolation property. Moreover, this logic is not H^*-complete.*

Proof. We consider two J-algebras **B** and **C** defined as follows. The universe of **B** consists of four elements $\{a, b, \bot, \top\}$, where $a < b < \bot < \top$. The algebra **C** consists of five elements $\{c, d, e, \bot, \top\}$, where $e < x < \bot < \top$ for $x \in \{c, d\}$ and the elements c and d are incomparable. We note that **B** and **C** have a common subalgebra **A**, which is a two-element boolean algebra consisting of $\bot, \top$.

Let a J-logic L_1 be a set of all formulas valid in the both algebras **B** and **C**. We note that the formula $(p \vee (p \to \bot))$ is valid in both algebras **B** and **C**. So the logic built in this theorem is an extension of Gl.

Define the formulas

$$A(x, y) = (x \to y) \& ((y \to x) \to x) \& (y \to \bot) \& ((\bot \to y) \to y),$$

$$B(u, w) = ((u \to w) \to w) \& ((w \to u) \to u) \& ((u \vee w) \leftrightarrow \bot).$$

We prove that

(1) $\quad A(x, y), B(u, w) \vdash_{L_1} \bot.$

Assume that the formula $(A(x, y) \& B(u, w) \to \bot)$ is not valid in one of the algebras **B** and **C**. We denote this algebra by **D**. There is a valuation v in **D** such that

$$v(A(x, y)) \& v(B(u, w)) \not\leq \bot.$$

It follows that

$$v(A(x, y)) \& v(B(u, w)) = \top.$$

Let $v(x) = x_0$, $v(y) = y_0$, $v(u) = v_0$, $v(w) = w_0$. Then we obtain:

$$x_0 \leq y_0, \ (y_0 \to x_0) \leq x_0, \ y_0 \leq \bot, \ (\bot \to y_0) \leq y_0,$$

$$(u_0 \to w_0) \leq w_0, \ (w_0 \to u_0) \leq u_0, \ u_0 \vee w_0 = \bot).$$

Therefore $u_0 < \top$, $w_0 < \top$. From $(u_0 \to w_0) \leq w_0$ we obtain $u_0 \not\leq w_0$; similarly, $w_0 \not\leq u_0$, that is, u_0 and w_0 are incomparable, so that the algebra **D** coincides with **C**.

On the other hand, $x_0 \leq y_0 \leq \bot < \top$. From the conditions $(y_0 \to x_0) \leq x_0$, $(\bot \to y_0) \leq y_0$ we obtain $y_0 \not\leq x_0$, $\bot \not\leq y_0$, hence $x_0 < y_0 < \bot$. Therefore $x_0 = e$, $y_0 \in \{c, d\}$. But in this case $y_0 \to x_0 \not\leq x_0$ – contradiction. So we proved (1).

Now we will prove that this formula has no interpolant in L_1.

Assume that there exists a variable-free formula C such that

(2) $\quad A(x, y) \vdash_{L_1} C$ and $C, B(u, w) \vdash_{L_1} \bot$.

Note that $J \vdash (C \leftrightarrow \top)$ or $J \vdash (C \leftrightarrow \bot)$.

Consider the following valuation v_1 in **B**: $v_1(x) = a$, $v_1(y) = b$. Then $v_1(A(x, y)) = \top$, and from (2) we obtain $v_1(C) = \top$. Since $v_1(C) \neq \bot$, we have $J \vdash (C \leftrightarrow \top)$.

We define the following valuation v_2 in **C**: $v_2(u) = c$, $v_2(w) = d$. Then $v_2(B(u, w)) = \top$, and from (2) we conclude $v_2(C) = \bot$, and so $J \vdash (C \leftrightarrow \bot)$ and $J \vdash (\top \leftrightarrow \bot)$. So we arrived at a contradiction.

To prove that L_1 is not H^*-complete, we note that the formula $(\neg A(x, y) \vee \neg B(u, w))$ is valid in L_1. Indeed, $\neg A(x, y)$ is valid in **C**, and $\neg B(u, w)$ is valid in **B**. So $(\neg A(x, y) \vee \neg B(u, w))$ is valid in both algebras **B** and **C**. On the other hand, neither $\neg\neg\neg A(x, y)$ nor $\neg\neg\neg B(u, v)$ belongs to L_1 because the former formula is refuted by **B**, and the latter by **C**. ∎

6 Algebraic equivalent of weak interpolation

In this section we find an algebraic equivalent of the weak interpolation property.

It is well known that the family of all J-algebras forms a variety, that is, can be defined by identities. There exists a one-to-one correspondence between the logics extending the logic J and the varieties of J-algebras. If A is a formula and **A** is an algebra, we say that A is valid in **A** and write $\mathbf{A} \models A$ if the identity $A = \top$ is satisfied in **A**. We write $\mathbf{A} \models L$ instead of $(\forall A \in L)(\mathbf{A} \models A)$.

To any logic $L \in E(J)$ there corresponds a variety

$$V(L) = \{\mathbf{A} | \mathbf{A} \models L\}.$$

Every logic L is characterized by the variety $V(L)$.

If $L \in E(\text{Int})$, then $V(L)$ is a variety of Heyting algebras, and if $L \in E(\text{Neg})$, then a variety of negative algebras.

Recall [9] that a J-logic has the Craig interpolation property if and only if $V(L)$ has the amalgamation property AP. We will recall now the necessary definitions.

Let V be a class of algebras invariant under isomorphisms. The class V has *the amalgamation property* if it satisfies the following condition AP for any algebras **A**,**B**,**C** in V:

AP. If **A** is a common subalgebra of **B** and **C**, then there exist **D** in V and monomorphisms $\delta : \mathbf{B} \to \mathbf{D}$, $\varepsilon : \mathbf{C} \to \mathbf{D}$ such that $\delta(x) = \varepsilon(x)$ for all $x \in \mathbf{A}$.

We will now review some definitions introduced in [10].

If $\mathbf{A} = \langle A; \&, \vee, \rightarrow, \bot, \top \rangle$ is a negative algebra and $\mathbf{B} = \langle B; \&, \vee, \rightarrow, \bot, \top \rangle$ is a Heyting algebra, then we define a new J-algebra $\mathbf{A} \uparrow \mathbf{B}$ as follows:

the universe of the new algebra is $C = A \cup B'$, where B' is isomorphic to B, $A \cap B' = \{\bot_\mathbf{A}\} = \{\bot_{\mathbf{B}'}\}$ and C is partially ordered by the relation

$$x \leq_\mathbf{C} y \Leftrightarrow$$

$[(x \in A \text{ and } y \in B') \text{ or } (x, y \in A \text{ and } x \leq_\mathbf{A} y) \text{ or } (x, y \in B' \text{ and } x \leq_{\mathbf{B}'} y)].$

As a consequence, $\bot_\mathbf{C} = \bot_\mathbf{A} = \bot_{\mathbf{B}'}$, $\top_\mathbf{C} = \top_{\mathbf{B}'}$.

We say that a J-algebra is *well-composed* if it has the form $\mathbf{A} \uparrow \mathbf{B}$ for a suitable negative algebra $\mathbf{A}$ and a Heyting algebra $\mathbf{B}$. In particular, any negative algebra $\mathbf{A}$ can be represented as $\mathbf{A} \uparrow \mathbf{E}$, and a Heyting algebra $\mathbf{B}$ as $\mathbf{E} \uparrow \mathbf{B}$, where $\mathbf{E}$ is the one-element J-algebra. If $\mathbf{A}$ is a negative algebra and B_0 is the two-element boolean algebra, then the algebra $\mathbf{A}' = \mathbf{A} \uparrow B_0$ is formed by adding a new greatest element $\top_{\mathbf{A}'}$ to $\mathbf{A}$, and $\bot_{\mathbf{A}'}$ is the opremum of $\mathbf{A}'$.

In the following theorem we formulate an algebraic equivalent of WIP in J-logics.

Theorem 6.1. *A consistent J-logic L has WIP if and only if the class of negative algebras $\mathbf{A}$ such that $(\mathbf{A} \uparrow B_0) \in V(L)$ has the amalgamation property.*

We give a proof in another paper. For modal logics an algebraic equivalent of WIP was found in [7].

For any non-trivial extension L_1 of Neg and any consistent superintuitionistic logic L_2 we defined a logic $(L_1 \uparrow L_2)$ [10]. It is characterized by all J-algebras of the form $\mathbf{A} \uparrow \mathbf{B}$, where $\mathbf{A} \in V(L_1)$ and $\mathbf{B} \in V(L_2)$. In [10] the following axiomatization was found:

$$L_1 \uparrow L_2 = (L_2 * L_1) + (\bot \rightarrow p) \vee (p \rightarrow \bot),$$

where

$$L * L' = J + \{I(A) |\ A \in L\} + \{\bot \rightarrow A |\ A \in L'\}$$

and $I(A)$ is a result of substitution of $p_i \vee \bot$ for any variable p_i in A.

It was proved in [10] that for any non-trivial L_1 and consistent L_2, a logic $(L_1 \uparrow L_2)$ has CIP if and only if both L_1 and L_2 have CIP. In [9] all non-trivial extensions of the logic Neg with CIP were found:

$$\text{Neg, NC} = \text{Neg} + (p \rightarrow q) \vee (q \rightarrow p),\ \text{NE} = \text{Neg} + p \vee (p \rightarrow q).$$

We can prove the following fact:

Proposition 6.2. *Let L_1 be a non-trivial extension of Neg and L_2 a consistent superintuitionistic logic. Then the following conditions are equivalent:*

1. $(L_1 \uparrow L_2)$ has WIP;

2. $(L_1 \uparrow \text{Cl})$ has WIP;

3. $(L_1 \uparrow \text{Cl})$ has CIP;

4. L_1 has CIP.

The logic Gl = J + $(p \vee \neg p)$ considered in Section 4 coincides with Neg $\uparrow$ Cl; it is complete with respect to the class of all algebras of the form $\mathbf{A} \uparrow B_0$ [10].

In conclusion we recall that there are only finitely many propositional superintuitionistic logics with CIP. In addition, CIP is decidable over intuitionistic logic Int, that is, there exists an algorithm for recognizing CIP in any calculus obtained from Int by adding finitely many axiom schemas [8]. As we have seen, WIP is trivial over Int because all superintuitionistic logics possess this property. We have shown that WIP is not trivial over the minimal logic J: in $E(\text{J})$ there is a continuum of logics with WIP and a continuum of logics without WIP. The following problems are still open:
1. Are WIP or CIP decidable over J?
2. How many J-logics possess CIP?

BIBLIOGRAPHY

[1] J.Barwise, S.Feferman, eds. Model-Theoretic Logics. New York: Springer-Verlag, 1985.

[2] W.Craig. Three uses of Herbrand-Gentzen theorem in relating model theory. J. Symbolic Logic, 22 (1957), 269-285.

[3] D.M.Gabbay. Semantical Investigations in Heyting's Intuitionistic Logic, D.Reidel Publ. Co., Dordrecht, 1981.

[4] D.M.Gabbay, L.Maksimova. Interpolation and Definability: Modal and Intuitionistic Logics. Clarendon Press, Oxford, 2005.

[5] V.I.Glivenko. Sur quelques points de la logique de M. Brouwer. Acad. Roy. Belgique. Bull. cl. sci., Ser. 5, 15 (1929), 183-188.

[6] I.Johansson. Der Minimalkalkül, ein reduzierter intuitionistic Formalismus. Compositio Mathematica 4 (1937), 119-136.

[7] L.Maksimova. Interpolation and joint consistency. In: We Will Show Them! Essays in Honour of Dov Gabbay. Volume 2, S. Artemov, H. Barringer, A. d'Avila Garcez, L. Lamb and J. Woods, eds. King's College Publications, London, 2005, pp. 293-305. Bogota,

[8] L.L.Maksimova. Craig's theorem in superintuitionistic logics and amalgamable varieties of pseudoboolean algebras. Algebra and Logic, 16, 6 (1977), 643-681.

[9] L.L.Maksimova. Implicit definability in positive logics. Algebra and Logic, 42, 1 (2003), 65-93.

[10] L.L.Maksimova. Interpolation and definability in extensions of the minimal logic. Algebra and Logic, 44, no. 6 (2005), 726-750.

[11] L.L.Maksimova. A method of proving interpolation in paraconsistent extensions of the minimal logic. Algebra and Logic, 46, no. 5 (2007), 627-648.

[12] L.L.Maksimova. Weak form of interpolation in equational logic. Algebra and Logic, 47, no. 8 (2008), 94-107.

[13] G.Mints. Modularisation and interpolation. Technical Report KES.U.98.4, Kestrel Institute, 1998.

[14] G.Mints. Interpolation theorems for intuitionistic predicate logic. Annals of Pure and Applied Logic, 113: 225-242, 2002.

[15] S.P.Odintsov. Logic of classical refutability and class of extensions of minimal logic. Logic and Logical Philosophy, 9 (2001), 91-107.

[16] A. Robinson. A result on consistency and its application to the theory of definition. Indagationes Mathematicae, 18 (1956), 47-58.

[17] K.Schütte. Der Interpolationssatz der intuitionistischen Prädikatenlogik. Mathematische Annalen, 148:192-200, 1962.

[18] K.Segerberg. Propositional logics related to Heyting's and Johansson's. Theoria, 34 (1968), 26-61.

[19] N.-Y.Suzuki. Hallden-completeness in super-intuitionistic predicate logics. Studia Logica 73 (2003), 113-130.

Decorating proofs[1]

DIANA RATIU AND HELMUT SCHWICHTENBERG

ABSTRACT. The programs synthesized from proofs are guaranteed to be correct, however at the cost of sometimes introducing irrelevant computations, as a consequence of the fact that the extracted code faithfully reflects the proof. In this paper we extend the work of Ulrich Berger [2], which introduces the concept of "non-computational universal quantifiers", and propose an algorithm by which we identify at the proof level the components - quantified variables, as well as premises of implications - that are computationally irrelevant and mark them as such. We illustrate the benefits of this (optimal) decorating algorithm in some case studies and present the results obtained with the proof assistant Minlog.

We consider proofs in minimal logic, written in natural deduction style. The only rules are introduction and elimination for implication and the universal quantifier. The logical connectives $\exists, \wedge$ are seen as special cases of inductively defined predicates, and hence are defined by the introduction and elimination schemes

$$\forall_x (A \to \exists_x A), \qquad\qquad A \to B \to A \wedge B,$$
$$\exists_x A \to \forall_x (A \to B) \to B \ (x \text{ not free in } B), \qquad A \wedge B \to (A \to B \to C) \to C.$$

Disjunction can be defined by $A \vee B := \exists_p((p \to A) \wedge (\neg p \to B))$ with p a boolean variable. When the computational content of a proof is of interest, it is appropriate to distinguish between computational and non-computational variants of $\to, \forall$, written $\to^c, \forall^c$ and $\to^{nc}, \forall^{nc}$, respectively; for $\forall^{nc}$ this was first done by Berger [2]. The introduction rules for $\to^{nc}, \forall^{nc}$ then need an additional restriction: the abstracted (assumption or object) variable is not allowed to be "computational", which can be defined to mean "not free in the extracted term of the premise proof". The insertion of such marks is called a "decoration" of the proof.

In the present paper we are interested in "fine-tuning" the computational content of proofs, by inserting decorations. After adapting in section 1 the standard theory of proof interpretation by (modified) realizability, in section 2 we define what a computational strengthening of a decorated formula is, and construct a derivation of $A_1 \to^c A_2$ for A_1 a computational strengthening of A_2. Here is an example (due to Robert Constable) of why this is of interest. Suppose that in a proof M of a formula C we have made use of a case distinction based on an auxiliary lemma stating a disjunction, say $L: A \vee B$. Then the extract $[\![M]\!]$ will contain the extract $[\![L]\!]$ of the proof of the auxiliary lemma,

---
[1] Dedicated to Grigori Mints on occasion of his 70th birthday

which may be large. Now suppose further that in the proof M of C, the only computationally relevant use of the lemma was which one of the two alternatives holds true, A or B. We can express this fact by using a weakened form of the lemma instead: $L'\colon A \vee^u B$. Since the extract $[\![L']\!]$ is a boolean, the extract of the modified proof has been "purified" in the sense that the (possibly large) extract $[\![L]\!]$ has disappeared.

In section 3 we consider the question of "optimal" decorations of proofs: suppose we are given an undecorated proof, and a decoration of its end formula. The task then is to find a decoration of the whole proof (including a further decoration of its end formula) in such a way that any other decoration "extends" this one. Here "extends" just means that some connectives have been changed into their more informative versions, disregarding polarities. We show that such an optimal decoration exists, and give an algorithm to construct it.

The final section 4 first takes up the example of list reversal used by Berger [3] to demonstrate that usage of $\forall^{nc}$ rather than $\forall^c$ can significantly reduce the complexity of extracted programs, in this case from quadratic to linear. Our implementation (cf. http://www.minlog-system.de) of the decoration algorithm from section 3 automatically finds the optimal decoration. A similar application of decoration occurs when one derives double induction (recurring to two predecessors) in continuation passing style, i.e., not directly, but using as an intermediate assertion (proved by induction)

$$\forall^c_{n,m}((Qn \to^c Q(Sn) \to^c Q(n+m)) \to^c Q0 \to^c Q1 \to^c Q(n+m)).$$

After decoration, the formula becomes

$$\forall^c_n \forall^{nc}_m ((Qn \to^c Q(Sn) \to^c Q(n+m)) \to^c Q0 \to^c Q1 \to^c Q(n+m)).$$

This is applied (as in [5]) to obtain a continuation based tail recursive definition of the Fibonacci function, from a proof of its totality. We finally give two more applications of a slightly extended version of the decoration algorithm involving a data base of proven theorems and proof transformations. The first one concerns Constable's example above, and the second one the maximal segment problem studied in [1].

1 Computational content of proofs

Computational content in proofs arises when the formula proved contains a strictly positive occurrence of an existential quantifier, as in $\forall_x \exists_y A(x,y)$. If $A(x,y)$ is quantifier-free, the computational content of the proof consists of a function assigning to every number n a number m such that $A(n,m)$ holds. One can view $\exists_y A(x,y)$ as a (somewhat degenerated) inductively defined predicate, whose single clause is $\forall_{x,y}(A(x,y) \to \exists_y A(x,y))$. More generally, the computational content of a proof of $I\vec{r}$ for an inductively defined predicate I is a "generation tree", witnessing how the arguments $\vec{r}$ were put into I. For example, consider the clauses Even(0) and $\forall_n(\text{Even}(n) \to \text{Even}(S(Sn)))$. A generation tree for Even(6) should consist of a single branch with nodes Even(0), Even(2), Even(4) and Even(6).

It is a tempting idea that one should be able in such cases to switch on or off the computational influence of a universally quantified variable (as in [2])

derivation	term
$u \colon A$	u^A
$[u \colon A]$ $\mid M$ $\dfrac{B}{A \to B} \to^+ u$	$(\lambda_{u^A} M^B)^{A \to B}$
$\mid M \quad\quad \mid N$ $\dfrac{A \to B \quad A}{B} \to^-$	$(M^{A \to B} N^A)^B$
$\mid M$ $\dfrac{A}{\forall_x A} \forall^+ x$ (with var.cond.)	$(\lambda_x M^A)^{\forall_x A}$ (with var.cond.)
$\mid M$ $\dfrac{\forall_x A(x) \quad r}{A(r)} \forall^-$	$(M^{\forall_x A(x)} r)^{A(r)}$

Figure 1. Derivation terms for $\to$ and $\forall$

or of the premise of an implication. For instance, in the clause $\forall_n(\text{Even}(n) \to \text{Even}(\text{S}(\text{S}n)))$ only the premise $\text{Even}(n)$ should be computationally relevant, not the quantifier $\forall_n$. We therefore "decorate" $\to$ and write $\to^c$, but leave the quantifier $\forall_n$ undecorated, i.e., consider it as "non-computational" and write $\forall^{nc}$. In the clause $\forall_x(A \to \exists_x A)$ for an existential quantifier $\exists_x A$ one may decorate the quantifier $\forall_x$ and the implication $\to$ independently, which gives four (only) computationally different variants $\exists^d, \exists^l, \exists^r, \exists^u$ (with d, l, r, u for "double", "left", "right", "uniform") of the existential quantifier, defined below.

Incorporating these two computationally different variants of implication and universal quantification into Gentzen's calculus of natural deduction is rather easy. Recall the introduction and elimination rules for $\to$ and $\forall$, written in the standard tree notation as well as in a (more condensed) linear term notation, in figure 1. For the universal quantifier $\forall$ there is an introduction rule $\forall^+ x$ and an elimination rule $\forall^-$, whose right premise is the term r to be substituted. The rule $\forall^+ x$ is subject to an *(Eigen-) variable condition*: The derivation term

M of the premise A should not contain any open assumption with x as a free variable.

For $\to^c$ and $\forall^c$ no changes are necessary, and for $\to^{nc}$ and $\forall^{nc}$ only the introduction rules need to be adapted: in "computational relevant" parts of the derivation the abstracted (assumption or object) variable needs to be "non-computational". This will be defined later, simultaneously with the notion of an "extracted term" of a derivation.

For $\exists, \wedge$ we have a whole zoo of (only) computationally different variants: $\exists^d, \exists^l, \exists^r, \exists^u, \wedge^d, \wedge^l, \wedge^r, \wedge^u$ ($\exists^r$ already appeared in [3]). They are defined by their introduction and elimination axioms, which involve both $\to^c$, $\forall^c$ and $\to^{nc}$, $\forall^{nc}$. For the first four these are

$$\forall_x^c (A \to^c \exists_x^d A), \qquad \exists_x^d A \to^c \forall_x^c (A \to^c B) \to^c B,$$
$$\forall_x^c (A \to^{nc} \exists_x^l A), \qquad \exists_x^l A \to^c \forall_x^c (A \to^{nc} B) \to^c B,$$
$$\forall_x^{nc} (A \to^c \exists_x^r A), \qquad \exists_x^r A \to^c \forall_x^{nc} (A \to^c B) \to^c B,$$
$$\forall_x^{nc} (A \to^{nc} \exists_x^u A), \qquad \exists_x^u A \to^c \forall_x^{nc} (A \to^{nc} B) \to^c B$$

where x is not free in B, and similar for $\wedge$:

$$A \to^c B \to^c A \wedge^d B, \qquad A \wedge^d B \to^c (A \to^c B \to^c C) \to^c C,$$
$$A \to^c B \to^{nc} A \wedge^l B, \qquad A \wedge^l B \to^c (A \to^c B \to^{nc} C) \to^c C,$$
$$A \to^{nc} B \to^c A \wedge^r B, \qquad A \wedge^r B \to^c (A \to^{nc} B \to^c C) \to^c C,$$
$$A \to^{nc} B \to^{nc} A \wedge^u B, \qquad A \wedge^u B \to^c (A \to^{nc} B \to^{nc} C) \to^c C.$$

Types ρ, σ, τ are generated from ground types like the types $\mathbf{N}$ of natural numbers and $\mathbf{B}$ of booleans by forming list types $\mathbf{L}(\rho)$, product types $\rho \times \sigma$ and function types $\rho \to \sigma$. The types $\mathbf{N}$, $\mathbf{B}$, $\mathbf{L}(\rho)$ and $\rho \times \sigma$ can be viewed as special cases of the formation of free algebras (with parameters); however, for simplicity we only consider these cases here.

Terms r, s, t of a given type are built from (typed) variables and (typed) constants by (type-correct) abstraction and application. The *projections* of a term r of product type $\rho \times \sigma$ are written as $r0$ and $r1$.

Kolmogorov [9] proposed to view a formula A as a "computational problem", of type $\tau(A)$, the type of a potential solution or "realizer" of A. More precisely, $\tau(A)$ should be the type of the term (or "program") to be extracted from a proof of A. Formally, we assign to every formula A an object $\tau(A)$ (a type or the "nulltype" symbol $\circ$). In case $\tau(A) = \circ$ proofs of A have no computational content; such formulas A are called *computationally irrelevant* (c.i.) (or *Harrop* formulas); the other ones are called *computationally relevant* (c.r.). The definition can be conveniently written if we extend the use of $\rho \to \sigma$ and $\rho \times \sigma$ to the nulltype symbol $\circ$:

$$(\rho \to \circ) := \circ, \quad (\circ \to \sigma) := \sigma, \quad (\circ \to \circ) := \circ,$$
$$(\rho \times \circ) := \rho, \quad (\circ \times \sigma) := \sigma, \quad (\circ \times \circ) := \circ.$$

With this understanding of $\rho \to \sigma$ and $\rho \times \sigma$ we can simply write

$$\tau(P) := \circ \quad \text{for } P \text{ a decidable prime formula (given by a boolean term)},$$

$\tau(A \to^c B) := \tau(A) \to \tau(B), \quad \tau(A \to^{nc} B) := \tau(B),$
$\tau(\forall^c_{x^\rho} A) := \rho \to \tau(A), \quad \tau(\forall^{nc}_{x^\rho} A) := \tau(A),$
$\tau(\exists^d_{x^\rho} A) := \rho \times \tau(A), \quad \tau(\exists^l_{x^\rho} A) := \rho, \quad \tau(\exists^r_{x^\rho} A) := \tau(A), \quad \tau(\exists^u_{x^\rho} A) := \circ,$
$\tau(A \wedge^d B) := \tau(A) \times \tau(B), \quad \tau(A \wedge^l B) := \tau(A), \quad \tau(A \wedge^r B) := \tau(B),$
$\tau(A \wedge^u B) := \circ.$

We now define *realizability*. It will be convenient to introduce a special "null-term" symbol ε to be used as a "realizer" for c.i. formulas. We extend term application to the nullterm symbol by

$$\varepsilon t := \varepsilon, \quad t\varepsilon := t, \quad \varepsilon\varepsilon := \varepsilon.$$

DEFINITION 1 (*t realizes A*). Let A be a formula and t either a term of type $\tau(A)$ if the latter is a type, or the nullterm symbol ε if for c.i. A. We use P for decidable prime formulas (given by a boolean term).

$$\begin{aligned}
\varepsilon \mathbf{r}\, P &:= P \\
t \mathbf{r}\, (A \to^c B) &:= \forall^{nc}_x (x \mathbf{r}\, A \to^{nc} tx \mathbf{r}\, B), \\
t \mathbf{r}\, (A \to^{nc} B) &:= \forall^{nc}_x (x \mathbf{r}\, A \to^{nc} t \mathbf{r}\, B), \\
t \mathbf{r}\, (\forall^c_x A) &:= \forall^{nc}_x (tx \mathbf{r}\, A), \\
t \mathbf{r}\, (\forall^{nc}_x A) &:= \forall^{nc}_x (t \mathbf{r}\, A).
\end{aligned}$$

In case A is c.i., $\forall^{nc}_x (x \mathbf{r}\, A \to^{nc} B(x))$ means $\varepsilon \mathbf{r}\, A \to^{nc} B(\varepsilon)$. For the other connectives realizability is defined by

$$t \mathbf{r}\, \exists^d_x A(x) := \begin{cases} t1 \mathbf{r}\, A(t0) & \text{if } A \text{ is c.r.} \\ \varepsilon \mathbf{r}\, A(t) & \text{otherwise,} \end{cases} \quad t \mathbf{r}\, \exists^l_x A(x) := \exists^u_y (y \mathbf{r}\, A(t)),$$

$t \mathbf{r}\, \exists^r_x A(x) := \exists^u_x (t \mathbf{r}\, A(x)), \quad \varepsilon \mathbf{r}\, \exists^u_x A(x) := \exists^u_{x,y}(y \mathbf{r}\, A(x)),$
$t \mathbf{r}\, (A \wedge^d B) := t0 \mathbf{r}\, A \wedge t1 \mathbf{r}\, B, \quad t \mathbf{r}\, (A \wedge^l B) := t \mathbf{r}\, A \wedge \exists^u_y (y \mathbf{r}\, B),$
$t \mathbf{r}\, (A \wedge^r B) := \exists^u_y (y \mathbf{r}\, A) \wedge t \mathbf{r}\, B,$
$\varepsilon \mathbf{r}\, (A \wedge^u B) := \exists^u_y (y \mathbf{r}\, A) \wedge \exists^u_z (z \mathbf{r}\, B).$

Call two formulas A and A' *computationally equivalent* if each of them computationally implies the other, and in addition the identity realizes each of the two derivations of $A' \to^c A$ and of $A \to^c A'$. It is an easy exercise to verify that for c.i. A, the formulas $A \to^c B$ and $A \to^{nc} B$ are computationally equivalent, and hence can be identified. In the sequel we shall simply write $A \to B$ for either of them. Similarly, for c.i. A the two formulas $\forall^c_x A$ and $\forall^{nc}_x A$ are c.i., and both $\varepsilon \mathbf{r}\, \forall^c_x A$ and $\varepsilon \mathbf{r}\, \forall^{nc}_x A$ are defined to be $\forall^{nc}_x (\varepsilon \mathbf{r}\, A)$. Hence they can be identified as well, and we shall simply write $\forall_x A$ for either of them. Since the formula $t \mathbf{r}\, A$ is c.i., under this convention the $\to, \forall$-cases in the definition of realizability can be written

$$\begin{aligned}
t \mathbf{r}\, (A \to^c B) &:= \forall_x (x \mathbf{r}\, A \to tx \mathbf{r}\, B), \\
t \mathbf{r}\, (A \to^{nc} B) &:= \forall_x (x \mathbf{r}\, A \to t \mathbf{r}\, B),
\end{aligned}$$

$$t \mathbf{r} (\forall_x^c A) := \forall_x (tx \mathbf{r} A),$$
$$t \mathbf{r} (\forall_x^{nc} A) := \forall_x (t \mathbf{r} A).$$

Notice that the formula $t \mathbf{r} A$ is "invariant" in the sense that $\varepsilon \mathbf{r} (t \mathbf{r} A)$ and $t \mathbf{r} A$ are identical formulas.

We simultaneously define (1) *derivations* M, and (2) the *extracted term* $[\![M]\!]$ of type $\tau(A)$ of a derivation M with endformula A. As already mentioned, this simultaneous definition is needed because the introduction rules for $\to^{nc}$ and $\forall^{nc}$ must be restricted: in "computational relevant" parts of the derivation the abstracted (assumption or object) variable needs to be "non-computational".

For derivations M^A with A c.i. let $[\![M^A]\!] := \varepsilon$; there is no condition on the usage of introduction rules for $\to^{nc}$ and $\forall^{nc}$ in such derivations. Now assume that M derives a c.r. formula. Then

$$[\![u^A]\!] := x_u^{\tau(A)} \quad (x_u^{\tau(A)} \text{ uniquely associated with } u^A),$$
$$[\![(\lambda_{u^A} M^B)^{A \to^c B}]\!] := \lambda_{x_u^{\tau(A)}}[\![M]\!],$$
$$[\![(M^{A \to^c B} N^A)^B]\!] := [\![M]\!][\![N]\!],$$
$$[\![(\lambda_{x^\rho} M^A)^{\forall_x^c A}]\!] := \lambda_{x^\rho}[\![M]\!],$$
$$[\![(M^{\forall_x^c A(x)} r)^{A(r)}]\!] := [\![M]\!]r,$$
$$[\![(\lambda_{u^A} M^B)^{A \to^{nc} B}]\!] := [\![(M^{A \to^{nc} B} N^A)^B]\!] := [\![(\lambda_{x^\rho} M^A)^{\forall_x^{nc} A}]\!]$$
$$:= [\![(M^{\forall_x^{nc} A(x)} r)^{A(r)}]\!] := [\![M]\!].$$

Here $\lambda_{x_u^{\tau(A)}}[\![M]\!]$ means just $[\![M]\!]$ if A is c.i.

In all these cases the correctness of the derivation is preserved, except possibly in those involving λ: $(\lambda_{u^A} M^B)^{A \to^{nc} B}$ is correct if M is and in case A is c.r., $x_u \notin \mathrm{FV}([\![M]\!])$. $(\lambda_{x^\rho} M^A)^{\forall_x^{nc} A}$ is correct if M is and – in addition to the usual variable condition – $x \notin \mathrm{FV}([\![M]\!])$.

It remains to define extracted terms for the axioms. Consider for example the induction schema for the type $\mathbf{L}(\rho)$ of lists of type ρ objects. It is written here in the form

$$\forall_v^c (A(\mathrm{nil}) \to^c \forall_{x,v}^c (A(v) \to^c A(xv)) \to^c A(v)),$$

with x, v variables of type $\rho, \mathbf{L}(\rho)$ and xv denoting $\mathrm{cons}(x, v)$; clearly the decoration is appropriate given the computational meaning of induction. The extracted term is the corresponding recursion operator $\mathcal{R}_{\mathbf{L}(\rho)}^\tau$ (in the sense of Gödel [8]), of type

$$\mathbf{L}(\rho) \to \tau \to (\rho \to \mathbf{L}(\rho) \to \tau \to \tau) \to \tau.$$

Notice that a different decoration of induction is possible:

$$\forall_v^c (A(\mathrm{nil}) \to^c \forall_{x,v}^{nc} (A(v) \to^{nc} A(xv)) \to^c A(v)).$$

The computational meaning then is case distinction (whether a list is empty or not) rather than recursion. This arises naturally if induction is introduced as elimination scheme for an inductive definition of totality, from a different decoration of the clauses.

For the introduction and elimination axioms for $\exists^d, \exists^l, \exists^r, \exists^u, \wedge^d, \wedge^l, \wedge^r, \wedge^u$ the extracted terms are rather obvious and not spelled out here.

THEOREM 2 (Soundness). *Let M be a derivation of A from assumptions $u_i \colon C_i$ ($i < n$). Then we can derive $[\![M]\!]$ **r** A from assumptions x_{u_i} **r** C_i (with $x_{u_i} := \varepsilon$ in case C_i is c.i.).*

2 Computational strengthening

Formulas can be decorated in many different ways, and it is a natural question to ask when one such decoration A' is "stronger" than another one A, in the sense that the former computationally implies the latter, i.e., $\vdash A' \to^c A$. We give a partial answer to this question in the proposition below.

We define a relation $A' \sqsupseteq A$ (A' is a *computational strengthening* of A) between c.r. formulas A', A inductively. It is reflexive, transitive and satisfies

$$(A \to^{nc} B) \sqsupseteq (A \to^c B),$$
$$(A \to^c B) \sqsupseteq (A \to^{nc} B) \quad \text{if } A \text{ is c.i.,}$$
$$(A \to B') \sqsupseteq (A \to B) \quad \text{if } B' \sqsupseteq B, \text{ with } \to \in \{\to^c, \to^{nc}\},$$
$$(A \to B) \sqsupseteq (A' \to B) \quad \text{if } A' \sqsupseteq A, \text{ with } \to \in \{\to^c, \to^{nc}\},$$
$$\forall_x^{nc} A \sqsupseteq \forall_x^c A,$$
$$\forall_x A' \sqsupseteq \forall_x A \quad \text{if } A' \sqsupseteq A, \text{ with } \forall \in \{\forall^c, \forall^{nc}\}.$$

Moreover, for the defined $\exists^d, \exists^l, \exists^r, \wedge^d, \wedge^l, \wedge^r$ it satisfies

$$\exists_x^d A \sqsupseteq \exists_x^l A, \exists_x^r A,$$
$$\exists_x A' \sqsupseteq \exists_x A \quad \text{if } A' \sqsupseteq A, \text{ with } \exists \in \{\exists^d, \exists^l, \exists^r\},$$
$$(A \wedge^d B) \sqsupseteq (A \wedge^l B), (A \wedge^r B),$$
$$(A' \wedge B) \sqsupseteq (A \wedge B) \quad \text{if } A' \sqsupseteq A, \text{ with } \wedge \in \{\wedge^d, \wedge^l, \wedge^r\},$$
$$(A \wedge B') \sqsupseteq (A \wedge B) \quad \text{if } B' \sqsupseteq B, \text{ with } \wedge \in \{\wedge^d, \wedge^l, \wedge^r\}.$$

PROPOSITION 3. *If $A' \sqsupseteq A$, then $\vdash A' \to^c A$.*

Proof. We show that the relation "$\vdash A' \to^c A$" has the same closure properties as "$A' \sqsupseteq A$". For reflexivity and transitivity this is clear. For the rest we give some sample derivations.

$$\cfrac{\cfrac{A \to^{nc} B \quad u \colon A}{B}}{A \to^c B}\,(\to^c)^+, u \qquad \cfrac{\cfrac{\mid \text{assumed}}{B' \to^c B} \quad \cfrac{A \to^{nc} B' \quad u \colon A}{B'}}{\cfrac{B}{A \to^{nc} B}\,(\to^{nc})^+, u}$$

where in the last derivation the final $(\to^{nc})^+$-application is correct since u is not a computational assumption variable in the premise derivation of B.

$$\dfrac{A \to^{nc} B \qquad \dfrac{\dfrac{A' \to^{c} A \qquad \overset{\mid \text{assumed}}{u\colon A'}}{A}}{\dfrac{B}{A' \to^{nc} B} (\to^{nc})^+, u}$$

where for the same reason the final $(\to^{nc})^+$-application is correct. ∎

3 Optimal decorations

We denote the *sequent* of a proof M by $\mathrm{Seq}(M)$; it consists of its *context* and *end formula*.

The *proof pattern* $\mathrm{P}(M)$ of a proof M is the result of marking in c.r. formulas of M (i.e., those not above a c.i. formula) all occurrences of implications and universal quantifiers as non-computational, except the "uninstantiated" formulas of axioms and theorems. For instance, the induction axiom for $\mathbf{N}$ consists of the uninstantiated formula $\forall_n^c (P0 \to^c \forall_n^c (Pn \to^c P(Sn)) \to^c Pn^{\mathbf{N}})$ with a unary predicate variable P and a predicate substitution $P \mapsto \{\, x \mid A(x) \,\}$. Notice that a proof pattern in most cases is not a correct proof, because at axioms formulas may not fit.

We say that a formula D *extends* C if D is obtained from C by changing some (possibly zero) of its occurrences of non-computational implications and universal quantifiers into their computational variants $\to^c$ and $\forall^c$.

A proof N *extends* M if (i) N and M are the same up to variants of implications and universal quantifiers in their formulas, and (ii) every c.r. formula of M is extended by the corresponding one in N. Every proof M whose proof pattern $\mathrm{P}(M)$ is U is called a *decoration* of U.

REMARK 4. Notice that if a proof N extends another one M, then $\mathrm{FV}(\llbracket N \rrbracket)$ is essentially (that is, up to extensions of assumption formulas) a superset of $\mathrm{FV}(\llbracket M \rrbracket)$. This can be proven by induction on N.

In the sequel we assume that every axiom has the property that for every extension of its formula we can find a further extension which is an instance of an axiom, and which is the least one under all further extensions that are instances of axioms. This property clearly holds for axioms whose uninstantiated formula only has the decorated $\to^c$ and $\forall^c$, for instance induction. However, in $\forall_n^c (A(0) \to^c \forall_n^c (A(n) \to^c A(Sn)) \to^c A(n^{\mathbf{N}}))$ the given extension of the four A's might be different. One needs to pick their "least upper bound" as further extension. To make this assumption true for the other axioms listed above (introduction and elimination axioms for $\exists^d, \exists^l, \exists^r, \exists^u, \wedge^d, \wedge^l, \wedge^r, \wedge^u$) we also add all their extensions as axioms.

We will define a *decoration algorithm*, assigning to every proof pattern U and every extension of its sequent an "optimal" decoration M_∞ of U, which further extends the given extension of its sequent.

THEOREM 5. *Under the assumption above, for every proof pattern U and every extension of its sequent $\mathrm{Seq}(U)$ we can find a decoration M_∞ of U such that*

(a) $\mathrm{Seq}(M_\infty)$ extends the given extension of $\mathrm{Seq}(U)$, and

(b) M_∞ is optimal *in the sense that any other decoration M of U whose sequent* $\mathrm{Seq}(M)$ *extends the given extension of* $\mathrm{Seq}(U)$ *has the property that M also extends* M_∞.

Proof. By induction on derivations. It suffices to consider derivations with a c.r. endformula. For axioms the validity of the claim was assumed, and for assumption variables it is clear.

Case $(\to^{nc})^+$. Consider the proof pattern

$$\frac{\begin{array}{c}\Gamma, u\colon A\\ |\, U\\ B\end{array}}{A \to^{nc} B}\; (\to^{nc})^+, u$$

with a given extension $\Delta \Rightarrow C \to^{nc} D$ or $\Delta \Rightarrow C \to^c D$ of its sequent $\Gamma \Rightarrow A \to^{nc} B$. Applying the induction hypothesis for U with sequent $\Delta, C \Rightarrow D$, one obtains a decoration M_∞ of U whose sequent $\Delta_1, C_1 \Rightarrow D_1$ extends $\Delta, C \Rightarrow D$. Now apply $(\to^{nc})^+$ in case the given extension is $\Delta \Rightarrow C \to^{nc} D$ and $x_u \notin \mathrm{FV}(\llbracket M_\infty \rrbracket)$, and $(\to^c)^+$ otherwise.

For (b) consider a decoration $\lambda_u M$ of $\lambda_u U$ whose sequent extends the given extended sequent $\Delta \Rightarrow C \to^{nc} D$ or $\Delta \Rightarrow C \to^c D$. Clearly the sequent $\mathrm{Seq}(M)$ of its premise extends $\Delta, C \Rightarrow D$. Then M extends M_∞ by induction hypothesis for U. If $\lambda_u M$ derives a non-computational implication then the given extended sequent must be of the form $\Delta \Rightarrow C \to^{nc} D$ and $x_u \notin \mathrm{FV}(\llbracket M \rrbracket)$, hence $x_u \notin \mathrm{FV}(\llbracket M_\infty \rrbracket)$. But then by construction we have applied $(\to^{nc})^+$ to obtain $\lambda_u M_\infty$. Hence $\lambda_u M$ extends $\lambda_u M_\infty$. If $\lambda_u M$ does not derive a non-computational implication, the claim follows immediately.

Case $(\to^{nc})^-$. Consider a proof pattern

$$\frac{\begin{array}{cc}\Phi, \Gamma & \Gamma, \Psi\\ |\, U & |\, V\\ A \to^{nc} B & A\end{array}}{B}\;(\to^{nc})^-$$

We are given an extension $\Pi, \Delta, \Sigma \Rightarrow D$ of $\Phi, \Gamma, \Psi \Rightarrow B$. Then we proceed in alternating steps, applying the induction hypothesis to U and V.

(1) The induction hypothesis for U for the extension $\Pi, \Delta \Rightarrow A \to^{nc} D$ of its sequent gives a decoration M_1 of U whose sequent $\Pi_1, \Delta_1 \Rightarrow C_1 \to D_1$ extends $\Pi, \Delta \Rightarrow A \to^{nc} D$, where $\to$ means $\to^{nc}$ or $\to^c$. This already suffices if A is c.i., since then the extension $\Delta_1, \Sigma \Rightarrow C_1$ of V is a correct proof (recall that in c.i. parts of a proof decorations of implications and universal quantifiers can be ignored). If A is c.r.:

(2) The induction hypothesis for V for the extension $\Delta_1, \Sigma \Rightarrow C_1$ of its sequent gives a decoration N_2 of V whose sequent $\Delta_2, \Sigma_2 \Rightarrow C_2$ extends $\Delta_1, \Sigma \Rightarrow C_1$.

(3) The induction hypothesis for U for the extension $\Pi_1, \Delta_2 \Rightarrow C_2 \to D_1$ of its sequent gives a decoration M_3 of U whose sequent $\Pi_3, \Delta_3 \Rightarrow C_3 \to D_3$ extends $\Pi_1, \Delta_2 \Rightarrow C_2 \to D_1$.

(4) The induction hypothesis for V for the extension $\Delta_3, \Sigma_2 \Rightarrow C_3$ of its sequent gives a decoration N_4 of V whose sequent $\Delta_4, \Sigma_4 \Rightarrow C_4$ extends $\Delta_3, \Sigma_2 \Rightarrow C_3$. This process is repeated until in V no further proper extension of Δ_3 and C_3 is returned. Such a situation will always be reached since there is a maximal extension, where all connectives are maximally decorated. But then we easily obtain (a): Assume that in (4) we have $\Delta_4 = \Delta_3$ and $C_4 = C_3$. Then the decoration

$$\cfrac{\begin{array}{cc} \Pi_3, \Delta_3 & \Delta_4, \Sigma_4 \\ \mid M_3 & \mid N_4 \\ C_3 \to D_3 & C_4 \end{array}}{D_3} \to^-$$

of UV derives a sequent $\Pi_3, \Delta_3, \Sigma_4 \Rightarrow D_3$ extending $\Pi, \Delta, \Sigma \Rightarrow D$.

For (b) we need to consider a decoration MN of UV whose sequent $\mathrm{Seq}(MN)$ extends the given extension $\Pi, \Delta, \Sigma \Rightarrow D$ of $\Phi, \Gamma, \Psi \Rightarrow B$. We must show that MN extends $M_3 N_4$. To this end we go through the alternating steps again.

(1) Since the sequent $\mathrm{Seq}(M)$ extends $\Pi, \Delta \Rightarrow A \to^{nc} D$, the induction hypothesis for U for the extension $\Delta \Rightarrow A \to^{nc} D$ of its sequent ensures that M extends M_1.

(2) Since then the sequent $\mathrm{Seq}(N)$ extends $\Delta_1, \Sigma \Rightarrow C_1$, the induction hypothesis for V for the extension $\Delta_1, \Sigma \Rightarrow C_1$ of its sequent ensures that N extends N_2.

(3) Therefore $\mathrm{Seq}(M)$ extends the sequent $\Pi_1, \Delta_2 \Rightarrow C_2 \to D_1$, and the induction hypothesis for U for the extension $\Pi_1, \Delta_2 \Rightarrow C_2 \to D_1$ of U's sequent ensures that M extends M_3.

(4) Therefore $\mathrm{Seq}(N)$ extends $\Delta_3, \Sigma_2 \Rightarrow C_3$, and induction hypothesis for V for the extension $\Delta_3, \Sigma_2 \Rightarrow C_3$ of V's sequent ensures that N also extends N_4.

But since $\Delta_4 = \Delta_3$ and $C_4 = C_3$ by assumption, MN extends the decoration $M_3 N_4$ of UV constructed above.

Case $(\forall^{nc})^+$. Consider a proof pattern

$$\cfrac{\begin{array}{c} \Gamma \\ \mid U \\ A \end{array}}{\forall_x^{nc} A}\,(\forall^{nc})^+$$

with a given extension $\Delta \Rightarrow \forall_x^{nc} C$ or $\Delta \Rightarrow \forall_x^{c} C$ of its sequent. Applying the induction hypothesis for U with sequent $\Delta \Rightarrow C$, one obtains a decoration M_∞ of U whose sequent $\Delta_1 \Rightarrow C_1$ extends $\Delta \Rightarrow C$. Now apply $(\forall^{nc})^+$ in case the given extension is $\Delta \Rightarrow \forall_x^{nc} C$ and $x \notin \mathrm{FV}(\llbracket M_\infty \rrbracket)$, and $(\forall^c)^+$ otherwise.

For (b) consider a decoration $\lambda_x M$ of $\lambda_x U$ whose sequent extends the given extended sequent $\Delta \Rightarrow \forall_x^{nc} C$ or $\Delta \Rightarrow \forall_x^{c} C$. Clearly the sequent $\mathrm{Seq}(M)$ of its premise extends $\Delta \Rightarrow C$. Then M extends M_∞ by induction hypothesis for U. If $\lambda_x M$ derives a non-computational generalization, then the given extended sequent must be of the form $\Delta \Rightarrow \forall_x^{nc} C$ and $x \notin \mathrm{FV}(\llbracket M \rrbracket)$, hence $x \notin \mathrm{FV}(\llbracket M_\infty \rrbracket)$ (by the remark above). But then by construction we have applied $(\forall^{nc})^+$ to obtain $\lambda_x M_\infty$. Hence $\lambda_x M$ extends $\lambda_x M_\infty$. If $\lambda_x M$ does not derive a non-computational generalization, the claim follows immediately.

Case $(\forall^{nc})^-$. Consider a proof pattern

$$\dfrac{\begin{array}{c}\Gamma \\ \mid U \\ \forall_x^{nc} A(x) \quad r\end{array}}{A(r)}\;(\forall^{nc})^-$$

and let $\Delta \Rightarrow C(r)$ be any extension of its sequent $\Gamma \Rightarrow A(r)$. The induction hypothesis for U for the extension $\Delta \Rightarrow \forall_x^{nc} C(x)$ produces a decoration M_∞ of U whose sequent extends $\Delta \Rightarrow \forall_x^{nc} C(x)$. Then apply $(\forall^{nc})^-$ or $(\forall^c)^-$, whichever is appropriate, to obtain the required $M_\infty r$.

For (b) consider a decoration Mr of Ur whose sequent $\mathrm{Seq}(Mr)$ extends the given extension $\Delta \Rightarrow C(r)$ of $\Gamma \Rightarrow A(r)$. Then M extends M_∞ by induction hypothesis for U, and hence Mr extends $M_\infty r$. ∎

4 Applications

4.1 Decoration of implication and conjunction

We illustrate the effects of decoration on a simple example involving implications. Consider $A \to B \to A$ with the trivial proof $M := \lambda_{u_1}^A \lambda_{u_2}^B u_1$. Clearly, one does not need the decoration of the second implication, so we apply the decoration algorithm and specify as extension of $\mathrm{Seq}(\mathrm{P}(M))$ the formula $A \to^{nc} B \to^{nc} A$. The algorithm detects that the first implication needs to be decorated, since the abstracted assumption variable is computational. Since the second implication can be left undecorated, a proof of $A \to^c B \to^{nc} A$ is constructed from M.

A similar phenomenon occurs for $A \wedge^d B \to B$. Let M be its proof and $U := \mathrm{P}(M)$ its proof pattern. When given the extension $A \wedge^u B \to^{nc} B$ for $\mathrm{Seq}(U)$, the decoration algorithm constructs a correct proof of $A \wedge^r B \to^c B$.

4.2 List reversal

We work in Heyting Arithmetic HA^ω for a language based on Gödel's T [8], which is finitely typed (cf. Troelstra [10] for general background). We call the formulas built from terms r of type $\mathbf{B}$ by means of a special operator $\mathrm{atom}(r^{\mathbf{B}})$ *decidable* prime formulas. They include for instance equations between terms of type $\mathbf{N}$, since the boolean-valued binary equality function $=_{\mathbf{N}}: \mathbf{N} \to \mathbf{N} \to \mathbf{B}$ can be defined by

$$\begin{array}{ll}(0 =_{\mathbf{N}} 0) := \mathrm{tt}, & (Sn =_{\mathbf{N}} 0) := \mathrm{ff}, \\ (0 =_{\mathbf{N}} Sm) := \mathrm{ff}, & (Sn =_{\mathbf{N}} Sm) := (n =_{\mathbf{N}} m).\end{array}$$

For falsity we can take the atomic formula $\mathbf{F} := \mathrm{atom}(\mathrm{ff})$ – called *arithmetical falsity* – built from the boolean constant ff. Since in the A-translation below we need to substitute a formula for falsity, we alternatively use a special predicate

variable $\bot$ to mark such occurrences of falsity. The *formulas* of HA$^\omega$ are built from prime formulas by the connectives $\to$ and $\forall$. We define *negation* $\neg A$ by $A \to \mathbf{F}$ or $A \to \bot$ (depending on the context), and the *weak* (or "classical") existential quantifier by
$$\tilde{\exists}_x A := \neg\forall_x \neg A.$$

We first give an informal weak existence proof for list reversal. Write vw for the result $v * w$ of appending the list w to the list v, vx for the result $v * x$: of appending the one element list x: to the list v, and xv for the result $x :: v$ of constructing a list by writing an element x in front of a list v, and omit the parentheses in $R(v,w)$ for (typographically) simple arguments. Assuming

$$\text{InitRev:} \quad R(\text{nil}, \text{nil}), \tag{1}$$
$$\text{GenRev:} \quad \forall_{v,w,x}(Rvw \to R(vx, xw)) \tag{2}$$

we prove

(3) $\quad \forall_v \tilde{\exists}_w Rvw \qquad (:= \forall_v(\forall_w(Rvw \to \bot) \to \bot))$.

Fix v and assume $u\colon \forall_w \neg Rvw$; we need to derive a contradiction. To this end we prove that all initial segments of v are non-revertible, which contradicts (1). More precisely, from u and (2) we prove

$$\forall_{v_2} A(v_2) \quad \text{with } A(v_2) := \forall_{v_1}(v_1 v_2 = v \to \forall_w \neg R v_1 w)$$

by induction on v_2. For $v_2 = \text{nil}$ this follows from our initial assumption u. For the step case, assume $v_1(xv_2) = v$, fix w and assume further $Rv_1 w$. We must derive a contradiction. By (2) we conclude that $R(v_1 x, xw)$. On the other hand, properties of the append function imply that $(v_1 x)v_2 = v$. The induction for $v_1 x$ gives $\forall_w \neg R(v_1 x, w)$. Taking xw for w leads to the desired contradiction.

We formalize this proof, to prepare it for decoration. The following lemmata will be used.

$$\text{Compat:} \quad \forall_P \forall_{v_1,v_2}(v_1 = v_2 \to Pv_1 \to Pv_2),$$
$$\text{Symm:} \quad \forall_{v_1,v_2}(v_1 = v_2 \to v_2 = v_1),$$
$$\text{Trans:} \quad \forall_{v_1,v_2,v_3}(v_1 = v_2 \to v_2 = v_3 \to v_1 = v_3),$$
$$L_1: \quad \forall_v(v = v\,\text{nil}),$$
$$L_2: \quad \forall_{v_1,x,v_2}((v_1 x)v_2 = v_1(xv_2)),$$

The proof term is

$$M := \lambda_v \lambda_u^{\forall_w \neg Rvw}(\text{Ind}_{v_2, A(v_2)} vv M_{\text{Base}} M_{\text{Step}} \text{ nil } T^{\text{nil } v = v} \text{ nil InitRev})$$

with

$$M_{\text{Base}} := \lambda_{v_1} \lambda_{u_1}^{v_1 \text{nil} = v}(\text{Compat } \{\, v \mid \forall_w \neg Rvw \,\} vv_1$$
$$\qquad (\text{Symm } v_1 v(\text{Trans } v_1 (v_1 \text{ nil}) v(L_1 v_1) u_1)) u),$$
$$M_{\text{Step}} := \lambda_{x,v_2} \lambda_{u_0}^{A(v_2)} \lambda_{v_1} \lambda_{u_1}^{v_1(xv_2) = v} \lambda_w \lambda_{u_2}^{Rv_1 w}($$
$$\qquad u_0(v_1 x)(\text{Trans }((v_1 x)v_2)(v_1(xv_2)) v(L_2 v_1 x v_2) u_1)$$
$$\qquad (xw)(\text{GenRev } v_1 wxu_2)).$$

We now have a proof M of $\forall_v \tilde{\exists}_w Rvw$ from the clauses InitRev: D_1 and GenRev: D_2, with $D_1 := R(\text{nil}, \text{nil})$ and $D_2 := \forall_{v,w,x}(Rvw \to R(vx, xw))$. Using the Dragalin/Friedman [6, 7] A-translation (in its refined form of [4]) we can replace $\bot$ throughout by $\exists_w Rvw$. The end formula $\forall_v \tilde{\exists}_w Rvw := \forall_v \neg \forall_w \neg Rvw := \forall_v(\forall_w(Rvw \to \bot) \to \bot)$ is turned into $\forall_v(\forall_w(Rvw \to \exists_w Rvw) \to \exists_w Rvw)$. Since its premise is an instance of existence introduction we obtain a derivation $M^\exists$ of $\forall_v \exists_w Rvw$. Moreover, in this case neither the D_i nor any of the axioms used involves $\bot$ in its uninstantiated formulas, and hence the correctness of the proof is not affected by the substitution. The term `neterm` extracted in Minlog from a formalization of the proof above is (after "animating" Compat)

```
[v0]
(Rec list nat=>list nat=>list nat=>list nat)v0([v1,v2]v2)
([x1,v2,g3,v4,v5]g3(v4:+:x1:)(x1::v5))
(Nil nat)
(Nil nat)
```

with g a variable for binary functions on lists. In fact, the underlying algorithm defines an auxiliary function h by

$$h(\text{nil}, v_2, v_3) := v_3, \qquad h(xv_1, v_2, v_3) := h(v_1, v_2 x, xv_3)$$

and gives the result by applying h to the original list and twice nil.

Notice that the second argument of h is not needed. However, its presence makes the algorithm quadratic rather than linear, because in each recursion step $v_2 x$ is computed, and the list append function is defined by recursion on its first argument. We will be able to get rid of this superfluous second argument by decorating the proof. It will turn out that in the proof (by induction on v_2) of the auxiliary formula $A(v_2) := \forall_{v_1}(v_1 v_2 = v \to \forall_w \neg Rv_1 w))$, the variable v_1 is not used computationally. Hence, in the decorated version of the proof, we can use $\forall_{v_1}^{nc}$.

Let us now apply the general method of decorating proofs to the example of list reversal. To this end, we present our proof in more detail, particularly by writing proof trees with formulas. The decoration algorithm then is applied to its proof pattern with the sequent consisting of the context $R(\text{nil}, \text{nil})$ and $\forall_{v,w,x}^{nc}(Rvw \to^{nc} R(vx, xw))$ and the end formula $\forall_v^{nc} \exists_w^l Rvw$.

Rather than describing the algorithm step by step we only display the end result. Among the axioms used, the only ones in c.r. parts are Compat and list induction. They appear in the decorated proof in the form

Compat: $\forall_P \forall_{v_1, v_2}^{nc}(v_1 = v_2 \to Pv_1 \to^c Pv_2)$,

Ind: $\quad \forall_{v_2}^c(A(\text{nil}) \to^c \forall_{x, v_2}^c(A(v_2) \to^c A(xv_2)) \to^c A(v_2))$

with $A(v_2) := \forall_{v_1}^{nc}(v_1 v_2 = v \to \forall_w^c \neg^\exists Rv_1 w)$ and $\neg^\exists Rv_1 w := Rv_1 w \to \exists_w^l Rvw$. $M_{\text{Base}}^\exists$ is the derivation in Figure 2, where N is a derivation involving L_1 with a free assumption $u_1: v_1 \text{nil}=v$. $M_{\text{Step}}^\exists$ is the derivation in Figure 3, where N_1 is a derivation involving L_2 with free assumption $u_1: v_1(xv_2)=v$, and N_2 is one involving GenRev with the free assumption $u_2: Rv_1 w$.

The extracted term `neterm` then is

$$
\dfrac{\dfrac{\dfrac{\dfrac{\dfrac{\text{Compat}\quad \{v \mid \forall^{\mathrm{c}}_{w}\neg^{\exists}Rvw\}\quad v\quad v_1}{v{=}v_1 \to \forall^{\mathrm{c}}_{w}\neg^{\exists}Rvw \to^{\mathrm{c}} \forall^{\mathrm{c}}_{w}\neg^{\exists}Rv_1w}\qquad \begin{matrix}[u_1\colon v_1\,\mathrm{nil}{=}v]\\ \mid N\\ v{=}v_1\end{matrix}}{\forall^{\mathrm{c}}_{w}\neg^{\exists}Rvw \to^{\mathrm{c}} \forall^{\mathrm{c}}_{w}\neg^{\exists}Rv_1w}\qquad \exists^{+}\colon \forall^{\mathrm{c}}_{w}\neg^{\exists}Rvw}{\forall^{\mathrm{c}}_{w}\neg^{\exists}Rv_1w}\,(\to^{\mathrm{nc}})^{+}u_1}{v_1\,\mathrm{nil} = v \to \forall^{\mathrm{c}}_{w}\neg^{\exists}Rv_1w}}{\forall^{\mathrm{nc}}_{v_1}(v_1\,\mathrm{nil} = v \to \forall^{\mathrm{c}}_{w}\neg^{\exists}Rv_1w)\quad (= A(\mathrm{nil}))}
$$

Figure 2. The decorated base derivation

$$
\dfrac{\dfrac{\dfrac{\dfrac{\dfrac{\dfrac{\dfrac{\dfrac{\dfrac{\dfrac{[u_0\colon A(v_2)]\quad v_1x}{(v_1x)v_2{=}v \to \forall^{\mathrm{c}}_{w}\neg^{\exists}R(v_1x,w)}\qquad \begin{matrix}[u_1\colon v_1(xv_2){=}v]\\ \mid N_1\\ (v_1x)v_2{=}v\end{matrix}}{\forall^{\mathrm{c}}_{w}\neg^{\exists}R(v_1x,w)}\qquad xw}{\neg^{\exists}R(v_1x,xw)}\qquad \begin{matrix}[u_2\colon Rv_1w]\\ \mid N_2\\ R(v_1x,xw)\end{matrix}}{\exists^{\mathrm{l}}_{w}Rvw}\,(\to^{\mathrm{nc}})^{+}u_2}{\neg^{\exists}Rv_1w}}{\forall^{\mathrm{c}}_{w}\neg^{\exists}Rv_1w}\,(\to^{\mathrm{nc}})^{+}u_1}{v_1(xv_2){=}v \to \forall^{\mathrm{c}}_{w}\neg^{\exists}Rv_1w}}{\forall^{\mathrm{nc}}_{v_1}(v_1(xv_2){=}v \to \forall^{\mathrm{c}}_{w}\neg^{\exists}Rv_1w)\ \ ({=}A(xv_2))}\,(\to^{\mathrm{c}})^{+}u_0}{A(v_2) \to^{\mathrm{c}} A(xv_2)}}{\forall^{\mathrm{c}}_{x,v_2}(A(v_2) \to^{\mathrm{c}} A(xv_2))}
$$

Figure 3. The decorated step derivation

```
[v0]
(Rec list nat=>list nat=>list nat)v0([v1]v1)
([x1,v2,f3,v4]f3(x1::v4))
(Nil nat)
```

with f a variable for unary functions on lists. To run this algorithm one has to normalize the term obtained by applying neterm to a list:

```
(pp (nt (mk-term-in-app-form neterm (pt "1::2::3::4:"))))
; 4::3::2::1:
```

This time, the underlying algorithm defines an auxiliary function g by

$$g(\mathrm{nil}, w) := w, \qquad g(x :: v, w) := g(v, x :: w)$$

and gives the result by applying g to the original list and nil. In conclusion, we have obtained (by machine extraction from an automated decoration of a weak existence proof) the standard linear algorithm for list reversal, with its use of an accumulator.

4.3 Passing continuations

A similar application of decoration occurs when one derives double induction

$$\forall_n^c (Qn \to^c Q(Sn) \to^c Q(S(Sn))) \to^c \forall_n^c (Q0 \to^c Q1 \to^c Qn)$$

in continuation passing style, i.e., not directly, but using as an intermediate assertion (proved by induction)

$$\forall_{n,m}^c ((Qn \to^c Q(Sn) \to^c Q(n+m)) \to^c Q0 \to^c Q1 \to^c Q(n+m)).$$

After decoration, the formula becomes

$$\forall_n^c \forall_m^{nc} ((Qn \to^c Q(Sn) \to^c Q(n+m)) \to^c Q0 \to^c Q1 \to^c Q(n+m)).$$

This can be applied to obtain a continuation based tail recursive definition of the Fibonacci function, from a proof of its totality. Let G be the graph of the Fibonacci function, defined by the clauses

$$G(0,0), \quad G(1,1),$$
$$\forall_{n,v,w}^{nc} (G(n,v) \to^{nc} G(Sn, w) \to^{nc} G(S(Sn), v+w)).$$

¿From these assumptions one can easily derive

$$\forall_n^c \exists_v G(n, v),$$

using double induction (proved in continuation passing style). The term extracted from this proof is

```
[n0]
(Rec nat=>nat=>(nat=>nat=>nat)=>nat=>nat=>nat)
n0([n1,k2]k2)
([n1,p2,n3,k4]p2(Succ n3)([n7,n8]k4 n8(n7+n8)))
```

applied to 0, ([n1,n2]n1), 0 and 1. An unclean aspect of this term is that the recursion operator has value type

$$\mathtt{nat => (nat => nat => nat) => nat => nat => nat}$$

rather than `(nat=>nat=>nat)=>nat=>nat=>nat`, which would correspond to an iteration. However, we can repair this by decoration. After (automatic) decoration of the proof, the extracted term becomes

```
[n0]
(Rec nat=>(nat=>nat=>nat)=>nat=>nat=>nat)
n0([k1]k1)
([n1,p2,k3]p2([n6,n7]k3 n7(n6+n7)))
```

applied to ([n1,n2]n1), 0 and 1. This indeed is iteration in continuation passing style.

4.4 Proof transformations

In the next two examples we allow the decoration algorithm to substitute an auxiliary lemma used in the proof by a lemma that we specify explicitly. The algorithm will verify if the lemma passed to it as an argument is fitting and if this is the case, it will replace the lemma used in the original proof by the specified one. If not, the initial lemma is kept. This will allow for a certain control over the computational content, as shown by the following examples.

Our first example is an elaboration of Constable's idea described in the introduction. Let Pn mean "n is prime". Consider

$$\forall_n^c (Pn \vee^r \exists_{m,k>1}^d (n = mk)) \quad \text{factorization,}$$
$$\forall_n^c (Pn \vee^u \exists_{m,k>1}^d (n = mk)) \quad \text{prime number test.}$$

Euler's φ-function has the properties

$$\begin{cases} \varphi(n) = n-1 & \text{if } Pn, \\ \varphi(n) < n-1 & \text{if } n \text{ is composed.} \end{cases}$$

Suppose that somewhat foolishly we have used factorization and these properties to obtain a proof of

$$\forall_n^c (\varphi(n) = n-1 \vee^u \varphi(n) < n-1).$$

Our goal is to get rid of the expensive factorization algorithm in the computational content, via decoration.

The decoration algorithm arrives at the factorization theorem

$$\forall_n^c (Pn \vee^r \exists_{m,k>1}^d (n = mk))$$

with the decorated formula

$$\forall_n^c (Pn \vee^u \exists_{m,k>1}^d (n = mk)).$$

Since the prime number test can be considered instead of the factorization lemma, we can specify that the decoration algorithm should try to replace the former by the latter. In case this is possible, a new proof is constructed, using the the prime number test lemma. Should this fail, the factorization lemma is kept. As it turns out in this case, the replacement is possible.

In the Minlog implementation the difference is clearly visible. cL denotes the computational content of the factorization lemma L (i.e., the factorization algorithm), and cLU the computational content of the lemma LU expressing the prime number test. The extract from the original proof involves computing (cL n0), i.e., factorizing the argument, whereas after decoration the prime number test cLU suffices.

```
(pp (nt (proof-to-extracted-term proof)))
; [n0][if (cL n0) cInlOrU ([algC1]cInrOrU)]

(pp (nt (proof-to-extracted-term (decorate proof))))
; cLU
```

The second example is due to Bates and Constable [1], and deals with the "maximal segment problem". Let X be a set with a linear ordering $\leq$, and consider an infinite sequence $f \colon \mathbf{N} \to X$ of elements of X. Assume further that we have a function $M \colon (\mathbf{N} \to X) \to \mathbf{N} \to \mathbf{N} \to X$ such that $M(f, i, k)$ "measures" the segment $f(i), \ldots, f(k)$. The task is to find a segment determined by $i \leq k \leq n$ such that its measure is maximal. To simplify the formalization let us consider M and f fixed and define $\mathrm{seg}(i, k) := M(f, i, k)$.

Of course we can simply solve this problem by trying all possibilities; these are $O(n^2)$ many. The first proof to be given below corresponds to this general claim. Then we will show that for a more concrete problem with the sum $x_i + \cdots + x_k$ as measure the proof can be simplified, using monotonicity of the sum at an appropriate place. From this simplified proof one can extract a better algorithm, which is linear rather than quadratic. Our goal is to achieve this effect by decoration.

Let us be more concrete. The original specification is to find a maximal segment $x_i, \ldots, x_k$, i.e.,

$$\forall_n^c \exists_{i \leq k \leq n}^d \forall_{i' \leq k' \leq n}(\mathrm{seg}(i', k') \leq \mathrm{seg}(i, k)).$$

A special case is to find the maximal end segment

$$\forall_n^c \exists_{j \leq n}^1 \forall_{j' \leq n}(\mathrm{seg}(j', n) \leq \mathrm{seg}(j, n)).$$

We first provide two proofs of the existence of a maximal end segment for $n + 1$

$$\forall_n^c \exists_{j \leq n+1}^1 \forall_{j' \leq n+1}(\mathrm{seg}(j', n+1) \leq \mathrm{seg}(j, n+1)).$$

The first proof introduces an auxiliary variable m and proceeds by induction on m, with n a parameter:

$$\forall_n^{nc} \forall_{m \leq n+1}^c \exists_{j \leq n+1}^1 \forall_{j' \leq m}(\mathrm{seg}(j', n+1) \leq \mathrm{seg}(j, n+1)).$$

The second proof uses as assumptions an "induction hypothesis"

$$\text{IH}_n: \exists^l_{j \leq n} \forall_{j' \leq n}(\text{seg}(j', n) \leq \text{seg}(j, n))$$

and the additional assumption of monotonicity of seg

$$\text{Mon}: \text{seg}(i, k) \leq \text{seg}(j, k) \to \text{seg}(i, k+1) \leq \text{seg}(j, k+1).$$

It proceeds by cases on $\text{seg}(j, n+1) \leq \text{seg}(n+1, n+1)$. If $\leq$ holds, take $n+1$, else the previous j.

We now prove the existence of a maximal segment by induction on n, simultaneously with the existence of a maximal end segment.

$$\forall^c_n(\exists^d_{i \leq k \leq n} \forall_{i' \leq k' \leq n}(\text{seg}(i', k') \leq \text{seg}(i, k)) \wedge^d$$
$$\exists^l_{j \leq n} \forall_{j' \leq n}(\text{seg}(j', n) \leq \text{seg}(j, n)))$$

In the step, we compare the maximal segment i, k for n with the maximal end segment $j, n+1$ provided separately. If $\leq$ holds, take the new i, k to be $j, n+1$. Else take the old i, k.

Depending on how the existence of a maximal end segment was proved, we obtain a quadratic or a linear algorithm. For this reason, we can use the option of specifying which lemma the decoration algorithm should use. We have

L1: $\forall^c_n \exists^l_{j \leq n+1} \forall_{j' \leq n+1}(\text{seg}(j', n+1) \leq \text{seg}(j, n+1))$,

L2: $\forall^{nc}_n (\text{IH}_n \to^c \text{Mon} \to \exists^l_{j \leq n+1} \forall_{j' \leq n+1}(\text{seg}(j', n+1) \leq \text{seg}(j, n+1)))$,

so we can try to replace **L1** by **L2**. Since this is possible, the decoration algorithm constructs the correct proof from **L2** and produces the proof resulting in the linear algorithm.

BIBLIOGRAPHY

[1] Joseph L. Bates and Robert L. Constable. Proofs as programs. *ACM Transactions on Programming Languages and Systems*, 7(1):113–136, January 1985.

[2] Ulrich Berger. Program extraction from normalization proofs. In M. Bezem and J.F. Groote, editors, *Typed Lambda Calculi and Applications*, volume 664 of *LNCS*, pages 91–106. Springer Verlag, Berlin, Heidelberg, New York, 1993.

[3] Ulrich Berger. Uniform Heyting Arithmetic. *Annals of Pure and Applied Logic*, 133:125–148, 2005.

[4] Ulrich Berger, Wilfried Buchholz, and Helmut Schwichtenberg. Refined program extraction from classical proofs. *Annals of Pure and Applied Logic*, 114:3–25, 2002.

[5] Luca Chiarabini. *Program Development by Proof Transformation*. PhD thesis, Fakultät für Mathematik, Informatik und Statistik der LMU, München, 2009.

[6] Albert Dragalin. New kinds of realizability. In *Abstracts of the 6th International Congress of Logic, Methodology and Philosophy of Sciences*, pages 20–24, Hannover, Germany, 1979.

[7] Harvey Friedman. Classically and intuitionistically provably recursive functions. In D.S. Scott and G.H. Müller, editors, *Higher Set Theory*, volume 669 of *Lecture Notes in Mathematics*, pages 21–28. Springer Verlag, Berlin, Heidelberg, New York, 1978.

[8] Kurt Gödel. Über eine bisher noch nicht benützte Erweiterung des finiten Standpunkts. *Dialectica*, 12:280–287, 1958.

[9] Andrey N. Kolmogorov. Zur Deutung der intuitionistischen Logik. *Math. Zeitschr.*, 35:58–65, 1932.

[10] Anne S. Troelstra, editor. *Metamathematical Investigation of Intuitionistic Arithmetic and Analysis*, volume 344 of *Lecture Notes in Mathematics*. Springer Verlag, Berlin, Heidelberg, New York, 1973.

Searching for proofs (and uncovering capacities of the mathematical mind)[1]

WILFRIED SIEG

Abstract. What is it that shapes mathematical arguments into proofs that are intelligible to us, and what is it that allows us to find proofs efficiently? — This is the informal question I intend to address by investigating, on the one hand, the abstract ways of the axiomatic method in modern mathematics and, on the other hand, the concrete ways of proof construction suggested by modern proof theory. These theoretical investigations are complemented by experimentation with the proof search algorithm AProS. It searches for natural deduction proofs in pure logic; it can be extended directly to cover elementary parts of set theory and to find abstract proofs of Gödel's incompleteness theorems. The subtle interaction between understanding and reasoning, i.e., between *introducing concepts* and *proving theorems*, is crucial. It suggests principles for structuring proofs conceptually and brings out the dynamic role of *leading ideas*. Hilbert's work provides a perspective that allows us to weave these strands into a fascinating intellectual fabric and to connect, in novel and surprising ways, classical themes with deep contemporary problems. The connections reach from proof theory through computer science and cognitive psychology to the philosophy of mathematics and all the way back.

1 Historical perspective

It is definitely counter to the standard view of Hilbert's formalist perspective on mathematics that I associate his work with uncovering aspects of the mathematical mind; I hope you will see that he played indeed a pivotal role. He was deeply influenced by Dedekind and Kronecker; he connected these extraordinary mathematicians of the 19th century to two equally remarkable logicians of the 20th century, Gödel and Turing. The character of that connection is determined by Hilbert's focus on the *axiomatic method* and the associated *consistency problem*. What a remarkable path it is: emerging from the radical transformation of mathematics in the second half of the 19th century and leading to the dramatic development of metamathematics in the second half of the 20th century.

Examining that path allows us to appreciate Hilbert's perspective on the wide-open mathematical landscape. It also enriches our perspective on his

---

[1] This essay is dedicated to Grigori Mints on the occasion of his 70th birthday. Over the course of many years we have been discussing the fruitfulness of searching directly for natural deduction proofs. He and his Russian colleagues took already in 1965 a systematic and important step for propositional logic; see the co-authored paper (Shanin, et al. 1965), but also (Mints 1969) and the description of further work in (Maslov, Mints, and Orevkov 1983).

metamathematical work.[1] Some of Hilbert's considerations are, however, not well integrated into contemporary investigations. In particular, the *cognitive* side of proof theory has been neglected, and I intend to pursue it in this essay. It was most strongly, but perhaps somewhat misleadingly, expressed in Hilbert's Hamburg talk of 1927. He starts with a general remark about the "formula game" criticized by Brouwer:

> The formula game ... has, besides its mathematical value, an important general philosophical significance. For this formula game is carried out according to certain definite rules, in which the technique of our thinking is expressed. These rules form a closed system that can be discovered and definitively stated.

Then he continues with a provocative statement about the cognitive goal of proof theoretic investigations.

> The fundamental idea of my proof theory is none other than to describe the activity of our understanding, to make a protocol of the rules according to which our thinking actually proceeds.[2]

It is clear to us, and it was clear to Hilbert, that mathematical thinking does not proceed in the strictly regimented ways imposed by an austere formal theory. Though formal rigor is crucial, it is not sufficient to shape proofs intelligibly or to discover them efficiently, even in pure logic. Recalling the principle that mathematics should solve problems "by a minimum of blind calculation and a maximum of guiding thought", I will investigate the subtle interaction between understanding and reasoning, i.e., between *introducing concepts* and *proving theorems*. That suggests principles for structuring proofs conceptually and brings out the dynamic role of *leading ideas*.[3]

---

[1] In spite of the demise of the finitist program, proof theoretic work has been continued successfully along at least two dimensions. There is, first of all, the ever more refined formalization of mathematics with the novel mathematical end of extracting information from proofs. Formalizing mathematics was originally viewed as the basis for a mathematical treatment of foundational problems and, in particular, for obtaining consistency results. Gödel's theorems shifted the focus from absolute finitist to relative consistency proofs with the philosophical end of comparing foundational frameworks; that is the second dimension of continuing proof theoretic work. These two dimensions are represented by "proof mining" initiated by Kreisel and "reductive proof theory" pursued since Gödel and Gentzen's consistency proof of classical relative to intuitionistic number theory.

[2] (Hilbert 1927) in (van Heijenoort 1967, p. 475).

[3] The way in which I am pursuing matters is programmatically related to Wang's perspective in his (1970). In that paper Wang discusses, on p. 106, "the project of mechanizing mathematical arguments". The results that have been obtained so far, Wang asserts, are only "theoretical" ones, "which do not establish the strong conclusion that mathematical reasoning (or even a major part of it) is mechanical in nature". But the unestablished strong conclusion challenges us to address in novel ways "the perennial problem about mind and machine" — by dealing with mathematical activity in a systematic way. Wang continues: "Even though what is demanded is not mechanical simulation, the task requires a close examination of how mathematics is done in order to determine how informal methods can be replaced by mechanizable procedures and how the speed of computers can be employed to compensate for their inflexibility. The field is wide open, and like all good things, it is not easy. But one does expect and look for pleasant surprises in this enterprise which requires a novel combination of psychology, logic, mathematics and computer technology." Surprisingly, there is still no unified interdisciplinary approach; but see Appendix C below with the title "Confluence?".

In some sense, the development toward proof theory began in late 1917 when Hilbert gave a talk in Zürich, entitled *Axiomatisches Denken*. The talk was deeply rooted in the past and pointed decisively to the future. Hilbert suggested, in particular,

... we must — that is my conviction — take the concept of the specifically mathematical proof as an object of investigation, just as the astronomer has to consider the movement of his position, the physicist must study the theory of his apparatus, and the philosopher criticizes reason itself.

Hilbert recognized, in the very next sentence, that "the execution of this program is at present, to be sure, still an unsolved problem". Ironically, solving this problem was just a step in solving the most pressing issue with modern abstract mathematics as it had emerged in the second half of the 19th century. This development of mathematics is complemented by and connected to the dramatic expansion of logic facilitating steps toward full formalization.[4] Hilbert clearly hoped to address the issue he had already articulated in his Paris address of 1900 and had stated prominently as the second in his famous list of problems:

... I wish to designate the following as the most important among the numerous questions which can be asked with regard to the axioms [of arithmetic]: *To prove that they are not contradictory, that is, that a finite number of logical steps based upon them can never lead to contradictory results.*

As to the axioms of arithmetic, Hilbert points to his paper *Über den Zahlbegriff* delivered at the Munich meeting of the German Association of Mathematicians in September of 1899. The title alone indicates already its intellectual context: twelve years earlier, Kronecker had published a well-known paper with the very same title and had sketched a way of introducing irrational numbers without accepting the general notion. It is precisely to the general concept that Hilbert wants to give a proper foundation — using the axiomatic method and following Dedekind who represents most strikingly the development toward greater abstractness in mathematics.

2 Abstract concepts

Howard Stein analyzed philosophical aspects of the 19th century expansion and transformation of mathematics I just alluded to.[5] Underlying these develop-

---

[4]The deepest philosophical connection between the mathematical and logical developments is indicated by the fact that both Dedekind and Frege considered the concept of a "function" to be central; it is a dramatic break from traditional metaphysics. Cf. Cassirer's *Substanzbegriff und Funktionsbegriff*.

[5]Stein did so in his marvelous paper (Stein 1988). The key words of its title (logos, logic, and logistiké) structure the systematic progression of my essay that was presented as the Howard Stein Lecture at the University of Chicago on 15 May 2008; Part 2 is a discussion of logos, Part 3 of logic, and Part 4 of logistiké. Improved versions of that talk were presented on 8 October 2008 to a workshop on "Mathematics between the Natural Sciences and the Humanities" held in Göttingen, on 28 December 2008 to the Symposium on "Hilbert's Place in the Foundations and Philosophy of Mathematics" at the meeting of the American Philosophical Association in Philadelphia, on 27 February 2009 in the series "Formal Methods in the Humanities" at Stanford University, and on 16 April 2009 to the conference on "The Fundamental Idea of Proof Theory" in Paris. I am grateful to many

ments is for him the rediscovery of a *capacity of the human mind* that had been first discovered by the Greeks between the 6th and 4th century B.C.:

> The expansion [of mathematics in the 19th century] was effected by the very same capacity of thought that the Greeks discovered; but in the process, something new was learned about the nature of that capacity — what it is, and what it is not. I believe that what has been learned, when properly understood, constitutes one of the greatest advances in philosophy — although it, too, like the advance in mathematics itself, has a close relation to ancient ideas.[6]

The deep connections and striking differences between the two discoveries can be examined by comparing Eudoxos's theory of proportion with Dedekind's and Hilbert's theory of real numbers. Fundamental for articulating this difference is Dedekind's notion of *system* that is also used by Hilbert.

2.1 Systems

When discussing Kronecker's demand that proofs be constructive and that notions be decidable, Stein writes:

> I think the issue concerns definitions rather more crucially than proofs; but let me say, borrowing a usage from Plato, that it concerns the mathematical *logos*, in the sense both of 'discourse' generally, and of definition — i.e., the formation of concepts — in particular. (p. 251)

Logos refers to definitions not only as abbreviatory devices, but also as providing a frame for discourse, here the discourse concerning irrational numbers. Indeed, the frame is provided by a *structural* definition that concerns systems and that imposes relations between their elements. This methodological perspective shapes Dedekind's mathematical and foundational work, and Hilbert clearly stands in this Dedekindian tradition. The structural definitions of Euclidean space in Hilbert's (1899a) and of real numbers in his (1900b) start out with, *We think three systems of things...* , respectively with *We think a system of things; we call these things numbers and denote them by a, b, c ... We think these numbers in certain mutual relations, the precise and complete description of which is given by the following axioms: ...*[7] The last sentence is followed by the conditions characterizing real numbers, i.e., those of Dedekind's (1872c), except that continuity is postulated in a different, though deeply related way (see below). Hilbert and Bernays called this way of giving a structural definition, or formulating a mathematical theory, *existential axiomatics*.

---

remarks from the various audiences. The final version of this essay was influenced by very helpful comments from two anonymous referees and Sol Feferman. — Dawn McLaughlin prepared the L^ATEX version of this document; many thanks to her for her meticulous attention to detail.

[6](Stein 1988, pp. 238–239). Stein continues: "I also believe that, when properly understood, this philosophical advance should conduce to a certain modesty: one of the things we should have learned in the course of it is how much we do *not* yet understand about the nature of mathematics." — I could not agree more.

[7]The German texts are: "*Wir denken drei Systeme von Dingen ...*, respectively *Wir denken ein System von Dingen; wir nennen diese Dinge Zahlen und bezeichnen sie mit a, b, c ... Wir denken diese Zahlen in gewissen gegenseitigen Beziehungen, deren genaue und vollständige Beschreibung durch die folgenden Axiome geschieht: ...*"

The introduction of concepts "rendered necessary by the frequent recurrence of complex phenomena, which could be controlled only with difficulty by the old ones" is praised by Dedekind as the engine of progress in mathematics and other sciences.[8] The definition of *continuity* or *completeness* in his (1872c) is to be viewed in this light. The underlying complex phenomena are related to orderings. Dedekind emphasizes transitivity and density as central properties of an ordered system O, and adds the feature that every element in O generates a cut; a *cut* of O is simply a partition of O into two non-empty parts A and B, such that all the elements of A are smaller than all the elements of B. Two different interpretations are presented for these principles, namely, the rational numbers with the ordinary less-than relation and the geometric line with the to-the-right-of relation. On account of this fact the ordering phenomena for the rationals and the geometric line are viewed as *analogous*. Finally, the *continuity principle* is the converse of the last condition: every cut of the ordered system is produced by exactly one element. For Dedekind this principle expresses the essence of continuity and holds for the geometric line.[9]

In order to capture continuity arithmetically and to define a system of real numbers, Dedekind turns the analogy between the rationals and the geometric line into a *real correspondence* by embedding the rationals into the line (after having fixed an origin and a unit). This makes clear that the system of rationals is not continuous, and it motivates considering cuts of rationals as arithmetic counterparts to geometric points. Dedekind shows the system of these cuts to be an ordered field that is also continuous or complete.[10] The completeness of the system, its non-extendibility, points to the core of the difference with Eudoxos's definition of proportionality in Book V of Euclid's *Elements*. The ancient definition applies to many different kinds of geometric magnitudes without requiring that their respective systems be complete, as they may be open to new geometric constructions. Hilbert's completeness axiom expresses the condition of non-extendibility most directly as part of the structural definition. As a matter of fact, even in his (1922) Hilbert articulates Dedekind's structural way of thinking of the system of real numbers when describing the *axiomatische Begründungsmethode* for analysis (that is done still before finitist proof theory is given its proper methodological articulation in 1922):

The continuum of real numbers is a system of things, which are linked to one another by determinate relations, the so-called axioms. In particular, in place of the definition of real numbers by Dedekind cuts, we have the two axioms of continuity, namely, the

---

[8](Dedekind 1888, p. VI).

[9]Dedekind remarks on p. 11 of (1872c): "Die Annahme dieser Eigenschaft der Linie ist nichts als ein Axiom, durch welches wir erst der Linie ihre Stetigkeit zuerkennen, durch welches wir die Stetigkeit in die Linie hineindenken." Then he continues that the "really existent" space may or may not be continuous and that — even if it were not continuous — we could make it continuous in thought. On p. VII of (1888) he discusses a model for Euclid's *Elements* that is everywhere discontinuous.

[10]My interpretation of these considerations reflects Dedekind's methodological practice that is tangible in (1872c) and perfectly explicit five years later in his (1877) — with reference back to (1872c). Thus, Noether attributed the "axiomatische Auffassung" to Dedekind in her comments on (1872c). Notice that Dedekind does *not* identify real numbers with cuts of rationals; real numbers are associated with or determined by cuts, but are viewed as *new* objects. That is vigorously expressed in letters to Lipschitz.

Archimedean axiom and the so-called completeness axiom. To be sure, Dedekind cuts can then also be used to specify individual real numbers, but they do not provide the definition of the concept of real number. Rather, a real number is conceptually just a thing belonging to our system. ...

This standpoint is logically completely unobjectionable, and the only thing that remains to be decided is, whether a system of the requisite sort is thinkable, that is, whether the axioms do not, say, lead to a contradiction.[11]

The axioms serve, of course, also as starting-points for the systematic development of analysis; consistency is to ensure that not too much can be proved, namely, everything. This is one of the crucially important connections to provability. Dedekind also points repeatedly and polemically to the fact that we have finally a proof of $\sqrt{3}\sqrt{2} = \sqrt{6}$ and indicates how analysis can be developed; he shows the continuity principle to be *equivalent* to the basic analytic fact that bounded, monotonically increasing functions have a limit. That is methodology par excellence: The continuity principle is not only sufficient to prove the analytic fact, but indeed necessary.

2.2 Consistency

For both Dedekind and Hilbert, the coherence of their theories for real numbers was central. Dedekind had aimed for, and thought he had achieved in his (1888), "the purely logical construction of the science of numbers and the continuous realm of numbers gained in it."[12] Within the logical frame of that essay Dedekind defines simply infinite systems and provides also an "example" or "instance". The point of such an instantiation is articulated sharply and forcefully in his famous letter to Keferstein where he asks, whether simply infinite systems "exist at all in the realm of our thoughts". He supports the affirmative answer by a logical existence proof. Without such a proof, he explains, "it would remain doubtful, whether the concept of such a system does not perhaps contain internal contradictions". His *Gedankenwelt*, "the totality S of all things that can be object of my thinking", was crucial for obtaining a simply infinite system.[13]

Cantor recognized Dedekind's *Gedankenwelt* as an inconsistent system and communicated that fact to both Dedekind (in 1896) and to Hilbert (in 1897). When Hilbert formulated arithmetic in his (1900b), he reformulated the

---

[11] (Hilbert 1922) in (Ewald 1996, p. 1118). — That is fully in Dedekind's spirit: Hilbert's critical remark about the definition of real numbers as cuts do not apply to Dedekind, as should be clear from my discussion (in the previous note), and the issue of consistency was an explicit part of Dedekind's logicist program.

[12] The systematic build-up of the continuum envisioned in (1872c, pp. 5–6) is carried out in later manuscripts where integers and rationals are introduced as equivalence classes of pairs of natural numbers; they serve as models for subsystems of the axioms for the reals, in a completely modern way. — All of these developments as well as that towards the formulation of simply infinite systems are analyzed in (Sieg 2005).

[13] Let me support, by appeal to authority, the claim that Dedekind's thoughts are not psychological ideas: Frege asserts in his manuscript *Logik* from 1894 that he uses the word "Gedanke" in an unusual way and remarks that "Dedekind's usage agrees with mine". It is worthwhile noting that Frege, in this manuscript, approved of Dedekind's argument for the existence of an infinite system. — Note also that Hilbert formulated his existential axiomatics with the phrase "wir denken", so that the system is undoubtedly an object of our thought, indeed, "ein Gedanke".

problem of instantiating logoi as a quasi-syntactic problem: Show that no contradiction is provable from the axiomatic conditions in a finite number of logical steps. That is, of course, the second problem of his Paris address I discussed in section 1. He took for granted that consistency amounts to mathematical existence and assumed that the ordinary investigations of irrational numbers could be turned into a model theoretic consistency proof within a restricted logicist framework. This was crucial for the arithmetization of analysis and its logicist founding. It should be mentioned that Hilbert in *Grundlagen der Geometrie* also "geometrized" analysis by giving a geometric model via his "Streckenrechnung" for the axioms of arithmetic (with full continuity only in the second edition of the *Grundlagen* volume).

In his lecture (*1920b), Hilbert formulated the principles of Zermelo's set theory (in the language of first-order logic). He considered Zermelo's theory as providing the mathematical objects Dedekind had obtained through logicist principles; Hilbert remarked revealingly:

> The theory, which results from developing all the consequences of this axiom system, encompasses all mathematical theories (like number theory, analysis, geometry) in the following sense: the relations that hold between the objects of one of these mathematical disciplines are represented in a completely corresponding way by relations that obtain in a sub-domain of Zermelo's set theory.[14]

In spite of this perspective, Hilbert reconsidered at the end of the 1920-lecture his earlier attempt (published as (1905a)) to establish by mathematical proof that no contradiction can be proved in formalized elementary number theory. That had raised already then the issue, how proofs can be characterized and subjected to mathematical investigation. It was only after the study of *Principia Mathematica* that Hilbert had a properly general and precise concept of (formal) proof available.

3 Rigorous proofs

Proofs are essential for developing any mathematical subject, vide Euclid in the *Elements* or Dedekind in *Was sind und was sollen die Zahlen?*. In the introduction to his *Grundgesetze der Arithmetik*, Frege distinguished his systematic development from Euclid's by pointing to the list of explicit inference principles for obtaining gapless proofs. As to Dedekind's essay he remarked polemically that no proofs can be found in that work. Dedekind and Hilbert explicated the "science of (natural) number" and "arithmetic (of real numbers)" in similar ways; their theories start from the defining conditions for simply infinite systems, respectively complete ordered fields. Dedekind writes in (1888):

---

[14](Hilbert *1920b, p. 23). Here is the German text: "Die Theorie, welche sich aus der Entwicklung dieses Axiomensystems in seine Konsequenzen ergibt, schliesst alle mathematischen Theorien (wie Zahlentheorie, Analysis, Geometrie) in sich in dem Sinne, dass die Beziehungen, welche sich zwischen den Gegenständen einer dieser mathematischen Disziplinen finden, vollkommen entsprechend dargestellt werden durch die Beziehungen, welche in einem Teilgebiete der Zermeloschen Mengenlehre stattfinden."

The relations or laws which are derived exclusively from the conditions [for a simply infinite system] and are therefore always the same in all ordered simply infinite systems, ... form the next object of the *science of numbers* or *arithmetic*.[15]

The term "derive" is left informal; hence Frege's critique. Exactly at this point enters logic in the restricted modern sense as dealing with formal methods for correct, truth-preserving inference.

3.1 Natural deductions

Underlying Dedekind's and Hilbert's descriptions is an abstract concept of *logical consequence*. Hilbert stated in 1891 during a famous stop at a Berlin railway station that in a proper axiomatization of geometry "one must always be able to say 'tables, chairs, beer mugs' instead of 'points, straight lines, planes'." This remark has been taken as claiming that the basic terms must be meaningless, but it is more adequately understood if it is put side by side with a remark of Dedekind's in a letter to Lipschitz written fifteen years earlier: "All technical expressions [can be] replaced by arbitrary, newly invented (up to now meaningless) words; the edifice must not collapse, if it is correctly constructed, and I claim, for example, that my theory of real numbers withstands this test." Thus, logical arguments leading from principles to derived claims cannot be severed by a re-interpretation of the technical expressions or, to put it differently, there are no counterexamples to the arguments.

Dedekind's and Hilbert's presentations are detailed, reveal the logical form of arguments, and reflect features of the mathematical structures. In the very first sentence of the Preface to his (1888), Dedekind programmatically emphasizes that "in science nothing capable of proof should be accepted without proof" and claims that only common sense ("gesunder Menschenverstand") is needed to understand his essay. But he recognizes also that many readers will be discouraged, when asked to prove truths that seem obvious and certain by "the long sequence of simple inferences that corresponds to the nature of our step-by-step understanding" (Treppenverstand).[16] Dedekind believes that there are only a few such simple inferences, but he does not explicitly list them. Looking for an expressive formal language and powerful inferential tools, Hilbert moved slowly toward a presentation of proofs in logical calculi. He and his students started in 1913 to learn modern logic by studying *Principia Mathematica*. During the winter term 1917–18 he gave the first course in mathematical logic proper and sketched, toward the end of the term, how to develop analysis in ramified type theory with the axiom of reducibility.[17]

---

[15] (Dedekind 1888, sec. 73). In the letter to Keferstein, on p. 9, Dedekind reiterates this perspective and requires that every claim "must be derived completely abstractly from the logical definition of [the simply infinite system] N".

[16] (Dedekind 1888, p. IV). Dedekind continues: "Ich erblicke dagegen gerade in der Möglichkeit, solche Wahrheiten auf andere, einfachere zurückzuführen, mag die Reihe der Schlüsse noch so lang und scheinbar künstlich sein, einen überzeugenden Beweis dafür, daß ihr Besitz oder der Glaube an sie niemals unmittelbar durch innere Anschauung gegeben, sondern immer durch eine mehr oder weniger vollständige Wiederholung der einzelnen Schlüsse erworben ist."

[17] That is usually associated with the book (Hilbert and Ackermann 1928) that was published only in 1928; however, that book takes over the structure and much of the content from these earlier lecture notes. See my paper (1999) and the forthcoming third volume of

So there is finally (in Göttingen) a way of building up gapless proofs in Frege's sense. However, Hilbert aimed for a framework in which mathematics can be formalized in a natural and direct way. The calculus of *Principia Mathematica* did not lend itself to that task. In the winter term 1921–22 he presented a logical calculus that is especially interesting for sentential logic. He points to the parallelism with his axiomatization of geometry: groups of axioms are introduced there for each concept, and that is done here for each logical connective. Let me formulate the axioms for just conjunction and disjunction:

$A \& B \to A$ $((A \to C) \& (B \to C)) \to ((A \vee B) \to C)$

$A \& B \to B$ $A \to (A \vee B)$

$A \to (B \to A \& B)$ $B \to (A \vee B)$

The simplicity of this calculus and its directness for formalization inspired the work of Gentzen on natural reasoning. It should be pointed out that Bernays had proved the completeness of Russell's calculus in his *Habilitationsschrift* of 1918 and had investigated rule-based variants. The proof theoretic investigations of, essentially, primitive recursive arithmetic in the 1921–22 lectures also led to a tree-presentation of proofs, what Hilbert and Bernays called "the resolution of proofs into proof threads" (die Auflösung von Beweisen in Beweisfäden).[18] The full formulation of the calculus and the articulation of the methodological parallelism to *Grundlagen der Geometrie* are also found in (Hilbert and Bernays 1934, pp. 63–64).

3.2 Strategies

Gentzen formulated *natural deduction calculi* using Hilbert's axiomatic formulation as a starting point and called them calculi of *natural reasoning* (natürliches Schließen); he emphasized that making and discharging assumptions were their distinctive features. Here are the Elimination and Introduction rules for the connectives discussed above and as formulated in (Gentzen 1936); the configurations that are derived with their help are sequents of the form $\Gamma \supset \psi$ with Γ containing all the assumptions on which the proof of ψ depends:

$$\frac{\Gamma \supset A \& B}{\Gamma \supset A} \qquad \frac{\Gamma \supset A \& B}{\Gamma \supset B} \qquad \frac{\Gamma \supset A \vee B \quad \Gamma, A \supset C \quad \Gamma, B \supset C}{\Gamma \supset C}$$

$$\frac{\Gamma \supset A \quad \Gamma \supset B}{\Gamma \supset A \& B} \qquad \frac{\Gamma \supset A}{\Gamma \supset A \vee B} \qquad \frac{\Gamma \supset B}{\Gamma \supset A \vee B}$$

Hilbert's *Lectures on the Foundations of Mathematics and Physics*. — In the final section of his (2008), Wiedijk lists "three main revolutions" in mathematics: the introduction of *proof* in classical Greece (culminating in Euclid's *Elements*), that of *rigor* in the 19th century, and that of *formal mathematics* in the late 20th and early 21st centuries. The latter revolution, if it is one, took place in the 1920s.

[18] On account of this background, I assume, Gentzen emphasized in his dissertation and his first consistency proof for elementary number theory the dual character of introduction and elimination rules, but considered making and discharging assumptions as the most important feature of his calculi.

Gentzen and later Prawitz established normalization theorems for proofs in nd calculi.[19] As the calculi are complete, one obtains proof theoretically refined completeness theorems: if ψ is a logical consequence of Γ, then there is a *normal* proof of ψ from Γ. I reformulated the nd calculi as *intercalation calculi*[20] for which these refined completeness theorems can be proved semantically without appealing to a syntactic normalization procedure; see (Sieg and Byrnes 1998) for classical first-order logic as well as (Sieg and Cittadini 2005) for some non-classical logics, in particular, for intuitionist first-order logic.

The refined completeness results and their semantic proofs provide foundations to the systematic search for normal proofs in nd calculi. This is methodologically analogous to the use of completeness results for cut-free sequent calculi and was exploited in the pioneering work of Hao Wang.[21] The subformula property of normal and cut-free derivations is fundamental for mechanical search. The ic calculi enforce normality by applying the E-rules only on the left to premises and the I-rules only on the right to the goal. In the first case one really tries to "extract" a goal formula by a *sequence* of E-rules from an assumption in which it is contained as a strictly positive subformula. This feature is distinctive and makes search efficient, but it is in a certain sense just a natural systematization and logical deepening of the familiar forward and backward argumentation. Suitable strategies have been implemented and guide a complete search procedure for first-order logic, called AProS.[22] In Appendix A, I discuss examples of purely logical arguments.[23] The AProS strategies can be extended by E- and I-rules for definitions, so that the meanings of defined notions as well as those of logical connectives can be used to guide search. In this way we have developed quite efficiently the part of elementary set theory concerning Boolean operations, power sets, Cartesian products, etc. In Appendix B, the reader finds two examples of set theoretic arguments.

You might think, that is interesting, but what relevance do these considerations have for finding proofs in more complex parts of mathematics? To answer that question and put it into a broader context, let me first note that the history of such computational perspectives goes back at least to Leibniz, and that it can be illuminated by Poincaré's surprising view of Hilbert's *Grundlagen der Geometrie*. In his review of Hilbert's book, he suggested giving the axioms to a reasoning machine, like Jevons's logical piano, and observing whether all of geometry would be obtained. He wrote that such radical formalization might

---

[19] The first version of Gentzen's dissertation was recently discovered by Jan von Plato in the Bernays Nachlass of the ETH in Zürich. It contains a detailed proof of the normalization theorem for intuitionist predicate logic; see (von Plato 2008).

[20] I discovered only recently that Beth in his (1958) employs "intercalate" (on p. 87) when discussing the use of lemmata in the proofs of mathematical theorems.

[21] See the informative and retrospective discussion in his (1984) and, perhaps, also the programmatic (1970). — Cf. also my (2007).

[22] Nd calculi were considered as inappropriate for theorem proving because of the seemingly unlimited branching in a backward search afforded by modus ponens (conditional elimination). The global property of normality for nd proofs could not be directly exploited for a locally determined backward search; hence, the intercalation formulation of natural deduction. The implementation of AProS can be downloaded at http://caae.phil.cmu.edu/projects/apros/

[23] In (Sieg and Field 2005, pp. 334–5), the problem of proving that $\sqrt{2}$ is not rational is formulated as a logical problem, and AProS finds a proof directly; cf. the description of the difficulties of obtaining such a proof in (Wiedijk 2008).

seem "artificial and childish", were it not for the important question of "completeness":

> Is the list of axioms complete, or have some of them escaped us, namely those we use unconsciously? ... One has to find out whether geometry is a logical consequence of the explicitly stated axioms or, in other words, whether the axioms, when given to the reasoning machine, will make it possible to obtain the sequence of all theorems as output [of the machine].[24]

With respect to a sophisticated logical framework and under the assumption of the finite axiomatizability of mathematics, Poincaré's problem morphed into what Hilbert and others viewed in the 1920s as the *most important problem* of mathematical logic: the decision problem (*Entscheidungsproblem*) for predicate logic. Its special character was vividly described in a talk Hilbert's student Behmann gave in 1921:

> For the nature of the problem it is of fundamental significance that as auxiliary means ... only the completely mechanical reckoning according to a given prescription [Vorschrift] is admitted, i.e., without any thinking in the proper sense of the word. If one wanted to, one could speak of mechanical or machine-like thinking. (Perhaps it can later even be carried out by a machine.)

Johann von Neumann argued against the positive solvability of the decision problem, in spite of the fact that — as he formulated matters in 1924 — "... we have no idea how to prove the undecidability". It was only twelve years later that Turing provided the idea, i.e., introduced the appropriate concept, for proving the unsolvability of the *Entscheidungsproblem*.

The issue for Turing was, What are the procedures a human being can carry out when mechanically operating as a computer?[25] In his classical paper *On computable numbers with an application to the Entscheidungsproblem*, Turing isolated the basic steps underlying a computer's procedures as the operations of a Turing machine. He then proved: There is no procedure that can be executed by a Turing machine and solves the decision problem. Using the concepts of general recursive and λ-definable functions, Church had also established the undecidability of predicate logic. The core of Church's argument was presented in Supplement II of *Grundlagen der Mathematik, vol. II*. However, it was not only expanded by later considerations due to Church and Kleene, but also deepened by local axiomatic considerations for the concept of a *reckonable function*.[26]

---

[24](Poincaré 1902b, pp. 252–253).

[25]For Turing a "computer" is a human being carrying out a "calculation" and using only minimal cognitive capacities. The limitations of the human sensory apparatus motivate finiteness and locality conditions; Turing's supporting argument is not mathematically precise, and I don't think there is any hope of turning the analysis into a mathematical theorem. What one can do, however, is to exploit it as a starting point for formulating a general concept and establishing a representation theorem; cf. my paper (2008a).

[26]I distinguish local from global axiomatics. As an example of the former I discuss in part 4.1 an abstract proof of Gödel's incompleteness theorems. Other examples can be found in Hilbert's 1917-talk in Zürich, but also in contemporary discussions, e.g., Booker's report on L-functions in the Notices of the AMS, p. 1088. Booker remarks that many objects go by the name of L-function and that it is difficult to pin down exactly which ones are. He attributes then to A. Selberg an "axiomatic approach" consisting in "writing down the common properties of the known examples" — as axioms.

Hilbert and Bernays introduced reckonable functions informally as those number theoretic functions whose values can be determined in a "deductive formalism". They proved that, if the deductive formalism satisfies their *recursiveness conditions*, then the class of reckonable functions is co-extensional with that of the general recursive ones. (The crucial condition requires that the proof relation of the deductive formalism is primitive recursive.) Their concept is one way of capturing the "completely mechanical reckoning according to a given prescription" mentioned in the quotation from Behmann. Indeed, it generalizes Church's informal notion of calculable functions whose values can be determined in a logic and imposes the recursiveness condition in order to obtain a mathematically rigorous formulation. For us the questions are of course: Can a machine carry out this mechanical thinking? and, if a universal Turing machine in principle can, What is needed to *copy*, as Turing put it in 1948, aspects of mathematical thinking in such a machine? — Copying requires an original, i.e., that we have uncovered suitable aspects of the mathematical mind when trying to extend automated proof search from logic to mathematics.

4 Local axiomatics

At the end of his report on *Intelligent Machinery* from 1948, Turing suggested that machines might search for proofs of mathematical theorems in suitable formal systems. It was clear to Turing that one cannot just specify axioms and logical rules, state a theorem, and expect a machine to demonstrate the theorem. For a machine to exhibit the necessary intelligence it must "acquire both discipline and initiative". Discipline would be acquired by becoming (practically) a universal machine; Turing argued that "discipline is certainly not enough in itself to produce intelligence" and continued:

That which is required in addition we call initiative. This statement will have to serve as a definition. Our task is to discover the nature of this residue as it occurs in man, and try and copy it in machines. (p. 21)

The dynamic character of strategies constitutes but a partial and limited copy of human initiative. Nevertheless, local axiomatics that allows the expression of leading ideas together with a hierarchical organization that reflects the conceptual structure of a field can carry us a long way. Hilbert expressed his views in 1919 as follows, arguing against the logicists' view that mathematics consists of tautologies grounded in definitions:

If this view were correct, mathematics would be nothing but an accumulation of logical inferences piled on top of each other. There would be a random concatenation of inferences with logical reasoning as its sole driving force. But in fact there is no question of such arbitrariness; rather we see that the formation of concepts in mathematics is constantly guided by intuition and experience, so that mathematics on the whole forms a non-arbitrary, closed structure.[27]

---

[27] (Hilbert *1919, p. 5). Here is the German text: "Wäre die dargelegte Ansicht zutreffend, so müsste die Mathematik nichts anderes als eine Anhäufung von übereinander getürmten logischen Schlüssen sein. Es müsste ein wahlloses Aneinanderreihen von Folgerungen stattfinden, bei welchem das logische Schliessen allein die treibende Kraft wäre. Von einer solchen Willkür ist aber tatsächlich keine Rede; vielmehr zeigt sich, dass die Begriffsbildungen in der Mathematik beständig durch Anschauung und Erfahrung geleitet werden, sodass im grossen

Hilbert's grouping of the axioms for geometry in his (1899a) had the express purpose of organizing proofs and the subject in a conceptual way: parts of his development are marvelous instances of *local axiomatics*, analyzing which notions and principles are needed for which theorems.

4.1 Modern

The idea of local axiomatics can be used for individual mathematical theorems and asks, How can we prove this particular theorem or this particular group of theorems? Hilbert and Bernays used the technique in their *Grundlagen der Mathematik II* also outside a foundational axiomatic context: first for proving Gödel's incompleteness theorems and then, as I indicated at the end of section 3.2, for showing that the functions reckonable in formal deductive systems coincide with the general recursive ones. One crucial task has to be taken on for *local* as well as for *global axiomatics*, namely, isolating what is at the heart of an argument or uncovering its *leading (mathematical) idea*. That was proposed by Saunders MacLane in his Göttingen dissertation (of late 1933) and summarized in his (1935). MacLane emphasized that proofs are not "mere collections of atomic processes, but are rather complex combinations with a highly rational structure". When reviewing in 1979 this early logical work, he ended with the remark, "There remains the real question of the actual structure of mathematical proofs and their strategy. It is a topic long given up by mathematical logicians, but one which still — properly handled — might give us some real insight."[28] That is exactly the topic I am trying to explore.

As an illustration of the general point concerning the "rational structure" of mathematical arguments, I consider briefly the proofs of Gödel's incompleteness theorems. These proofs make use of the connection between the mathematics that is used to present a formal theory and the mathematics that can be formally developed in the theory. Three steps are crucial for obtaining the proofs, steps that go beyond the purely logical strategies and are merged into the search algorithm:

1. *Local axioms*: representability of the core syntactic notions, the diagonal lemma, and the Hilbert & Bernays derivability conditions.

2. *Proof-specific definitions*: formulating instances of existential claims, for example, the Gödel sentence for the first incompleteness theorem.

3. *Leading idea*: moving between object- and meta-theory, expressed by appropriate Elimination and Introduction rules (for example, if a proof of A has been obtained in the object-theory, then one is allowed to introduce the claim 'A is provable' in the meta-theory).

AProS finds the proofs efficiently and directly, even those that did not enter into the analysis of the leading idea, for example, the proof of Löb's theorem. All of this is found in (Sieg and Field 2005).

---

und ganzen die Mathematik ein willkürfreies, geschlossenes Gebilde darstellt."

[28]The first quotation is from MacLane's (1935, p. 130), the second from his (1979, p. 66). The processes by means of which MacLane tries to articulate the "rational structure" of proofs should be examined in greater detail.

It has been a long-standing tradition in mathematics to give and to analyze a variety of arguments for the same statement; the fundamental theorems of algebra and arithmetic are well-known examples. In this way we delimit conceptual contexts, provide contrasting *explanations* for the theorem at hand, and gain a deeper understanding by looking at it in different ways, e.g., from a topological or algebraic perspective.[29] An automated search requires obviously a sharp isolation of local axioms and leading ideas that underlie a proof. Such developments can be integrated into a global framework through a hierarchical organization, and that has been part and parcel of mathematical practice. Hilbert called it *Tieferlegung der Fundamente*!

These broad ideas are currently being explored in order to obtain an automated proof of the Cantor-Bernstein theorem from Zermelo's axioms for set theory.[30] The theorem claims that there is a bijection between two sets, in case there are injections from the first to the second and from the second to the first. The theorem is a crucial part of the investigations concerning the size of sets and guarantees the anti-symmetry of the partial ordering of sets by the "smaller-or-equal-size" relation.[31] We have begun to develop set theory from Zermelo's axioms and use three layers for the conceptual organization of the full proof:

A. Construction of sets, for example, empty set, power set, union, and pairs.
B. Introduction of functions as set theoretic objects.
C. The abstract proof.

The abstract proof is divided in the same schematic way as that of Gödel's theorems and is independent of the set theoretic definition of function. The *local axioms* are lemmata for injective, surjective, and bijective functions as well as a fixed-point theorem. The crucial *proof-specific definition* is that of the bijection claimed to exist in the theorem. Finally, the *leading idea* is simply to exploit the fixed-point property and verify that the defined function is indeed a bijection. — It is noteworthy that the differences between the standard proofs amount to different ways of obtaining the smallest fixed-point of an inductive definition.

4.2 Classical

Shaping a field and its proofs by concepts is classical; so is the deepening of its foundations. That can be beautifully illustrated by the developments in

---

[29] In the Introduction to the second edition of (Dirichlet 1863) Dedekind emphasized this aspect for the development of a whole branch of mathematics. In the tenth supplement to this edition of Dirichlet's lectures, he presented his general theory of ideals in order, as he put it, "to cast, from a higher standpoint, a new light on the main subject of the whole book". In German, "Endlich habe ich in dieses Supplement eine allgemeine Theorie der Ideale aufgenommen, um auf den Hauptgegenstand des ganzen Buches von einem höheren Standpunkte aus ein neues Licht zu werfen." He continues, "hierbei habe ich mich freilich auf die Darstellung der Grundlagen beschränken müssen, doch hoffe ich, daß das Streben nach charakteristischen Grundbegriffen, welches in anderen Teilen der Mathematik mit so schönem Erfolg gekrönt ist, mir nicht ganz mißglückt sein möge." (Dedekind 1932, pp. 396–7).

[30] My collaborators on this particular part of the AProS Project have been Ian Kash, Tyler Gibson, Michael Warren, and Alex Smith.

[31] On p. 209 of Cantor's (1932) *Gesammelte Abhandlungen*, Zermelo calls this theorem "one of the most important theorems of all of set theory".

the first two books of Euclid's *Elements* (and the related investigations at the beginning of Book XII). Proposition 47 of Book I, the Pythagorean theorem, is at the center of those developments. The broad mathematical context is given by the *quadrature problem*, i.e., determining the "size" or, in modern terms, the area of geometric figures in terms of squares. The problem is discussed in Book II for polygons. Polygons can be partitioned into triangles that can be transformed individually (by ruler and compass constructions) first into rectangles "of equal area" and then into equal squares.[32] The question is, how can we join these squares to obtain one single square that is equal to the polygon we started out with? It is precisely here that the Pythagorean theorem comes in and provides the most direct way of determining the larger square. Byrne's colorful diagram, displayed below, captures the construction and the abstract proof of the theorem. If one views the determination of the larger square as a geometric computation, then the proof straightforwardly verifies its correctness.[33]

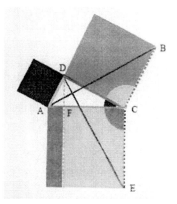

For the proof, Euclid has us first construct the squares on the triangle's sides and then make the observation that the extensions of the sides of the smaller squares by the contiguous sides of the original triangle constitute lines. In the next step a crucial auxiliary line is drawn, namely, the line that is perpendicular to the hypotenuse and that passes through the vertex opposite the hypotenuse. This auxiliary line partitions the big square into the blue and yellow rectangles. Two claims are now considered: the blue rectangle is equal to the black square, and the yellow rectangle is equal to the red square. Euclid uses three facts that are readily obtained from earlier propositions: (α) Triangles are equal when they have two equal sides and when the enclosed angles are equal (Proposition I.4); (β) Triangles are equal when they have the same base

---

[32] Euclid simply calls the geometric figures "equal". This is central and has been pursued throughout the evolution of geometry. In Hilbert's (1899a), a whole chapter is devoted to "Die Lehre von den Flächeninhalten in der Ebene", Chapter IV, making the implicit Euclidean assumptions concerning "area" explicit. See also Hartshorne's book, section 22, *Area in Euclid's geometry*.

[33] Hilbert remarked in his (*1899b), "Wir werden im folgenden häufig Gebrauch von Figuren machen, wir werden uns aber niemals auf sie verlassen. Stets müssen wir dafür sorgen, daß die an einer Figur vorgenommenen Operationen auch rein logisch gültig bleiben. Es kann dies garnicht genug betont werden; im richtigen Gebrauch der Figuren liegt eine Hauptschwierigkeit unserer Untersuchungen." (p. 303) Here I am obviously not so much interested in the (correct) use of diagrams as analyzed by Manders and for which Avigad e.a. have provided an informative formal framework. Manders's analysis led to the assertion that only topological features of diagrams are relevant for and appealed to in Euclidean proofs; the conceptual setting sketched above with its focus on "area" gives a reason for that assertion. There is also important work from the early part of the 20th century by Hans Brandes and, in particular, Paul Mahlo in his 1908 dissertation. This work tries to classify the "Zerlegungsbeweise" of the Pythagorean theorem and should be investigated carefully; (Bernstein 1924) reflects on those dissertations.

and when their third vertex lies on the same parallel to that base (Proposition I.37); (γ) A diagonal divides a rectangle into two equal triangles (Proposition I.41).[34]

Here is the proof based on (α) through (γ) for the red square and the yellow rectangle. (The common notions are implicitly appealed to; the argument for the equality of the black square and the blue rectangle is analogous.) The triangles ABC and DCE satisfy the conditions of (α) and are thus equal; on account of (β) they are equal to DBC and FCE, respectively. Finally, (γ) ensures the equality of the red square and yellow rectangle. Comparing the structure of this argument to that of the abstract proofs for the incompleteness theorem and the Cantor-Bernstein theorem, we can make the following general observations: (α) through (γ) are used as *local axioms*; the auxiliary line drawn through the vertex opposite of the hypotenuse and perpendicular to it is the central *proof-specific definition*; finally, the *leading idea* is the partitioning of squares and establishing that corresponding parts are equal.

The character of the "deepening of the foundations" is amusingly depicted by anecdotes concerning Hobbes and Newton: Hobbes started with Proposition 47 and was convinced of its truth only after having read its proof and all the (proofs of the) propositions supporting it; Newton, in contrast, started at the beginning and could not understand, why such evident propositions were being established — until he came to the Pythagorean theorem. Less historically, there is also a deeper parallelism with the overall structure of the proof of the Cantor-Bernstein theorem from Zermelo's axioms. The construction of figures like triangles and squares corresponds to A (in the list A–C concerning the Cantor-Bernstein theorem); the congruence criteria for such figures correspond to B; the abstract proof of the geometric theorem, finally, has the same conceptual organization as the set theoretic proof referenced in C.

The abstract proof of the Pythagorean theorem and its deepening are shaped by the mathematical context, here the quadrature problem. I want to end this discussion with two related observations. Recall that the Pythagorean theorem is used in Hippocrates's proof for the quadrature of the lune.[35] This is just one of its uses for solving quadrature problems, but it seems to be very special, as only the case for isosceles triangles is exploited. The crucial auxiliary line divides in half the square over the hypotenuse, and we have a perfectly symmetric configuration.[36] Here is the first observation, namely, the claim concerning the equality of the rectangles (into which the square over the hypoteneuse is divided) and squares (over the legs) is "necessary", and the proof idea is relatively straightforward. That leads me to the second observation that is speculative and formulated as a question: Isn't it plausible that the Euclidean proof is obtained by generalizing this special one?

---

[34] This is not *exactly* Euclid's proof. Euclid does not appeal to I.37, but just to I.41, which is really a combination of (β) and (γ) and applies directly to the diagram; I.37 is used in the proof of I.41. — The colorful diagram is from Byrne's edition of the first six books of the Euclidean *Elements*, London, 1847. [The labeling of points was added by me; WS.]

[35] See the very informative discussion in (Dunham 1990).

[36] In (Aumann 2009, pp. 64–65) knowledge of this geometric fact is attributed to the Babylonians, and it is the one Socrates extracts from the slave boy in Plato's *Meno*.

Karzel and Kroll, in their *Geschichte der Geometrie seit Hilbert* under the heading "Order and Topology", link the classical Greek considerations back (or rather forward) to modern developments:

> In Euclidean geometry triangles and also rectangles take on the role of elementary figures out of which more complex figures are thought to be composed. To these elementary figures one can assign in Euclidean geometry an area in a natural way. If one assumes in addition the axiom of continuity, then one arrives at the concept of an integral when striving to assign an area also to more complex figures.[37]

So we have returned to continuity and to Dedekind.

5 Cognitive aspects

In his (1888) Dedekind refers to his *Habilitationsrede* where he claimed that the need to introduce appropriate notions arises from the fact that our intellectual powers are imperfect. Their limitation leads us to frame the object of a science in different forms and introducing a concept means, in a certain sense, formulating a hypothesis on the inner nature of the science. How well the concept captures this inner nature is determined by its usefulness for the development of the science, and in mathematics that is mainly its usefulness for constructing proofs. Dedekind put the theories from his foundational essays to this test by showing that they allow the direct, stepwise development of analysis and number theory. Thus, Dedekind viewed general concepts and general forms of arguments as tools to overcome, at least partially, the imperfection of our intellectual powers. He remarked:

> Essentially, there would be no more science for a man gifted with an unbounded understanding — a man for whom the final conclusions, which we obtain through a long chain of inferences, would be immediately evident truths; and this would be so even if he stood in exactly the same relation to the objects of science as we do. (Ewald 1996, pp. 755–6)

The theme of bounded human understanding is sounded also in a remark from (Bernays 1954): "Though for differently built beings there might be a different kind of evidence, it is nevertheless our concern to find out what evidence is for us."[38] Bernays put forth the challenge of finding out what is evidence *for us*, not for some *differently built being*. Turing in his (1936) appealed crucially to *human* cognitive limitations to arrive at his notion of computability. Ten years later Gödel took the success of having given "an absolute definition of an interesting epistemological notion", i.e., of effective calculability, as encouragement to strive for "the same thing" with respect to demonstrability and mathematical definability. That was attempted in his (1946). Reflecting on a possible objection to his concept of ordinal definability, namely, that uncountably many sets are ordinal definable, Gödel considers as plausible the view

---

[37] (Karzel and Kroll 1988, p. 121). The German text is: "In der euklidischen Geometrie spielen neben den Dreiecksflächen noch die Rechtecksflächen ... die Rolle von Elementarflächen, aus denen man sich kompliziertere Flächen zusammengesetzt denkt. Diesen Elementarflächen kann man in der euklidischen Geometrie in natürlicher Weise einen Flächeninhalt zuweisen. Setzt man nunmehr noch das Stetigkeitsaxiom voraus, so gelangt man beim Bemühen, auch komplizierteren Flächen einen Inhalt zuzuweisen, zum Integralbegriff."

[38] (Bernays 1954, p. 18). The German text is: "Obwohl es für anders gebildete Wesen eine andere Evidenz geben könnte, so ist jedoch unser Anliegen festzustellen, was Evidenz für uns ist."

"that all things conceivable by us are denumerable". Indeed, he thinks that a concept of definability "satisfying the postulate of denumerability" is possible, but "that it would involve some extramathematical element concerning the psychology of the being who deals with mathematics".

Reflections on cognitive limitations motivated also the finitist program's goal of an absolute epistemological reduction. Bernays provides in his (1922) a view of the program in *statu nascendi* and connects it to the existential axiomatics discussed above in Part 2. When giving a rigorous foundation for arithmetic or analysis one proceeds axiomatically, according to Bernays, and assumes the existence of a system of objects satisfying the structural conditions expressed by the axioms. In the assumption of such a system "lies something transcendental for mathematics, and the question arises, which principled position is to be taken [towards that assumption]". An intuitive grasp of the completed sequence of natural numbers, for example, or even of the manifold of real numbers is not excluded outright. However, taking into account tendencies in the exact sciences, one might try "to give a foundation to these transcendental assumptions in such a way that only primitive intuitive knowledge is used". That is to be done by giving finitist consistency proofs for systems in which significant parts of mathematics can be formalized. The second incompleteness theorem implies, of course, that such an absolute epistemological reduction cannot be achieved. What then is evidence for principles that allow us to step beyond the finitist framework? — Bernays emphasized in his later writings that evidence is acquired by *intellectual experience* and through *experimentation* in an almost Dedekindian spirit. In his (1946) he wrote:

In this way we recognize the necessity of something like intelligence or reason that should not be regarded as a container of [items of] a priori knowledge, but as a mental activity that consists in reacting to given situations with the formation of experimentally applied categories.[39]

This intellectual experimentation in part supports the introduction of concepts to define abstract structures or to characterize accessible domains (obtained by general inductive definitions), and it is in part supported by using these concepts in proofs of central theorems.[40]

I intended to turn attention to those aspects of the mathematical mind that are central, if we want to grasp the subtle connection between reasoning and

---

[39] (Bernays 1946, p. 91). The German text is: "Wir erkennen so die Notwendigkeit von etwas wie Intelligenz oder Vernunft, die man nicht anzusehen hat als Behältnis von Erkenntnissen a priori, sondern als eine geistige Tätigkeit, die darin besteht, auf gegebene Situationen mit der Bildung von versuchsweise angesetzten Kategorien zu reagieren." — Unfortunately, "applied" is not capturing "angesetzten". The latter verb is related to "Ansatz". That noun has no adequate English rendering either, but is used (as in "Hilbertscher Ansatz") to express a particular approach to solving a problem that does however not guarantee a solution.

[40] Andrea Cantini expresses in his recent (2008) a similar perspective, emphasizing also the significance of "geistiges Experimentieren" in Bernays's reflections on mathematics; see pp. 34–37. In the very same volume in which Cantini's article is published, Carlo Cellucci describes a concept of *analytic* proof that incorporates many features of the experimentation both Cantini and I consider as important. However, Cellucci sharply contrasts that concept with that of an *axiomatic* proof. These two notions, it seems to me, stand in opposition only if one attaches to the latter concept a dogmatic foundationalist intention. — In (Sieg 2010b) I have compared Gödel's and Turing's approach to such intellectual experimentation.

understanding in mathematics, as well as the role of leading ideas in guiding proofs and of general concepts in providing explanations. Implicitly, I have been arguing for an *expansion of proof theory*: Let us take steps toward a theory that articulates principles for organizing proofs conceptually and for finding them dynamically. A good start is a thorough reconstruction of parts of the rich body of mathematical knowledge that *is* systematic, but is also structured for intelligibility and discovery, when viewed from the right perspective. Such an expanded proof theory should be called *structural* for two reasons. On the one hand one exploits the intricate internal structure of (normal) proofs, and on the other hand one appeals to the notions and principles characterizing mathematical structures. (Cf. also the very tentative parallel remarks in Appendix C.)

When focusing on formal methods and carrying out computations in support of proof search experiments, we have to isolate truly creative elements in proofs and thus come closer to an understanding of the technique of our mathematical thinking, be it mechanical or non-mechanical. Hilbert continued his remarks in (1927) about the formula game as follows:

Thinking, it so happens, parallels speaking and writing: we form statements and place them one behind another. If any totality of observations and phenomena deserves to be made the object of a serious and thorough investigation, it is this one — since, after all, it is part of the task of science to liberate us from arbitrariness, sentiment, and habit ... (p. 475)

I could not agree more (with the second sentence in this quotation) and share Hilbert's eternal optimism, "Wir müssen wissen! Wir werden wissen!"

Appendices

AProS' distinctive feature is its *goal-directed* search for *normal* proofs. It exploits an essential feature of normal proofs, i.e., the division of every branch in their representing tree into an E- and an I-part; see (Prawitz 1965, p. 41). This global property of nd proofs, far from being an obstacle to backward search, makes proof search both strategic and efficient. — Siekmann and Wrightson collected in their two volume *Automated Reasoning* classical papers that contain marvelous discussions of the broad methodology underlying different approaches in the emerging field from the late 1950s to the early 1970s. The papers by Beth, Kanger, Prawitz, Wang and the "Russian School" are of particular interest from my perspective, as we find in them serious attempts of searching for humanly intelligible proofs and of getting the logical framework right before building heuristics into the search. That was perhaps most clearly formulated by Kanger in his (1963, p. 364): "The introduction of heuristics may yield considerable simplifications of a given proof method, but I have the impression that it would be wise to postpone the heuristics until we have a satisfactory method to start with." The work with AProS and automated proof search support that view.

A. Purely logical arguments.

In the supplement to (Shanin, et al. 1965), one finds five propositional problems and their proofs; AProS solves them with just the basic rules whereas in this paper quite complex derived rules are used. I discuss one example in order to illustrate, how dramatically the search is impacted by "slight" reformulations of the problem to be solved or, what amounts to the same thing, by introducing specific heuristics. The problem in (Shanin, et al. 1965) is to derive

$$(\neg(K \to A) \vee (K \to B))$$

from the premises

$$(H \vee \neg(A \ \& \ K)) \text{ and } (H \to (\neg A \vee B)).$$

The pure AProS search procedure uses 277 search steps to find a proof of length 77. If one uses in a first step a derived rule to replace positive occurrences of $(\neg X \vee \Delta)$ or $(X \vee \neg \Delta)$ by $(X \to \Delta)$, respectively, $(\Delta \to X)$ then AProS uses 273 search steps for a proof of length 87 (having made the replacement in the first premise), 149 search steps for a proof of length 80 (having made the replacement also in the second premise), and finally 9 search steps to find a derivation of length 12 (having made the replacement also in the conclusion).

The Shanin-procedure introduces also instances of the law of excluded middle. In the above problem it does so for the left disjunct of the goal, i.e., it uses the instance $(\neg(K \to A) \vee (K \to A))$. If one adds that instance as an additional premise, then AProS takes 108 search steps to obtain a proof of 49 lines. If the conclusion $(\neg(K \to A) \vee (K \to B))$ is replaced by $((K \to A) \to (K \to B))$ then the instance of the law of excluded middle is not used when AProS obtains a proof of length 29 in 23 search steps. If only the goal is reformulated as a conditional, then the same proof is obtained with just 18 search steps.

The replacement step in the last proof amounts to using one of the available rules from (Shanin, et al. 1965) heuristically: if the goal is of the form $(\neg X \vee \Delta)$ or $(X \vee \neg \Delta)$ then prove instead $(X \to \Delta)$, respectively, $(\Delta \to X)$. Such a reformulation of a problem, or equivalently the strategic use of a derived rule, can thus have a dramatic consequence on the search and the resulting derivation. Let me discuss two additional examples and a motivated extension of this heuristic step:

(1) Prove from the premise $P \vee Q$ the disjunction

$$(P \ \& \ Q) \vee (P \ \& \ \neg Q) \vee (\neg P \ \& \ Q).$$

With its basic algorithm AProS uses 202 search steps to find a proof of length 58; however, if the goal is reformulated as the conditional

$$\neg(P \ \& \ Q) \to ((P \ \& \ \neg Q) \vee (\neg P \ \& \ Q))$$

then 28 steps lead to a proof of length 18.
(2) Prove $((P \vee Q) \to (P \vee R)) \to (P \vee (Q \to R))$.
103 steps in the basic search lead to a proof of length 47. If one considers instead

$$((P \vee Q) \to (P \vee R)) \to (\neg P \to (Q \to R))$$

AProS finds a proof of length 14 with 9 search steps.
These quasi-empirical observations can be used to articulate a heuristic for the purely logical search: if one encounters a disjunction $(X \vee \Delta)$ as the goal, prove instead the conditional $(\neg X \to \Delta)$ (and eliminate in the antecedent a double negation, in case X happens to be a negation).

B. Some elementary set theoretic arguments.

As I mentioned in sections 3.2 and 4.1, we have been extending the automated search procedure to elementary set theory. Though our goal is different from that of interactive theorem proving, there is a great deal of overlap: the hierarchical organization of the search can be viewed as reflecting and sharpening the interaction of a user with a proof assistant. After all, we start out by analyzing the structure of proofs, formalizing them, and then automating the proof search, i.e., completely eliminating interaction. The case of using computers as proof assistants is made in great detail in Harrison's paper (2008). For the case of automated proof search it is important, if not absolutely essential, that the logical calculus of choice is natural deduction.[41]

Natural deduction has been used for proof search in set theory; an informative description is found, for example, in (Bledsoe 1983). Pastre's (1976) dissertation, deeply influenced by Bledsoe's work, is mentioned in Bledsoe's paper. She has continued that early work, and her most recent paper (2007) addresses a variety of elementary set theoretic problems. Similar work, but in the context of the *Theorema* project, was done in (Windsteiger 2001) and (Windsteiger 2003). However, the term natural deduction is used here only in a very loose way: there is no search space that underlies the logical part and guarantees completeness of the search procedure. Rather, the search is guided in both logic and set theory by "natural heuristics" for the use of reduction rules that are not connected to a systematic logical search and, in Pastre's case, do not even allow for any backtracking.

Let me consider a couple of examples of APrOS proofs to show how the logical search is extended in a most natural way by exploiting the meaning of defined concepts by appropriate I- and E-rules. That has, in particular, the "side-effect" of articulating in a mathematically sensible way, at which point in the search definitions should be expanded. In each case, the reader should view the proof strategically, i.e., closing the gap between premises and conclusion by use of (inverted) I-rules and motivated E-rules.

Example 1: $a \in b$ proves $a \subseteq \bigcup(b)$

1.	$a \in b$	Premise
2.	$u \in a$	Assumption
3.	$(a \in b \ \& \ u \in a)$	&I 1, 2
4.	$(\exists z)(z \in b \ \& \ u \in z)$	$\exists$I 3
5.	$u \in \bigcup(b)$	Def.I (Union) 4
6.	$(u \in a \to u \in \bigcup(b))$	$\to$I 5
7.	$(\forall x)(x \in a \to x \in \bigcup(b))$	$\forall$I 6
8.	$a \subseteq \bigcup(b)$	Def.I (Subset) 7

---

[41] There have been attempts of using proofs by resolution or other "machine-oriented" procedures as starting points for obtaining natural deduction proofs; Peter Andrews and Frank Pfenning, but also more recently Xiaorong Huang did interesting work in that direction.

Example 2.1: $a \subseteq b$ proves $\wp(a) \subseteq \wp(b)$

1.	$a \subseteq b$	Premise
2.	$u \in \wp(a)$	Assumption
3.	$v \in u$	Assumption
4.	$(\forall x)(x \in a \to x \in b)$	Def.E (Subset) 1
5.	$(v \in a \to v \in b)$	$\forall$E 4
6.	$u \subseteq a$	Def.E (Power Set) 2
7.	$(\forall x)(x \in u \to x \in a)$	Def.E (Subset) 6
8.	$(v \in u \to v \in a)$	$\forall$E 7
9.	$v \in a$	$\to$E 8, 3
10.	$v \in b$	$\to$E 5, 9
11.	$(v \in u \to v \in b)$	$\to$I 10
12.	$(\forall x)(x \in u \to x \in b)$	$\forall$I 11
13.	$u \subseteq b$	Def.I (Subset) 12
14.	$u \in \wp(b)$	Def.I (Power Set) 13
15.	$(u \in \wp(a) \to u \in \wp(b))$	$\to$I 14
16.	$(\forall x)(x \in \wp(a) \to x \in \wp(b))$	$\forall$I 15
17.	$\wp(a) \subseteq \wp(b)$	Def.I (Subset) 16

Example 2.2: This is example 2.1 with the additional premise (lemma):

$$(\forall x)[(x \subseteq a \ \& \ a \subseteq b) \to x \subseteq b]$$

1.	$(\forall x)[(x \subseteq a \ \& \ a \subseteq b) \to x \subseteq b]$	Premise
2.	$a \subseteq b$	Premise
3.	$u \in \wp(a)$	Assumption
4.	$(u \subseteq a \ \& \ a \subseteq b) \to u \subseteq b$	$\forall$E 1
5.	$u \subseteq a$	Def.E (Power Set) 3
6.	$(u \subseteq a \ \& \ a \subseteq b)$	&I 5, 2
7.	$u \subseteq b$	$\to$E 4, 6
8.	$u \in \wp(b)$	Def.I (Power Set) 7
9.	$(u \in \wp(a) \to u \in \wp(b))$	$\to$I 8
10.	$(\forall x)(x \in \wp(a) \to x \in \wp(b))$	$\forall$I 9
11.	$\wp(a) \subseteq \wp(b)$	Def.I (Subset) 10

C. Confluence?

The AProS project intends also to throw some empirical light on the cognitive situation. With a number of collaborators I have been developing a web-based introduction to logic, called *Logic & Proofs*; it focuses on the strategically guided construction of proofs and includes dynamic tutoring via the search algorithm AProS. The course is an expansive Learning Laboratory, as students construct arguments in a virtual Proof Lab in which their every move is recorded. It allows the investigation of questions like:

- How do students go about constructing arguments?
- How do particular pedagogical interventions affect their learning?
- How efficient do students get in finding proofs with little backtracking?

- Does the skill of strategically looking for proofs transfer to informal considerations?

The last question hints at a broader and long-term issue I am particularly interested in, namely, to find out whether strategic-logical skills improve the ability of students to understand complex mathematics.

The practical educational aspects are deeply connected to a theoretical issue in cognitive science, namely, the stark opposition of "mental models" (Johnson-Laird) and "mental proofs" (Rips). I do not see an unbridgeable gulf, but consider the two views as complementary. Proofs as diagrams give rise to mental models, and the dynamic features of proof construction I emphasized are promoted by and reflect a broader structural, mathematical context; all of this is helping us to bridge the gap between premises and conclusion. The crucial question for me is: Can we make advances in isolating basic operations of the mind involved in constructing mathematical proofs or, in other words, can we develop a cognitive psychology of proofs that reflect logical and mathematical understanding?

There is deeply relevant work on analogical reasoning, e.g., Dedre Gentner's. In the (2008) manuscript with J. Colhoun they write, "Analogical processes are at the core of relational thinking, a crucial ability that, we suggest, is key to human cognitive prowess and separates us from other intelligent creatures. Our capacity for analogy ensures that every new encounter offers not only its own kernel of knowledge, but a potentially vast set of insights resulting from parallels past and future." Performance in particular tasks is enhanced when analogies, viewed as relational similarities, are strengthened by explicit comparisons and appropriate encodings. It seems that abstraction is here a crucial mental operation and builds on such comparisons. The underlying theoretical model of these investigations (structure mapping) is steeped in the language of the mathematics that evolved in the 19th century, in particular, through Dedekind's work. It was Dedekind who introduced *mappings* between arbitrary systems; he asserted in the strongest terms that without this *capacity of the mind* (to let a thing of one system correspond to a thing of another system) no thinking is possible at all. Modern abstract, structural mathematics, one can argue convincingly, makes analogies between different "structures" precise via appropriate axiomatic formulations. — All of this, so the rich psychological experimental work demonstrates, is important for learning. In the context of more sophisticated mathematics, *Kaminski e.a.* hypothesized (and confirmed) for example recently "that learning a single generic instantiation [i.e., a more abstract example of a structure or concept; WS] ... may result in better knowledge transfer than learning multiple concrete, contextualized instantiations." (p. 454)

There is a most plausible confluence of mathematical and psychological reflection that would get us closer to a better characterization of the "capacity of the human mind" that was discovered in Greek and rediscovered in 19th century mathematics; according to Stein, as quoted already at the beginning of Part 2, "what has been learned, when properly understood, constitutes one of the greatest advances in philosophy ..."

BIBLIOGRAPHY

The following abbreviations are being used: DMV for Deutsche Mathematiker Vereinigung; BSL for Bulletin of Symbolic Logic; AMS for American Mathematical Society.

Aspray, W. and Kitcher, P. (eds)

1988 *History and philosophy of modern mathematics*, Vol. XI of *Minnesota Studies in the Philosophy of Science*, University of Minnesota Press, Minneapolis.

Aumann, G.

2009 *Euklids Erbe — Ein Streifzug durch die Geometrie und ihre Geschichte*, third edition, Darmstadt.

Bernays, P.

1922 Über Hilberts Gedanken zur Grundlegung der Arithmetik, *Jahresbericht der DMV* **31**, 10–19. Translated in (Mancosu 1998, pp. 215–222).

1946 Gesichtspunkte zum Problem der Evidenz. Reprinted in (Bernays 1976, pp. 85–91).

1954 Zur Beurteilung der Situation in der beweistheoretischen Forschung, *Revue internationale de philosophie* **8**, 9–13. Discussion, pp. 15–21.

1976a *Abhandlungen zur Philosophie der Mathematik*, Wissenschaftliche Buchgesellschaft.

Bernstein, F.

1924 Der Pythagoräische Lehrsatz, *Zeitschrift für Mathematischen und Naturwissenschaftlichen Unterricht* **55**, 204–207.

Beth, E. W.

1958 On machines which prove theorems. Reprinted in (Siekmann and Wrightson 1983, pp. 79–90).

Bledsoe, W. W.

1983 Non-resolution theorem proving, *Artificial Intelligence* **9**, 1–35

Bledsoe, W. W. and Loveland, D. W. (eds)

1984 *Automated theorem proving: after 25 years*, Vol. 29 of *Contemporary Mathematics*, AMS.

Cantini, A.

2008 On formal proofs, in (Lupacchini and Corsi 2008, pp. 29–48).

Cantor, G.

1932 *Gesammelte Abhandlungen mathematischen und philosophischen Inhalts*, E. Zermelo (ed), Springer.

Davis, M.

1965 *The undecidable*, Raven Press.

Dedekind, R.

1872c *Stetigkeit und irrationale Zahlen*, Vieweg. Reprinted in (Dedekind 1932, pp. 315–324). Translated in (Ewald 1996, pp. 765–779).

1877 Sur la théorie des nombres entiers algébriques, *Bulletin des sciences mathématiques et astronomiques* **1**(XI), 2(I), pp. 1–121. Partially reprinted in (Dedekind 1932, pp. 262–296). Translated in (Dedekind 1996).

1888 *Was sind und was sollen die Zahlen?*, Vieweg. Reprinted in (Dedekind 1932, pp. 335–391). Translated in (Ewald 1996, pp. 787–833).

1932 *Gesammelte mathematische Werke*, Vol. 3. R. Fricke, E. Noether, and Ö. Ore (eds), Vieweg.

1996 *Theory of algebraic integers*, Cambridge University Press. Translated and introduced by J. Stillwell.

Dirichlet, P. G. L.

1863 *Vorlesungen über Zahlentheorie, Hrsg. und mit Zusätzen versehen von R. Dedekind*, Vieweg. 2nd edition 1871; 3rd 1879; 4th 1894.

Dunham, W.

1990 *Journey through genius: The great theorems of mathematics*, Wiley.

Ewald, W. B. (ed)

1996   *From Kant to Hilbert: A source book in the foundations of mathematics*, Oxford University Press. Two volumes.

Gentner, D.

2008   Analogical processes in human thinking and learning, in press.

Gentzen, G.

1936   Die Widerspruchsfreiheit der reinen Zahlentheorie, *Mathematische Annalen* **112**, 493–565. Translated in (Gentzen 1969).

1969   *The collected papers of Gerhard Gentzen*, North-Holland. Edited and translated by M. E. Szabo.

Gödel, K.

1946   Remarks before the Princeton bicentennial conference on problems in mathematics, in (Gödel 1990, pp. 150–153).

1990   *Collected Works*, Vol. II, Oxford University Press.

Harrison, J.

2008   Formal proof — theory and practice, *Notices of the AMS* **55**(11), 1395–1406.

Hilbert, D.

Unpublished Lecture Notes of Hilbert's are located in Göttingen in two different places, namely, the Staats- und Universitätsbibliothek and the Mathematisches Institut. The reference year of these notes is preceded by a "*"; their location is indicated by SUB xyz, repectively MI. Many of them are being prepared for publication in *David Hilbert's lectures on the foundations of mathematics and physics, 1891–1933*, Springer.

1899a   Grundlagen der Geometrie, in *Festschrift zur Feier der Enthüllung des Gauss-Weber-Denkmals in Göttingen*, Teubner, pp. 1–92.

*1899b   Zahlbegriff und Quadratur des Kreises, SUB 549 and also in SUB 557.

1900b   Über den Zahlbegriff, *Jahresbericht der DMV* **8**, 180–194. Reprinted in *Grundlagen der Geometrie*, third edition, Leipzig, 1909, pp. 256–262. Translated in (Ewald 1996, pp. 1089–1095).

1905a   Über die Grundlagen der Logik und der Arithmetik, in *Verhandlungen des Dritten Internationalen Mathematiker-Kongresses*, Teubner, pp. 174–185. Translated in (van Heijenoort 1967, pp. 129–138).

*1919   Natur und mathematisches Erkennen. Lecture notes by P. Bernays, MI. (These notes were edited by D. E. Rowe and published in 1992 by Birkhäuser.)

*1920b   Probleme der mathematischen Logik. Lecture notes by P. Bernays and M. Schönfinkel, MI.

1922   Neubegründung der Mathematik, *Abhandlungen aus dem mathematischen Seminar der Hamburgischen Universität* **1**, 157–177.

1927   Die Grundlagen der Mathematik, *Abhandlungen aus dem mathematischen Seminar der Hamburgischen Universität* **6**(1/2), 65–85. Translated in (van Heijenoort 1967, pp. 464–479).

Hilbert, D. and Ackermann, W.

1928   *Grundzüge der theoretischen Logik*, Springer.

Hilbert, D. and Bernays, P.

1934   *Grundlagen der Mathematik*, Vol. I, Springer. Second edition, 1968, with revisions detailed in foreword by Bernays.

Kanger, S.

1963   A simplified proof method for elementary logic. Reprinted in (Siekmann and Wrightson 1983, pp. 364–371).

Karzel, H. and Kroll, H.-J.

1988   *Geschichte der Geometrie seit Hilbert*, Wissenschaftliche Buchgesellschaft.

Lupacchini, R. and Corsi, G. (eds)

2008   *Deduction, computation, experiments — Exploring the effectiveness of proof*, Springer Italia.

Mac Lane, S.
    1935   A logical analysis of mathematical structure, *The Monist* **45**, 118–130.

    1979   A late return to a thesis in logic, *in* I. Kaplansky (ed), *Saunders MacLane — Selected papers*, Springer.

Mancosu, P.
    1998   *From Brouwer to Hilbert. The debate on the foundations of mathematics in the 1920s*, Oxford University Press.

Maslov, Y., Mints, G. E. and Orevkov, V. P.
    1983   Mechanical proof search and the theory of logical deduction in the USSR, in (Siekmann and Wrightson 1983, pp. 29–38).

Mints, G. E.
    1969   Variation in the deduction search tactics in sequential calculi, *Seminar in Mathematics V. A. Steklov Mathematical Institute* **4**, 52–59.

Pastre, D.
    1976   Démonstration automatique de théorèmes en théorie des ensembles, Dissertation, University of Paris.

    2007   Complementarity of a natural deduction knowledge-based prover and resolution-based provers in automated theorem proving. Manuscript, March 2007, 34 pages.

Poincaré, H.
    1902b   Review of (Hilbert 1899a), *Bulletin des sciences mathématiques* **26**, 249–272.

Prawitz, D.
    1965   *Natural deduction: A proof-theoretical study*, Stockholm.

Shanin, N. A., Davydov, G. E., Maslov, S. Yu., Mints, G. E., Orevkov, V. P. and Slisenk, A. O.
    1965   An algorithm for a machine search of a natural logical deduction in a propositional calculus. Translated in (Siekmann and Wrightson 1983, pp. 424–483).

Sieg, W.
    1999   Hilbert's programs: 1917–1922, *BSL* **5**, 1–44. In this volume, pp. 61–97.

    2005   Only two letters: The correspondence between Herbrand and Gödel, *BSL* **11**(2), 172–184. In this volume, pp. 125–136.

    2007   On mind & Turing's machines, *Natural Computing* **6**, 187–205.

    2008a   Church without dogma: axioms for computability, *in* S. B. Cooper, B. Löwe, and A. Sorbi (eds), *New computational paradigms — changing conceptions of what is computable*, Springer, pp. 139–152.

    2010b   Gödel's philosophical challenge (to Turing); to appear.

Sieg, W. and Byrnes, J.
    1998   Normal natural deduction proofs (in classical logic), *Studia Logica* **60**, 67–106.

Sieg, W. and Cittadini, S.
    2005   Normal natural deduction proofs (in non-classical logics), *in* D. Hutter and W. Stephan (eds), *Mechanizing mathematical reasoning*, Vol. 2605 of *Lecture Notes in Computer Science*, Springer, pp. 169–191.

Sieg, W. and Field, C.
    2005   Automated search for Gödel's proofs, *Annals of Pure and Applied Logic* **133**, 319–338. Reprinted in (Lupacchini and Corsi 2008, pp. 117–140).

Siekmann, J. and Wrightson, G. (eds)
    1983   *Automated Reasoning*, two volumes, Springer.

Stein, H.
    1988   Logos, logic, and logistiké: Some philosophical remarks on nineteenth-century transformation of mathematics, in (Aspray and Kitcher 1988, pp. 238–259).

Turing, A.
    1936   On computable numbers, with an application to the *Entscheidungsproblem*, *Proceedings of the London Mathematical Society* **42**, 230–265. Also in (Davis 1965, pp. 116–151).

van Heijenoort, J. (ed)
  1967 *From Frege to Gödel: A sourcebook of mathematical logic, 1879–1931*, Harvard University Press.

von Plato, J.
  2008 Gentzen's proof of normalization for natural deduction, *BSL* **14**, 240–257.

Wang, H.
  1970 On the long-range prospects of automatic theorem-proving, *in* M. Laudet et al. (eds), *Symposium on automated demonstration*, Vol. 125 in *Lecture Notes in Mathematics*, Springer, pp. 101–111.

  1984 Computer theorem proving and artificial intelligence, in (Bledsoe and Loveland 1984, pp. 49–70).

Wiedijk, F.
  2008 Formal proof — getting started, *Notices of the AMS* **55**(11), 1408–1414.

Windsteiger, W.
  2001 A set theory prover within *Theorema*, *in* R. Moreno-Diaz et al. (eds), *Eurocast 2001*, Vol. 2178 of *Lecture Notes in Computer Science*, Springer, pp. 525–539.

  2003 An automated prover for set theory in *Theorema*, Omega-Theorema Workshop, 14 pages.

On Varieties of Closed Categories and Dependency of Diagrams of Canonical Maps

A.EL KHOURY, S. SOLOVIEV, L.MEHATS, M.SPIVAKOVSKY

ABSTRACT. We present a series of diagrams D_n in Symmetric Monoidal Closed Categories such that there is infinitely many different varieties of SMCC (in the sense of universal algebra) defined by diagrams of this series as equations. Similar result will hold for weaker closed categories. We discuss the notion of dependency of diagrams in connection with this result.

1 Introduction

Canonical maps in closed categories may be seen as instances of morphisms of the free closed category generated by an infinite set of atoms.

There exist many types of closed categories, for example Cartesian Closed Categories (CCC), Symmetric Monoidal Closed Categories (SMCC) etc. Closed categories were first introduced and studied in the 1960ies and 1970ies (see [5]). G. Lambek was the first to notice and explicitly use a close connection with proof theory (see [7, 8, 9]).

Grigori Mints in the 1970ies (see [11, 12]) has shown that this connection extends to much deeper aspects of the structure of closed categories than was initially expected, for example, that the equality of morphisms in free closed categories can be faithfully represented using normalization in certain systems of natural deduction and lambda calculus.

His works have opened the way to the use of even more advanced proof-theoretic methods. For example, he suggested to one of the authors (his graduate student at the time) the idea of adapting a method of decreasing of the depth of formulas in proof theory to the study of commutativity of diagrams in closed categories. This approach helped to obtain many coherence theorems of category theory (see for example [13, 14]).

Typical examples of closed categories are the category of vector spaces over a field, the category of modules over a ring, the category of semi-modules over a semi-ring, the category of pointed sets, etc. If the ring is commutative with unit then the category is a SMCC. SMCCs will be our main interest in this paper.

A typical example of a CCC is the category of sets. In closed categories of certain types, for example CCCs, the "maximality" theorem holds (cf. [1]). If to the standard identities defining equality of morphisms in the free CCC is added a new identity[1] between morphisms that have the same domain and the same

---
[1] It is assumed that all identities are closed w.r.t. substitution.

codomain (i.e., form a diagram) then all the diagrams become commutative.

The situation is completely different in the case of SMCCs and closed categories with weaker theories. The "maximality theorem" does not hold for SMCCs. The negative result is even stronger. Unlike the case of CCC, in SMCC and closed categories with weaker structure, the graphs of naturality conditions (Kelly-Mac Lane graphs [5]) play very important role. There are diagrams $f, g : A \to B$ in the free SMCC where f, g have the same graph such that new identity $f \sim g$ added as a new axiom does not imply the commutativity of all diagrams with the same graph. Still, the commutativity of a diagram implies the commutativity of some other diagrams; we may say that the commutativity of these diagrams *depends* on the commutativity of the given one.

The main new result presented in this paper is the description of an infinite series of diagrams $D_1, ..., D_k, ...$ (with the same graph of naturality conditions) in the free SMCC such that for each k there exists a model K_k where $D_1, ..., D_k$ are non-commutative but for some $n, k < n, D_n, ...$ are commutative.

In universal algebra "variety" is a class of algebras defined by axioms that have the form of identities. Categories may be seen as partial many-sorted algebras. In this terminology, our new result is that there exist infinitely many different varieties of SMCC between the free category and the category of graphs[2].

This fact justifies the study of dependency of diagrams. Proof-theoretical methods have shown their strength in the study of commutativity and non-commutativity of diagrams in free closed categories [6, 11, 13, 14]. They provide, in particular, efficient deciding algorithms [15]. In all probability they will be also very useful in the study of dependency of diagrams in equationally defined subclasses of the class of closed categories and in concrete non-free models.

2 Algebra and Logic in Closed Categories: some Basic Facts

The connections between structural proof theory and categorical algebra are well known. In this section we follow roughly the schema introduced already in the works of Lambek [8, 9] and Mints [11, 12].

A SMCC is defined by the following data:

- A category K;

- An object $I \in Ob(K)$

- The bifunctors $\otimes : K \times K \to K$ (tensor) and $\multimap: K^{op} \times K \to K$ (internal hom-functor);

- the following families of maps (basic natural transformations): $1_A : A \to A$,

$$a_{ABC} : (A \otimes B) \otimes C \to A \otimes (B \otimes C), \quad a^{-1}_{ABC} : A \otimes (B \otimes C) \to (A \otimes B) \otimes C$$

---

[2] For example, if we add D_n as unique new axiom (declare that D_n is commutative) the resulting variety will be different from the variety defined by D_k as the new axiom. The result is valid also for all the weaker structures of closed category considered in this paper.

$$b_A : A \otimes I \to A, \quad b_A^{-1} : A \to A \otimes I, \quad c_{AB} : A \otimes B \to B \otimes A$$
$$e_{AB} : A \otimes (A \multimap B) \to B, \quad d_{AB} : A \to B \multimap A \otimes B \quad (A, B, C \in Ob(K)).$$

These data satisfy certain equations that we shall not describe in detail (see, e.g., [14]). The main groups of equations are:

- equations between components of basic natural transformations, such as $c_{AB} \circ c_{BA} = 1_{A \otimes B}$, Mac Lane's "pentagon" and "hexagon";
- naturality conditions, functoriality axioms for $\otimes$ and $\multimap$, general category axioms that involve arbitrary morphisms of K, the axioms (involving e, d, $f : A \otimes B \to C, g : A \to (B \multimap C)$) that make $\otimes$ left adjoint of $\multimap$.

The action of functors and composition on morphisms may be regarded as an application of the following *rules*:

$$\frac{f : A \to B \quad g : C \to D}{f \otimes g : A \otimes C \to B \otimes D}(\otimes) \qquad \frac{f : A \to B \quad g : C \to D}{f \multimap g : B \multimap C \to A \multimap D}(\multimap)$$

$$\frac{f : A \to B \quad g : B \to C}{g \circ f : A \to C}(cut).$$

The basic natural transformations correspond in this setting to axiom schemas ($1_A : A \to A, \quad c_{AB} : A \otimes B \to B \otimes A$ etc.).

The free SMCC $\mathbf{F}(\mathbf{A})$ over a set of atoms $\mathbf{A}$ may be built as follows.

- The objects are formulas built from atoms and the constant I using $\otimes$ and $\multimap$ as connectives.

- The morphisms are the expressions $f : A \to B$ derivable from axiom schemas corresponding to basic natural transformations by rules $\otimes$, $\multimap$ and *cut* above considered up to the smallest equivalence relation $\equiv$ such that all the equations of the SMCC mentioned above are satisfied.

The relation $\equiv$ is a congruence w.r.t. the application of $\otimes$, $\multimap$ and *cut* (composition). It is substitutive, i.e., $f \equiv g \Rightarrow \sigma f \equiv \sigma g$ for every substitution $\sigma = [A_1, ..., A_k/a_1, ..., a_k]$ because all the morphisms in $\mathbf{F}(\mathbf{A})$ are obtained from components of natural transformations (axiom schemes). If not stated otherwise we shall assume that $\mathbf{A}$ is infinite[3].

---

[3] A presentation of the free Cartesian Closed Category (CCC) along similar lines may be obtained if we add the following natural transformations (families of maps):

$$0_A : A \to I, \quad \delta_A : A \to A \otimes A,$$
$$l_{AB} : A \otimes B \to A, \quad r_{AB} : A \otimes B \to B$$

and the identities that will make I the terminal object and $\otimes$ the cartesian product with projections l, r. These new data may replace the transformations a, b, c because associativity, commutativity and the property of unit I can be obtained from $0, \delta, l, r$ and the new identities.

Maximality of the theory of CCC proved in [1] means that the only possible relation $\sim$ different from $\equiv$ in case of CCC is the relation that makes equivalent all the morphisms with the same source and target.

Let us mention some other types of closed categories:

- Monoidal Closed Categories, where the commutativity isomorphism c_{AB} is absent;

Let A be a formula built from atoms and I using $\otimes$ and $\multimap$ as connectives. From the categorical point of view, it represents a functor and every occurrence of an atom or I in A is co - or contravariant. In logic to categorical variance corresponds the notion of sign of an occurrence. It is defined by induction on the process of the construction of A.

DEFINITION 1.

- If $A = a$ or $A = I$ the occurrence of a (of I) in A is positive (covariant).

- In $A \otimes B$ the signs of occurrences are the same as the signs of corresponding occurrences in A and B. In $A \multimap B$ the signs of occurrences lying in B are the same as in B and the signs of occurrences lying in A are opposite to the signs in A.

- The signs of occurrences in the sequent $A \to B$ are the same as in $A \multimap B$.

DEFINITION 2. A sequent S is called balanced iff every atom has exactly two occurrences with opposite signs in S.

In the case of SMCC the following proposition holds [4].

THEOREM 3. Let $f, g : A \to B$. There exist $f', g' : A' \to B'$ where the sequent $A' \to B'$ is balanced such that f, g and $A \to B$ can be obtained from f', g' and $A' \to B'$ by identification of variables and $f' \equiv g'$ iff $f \equiv g$.

In this paper we shall consider other equivalence relations on morphisms of $\mathbf{F(A)}$. Below $\sim$ will denote any substitutive equivalence relation that contains $\equiv$, respects the graphs and is a congruence w.r.t. $\otimes$, $\multimap$ and cut. Obviously such a $\sim$ will define a structure of a SMCC on $\mathbf{F(A)}$. In connection with our main results we shall consider the relations $\sim_K$ generated by interpretations in certain SMCCs (models) K. More precisely, let us consider any function (valutaion) $v : \mathbf{A} \to Ob(K)$. Since $\mathbf{F(A)}$ is free, every v defines a unique structure-preserving functor (interpretation) $|-|_v : \mathbf{F(A)} \to K$ where $|a|_v = v(a)$. Assume that f, g have the same graph. The relation $\sim_K$ is defined by $f \sim g \iff |f|_v = |g|_v$ for every v in K. We shall say that the diagram $f', g' : A' \to B'$ depends on $f, g : A \to B$ if for every SMCC K the equivalence $f \sim_K g$ implies $f' \sim_K g'$.

- Symmetric Closed Categories (non-monoidal), where the functor $\otimes$ is absent and c_{AB} is replaced by the isomorphism $\xi_{ABC} : A \multimap (B \multimap C) \to B \multimap (A \multimap C)$;

- Closed Categories (without $\otimes$ and any symmetry isomorphism).

For all these types of categories a free category over a set of atoms $\mathbf{A}$ is constructed similarly to $\mathbf{F(A)}$. Each model for the theory of SMCC is at the same time a model for MCC, SCC and CC (but not of CCC). Our main results concerning SMCC will imply similar result for these closed categories.

[4]The pairs of occurrences of the same variable in balanced sequents correspond to the edges of the so called "graph" introduced by Kelly and Mac Lane, and also to "axiom links" in linear logic. The combinatorial proofs of a similar proposition were published as early as in [2, 5]. They use the fact that all the axioms that define the relation $\equiv$ may be written in a balanced form and the rather tricky definition of composition of graphs. The proof using Gentzen-style sequent calculus $\mathbf{L(A)}$ described below is much more straightforward. It uses cut-elimination and the properties of "linear" rules of $\mathbf{L(A)}$. There is no similar proposition that would hold in the CCCs case.

THEOREM 4. *Let $\sim$ be the smallest substitutive equivalence relation on the derivations of $\mathbf{F(A)}$ such that $f \sim g$, $\sim$ contains $\equiv$, respects the graphs and is a congruence w.r.t. $\otimes$, $\multimap$ and cut. The diagram $f', g' : A' \to B'$ depends on $f, g : A \to B$ iff $f' \sim g'$.*

This theorems shows that syntactic methods may be used for verification of dependency of diagrams, since the standard construction for the smallest equivalence relation $\sim$ uses certain syntactic calculus with pairs of derivations as derivable objects.

The interest of presentation of $\mathbf{F(A)}$ using "algebraic" axioms and rules above is that it opens a way to reformulations using different axiom and rules, already well studied in logics [5]. We shall consider in this paper one such reformulation, the sequent calculus for Intuitionistic Multiplicative Linear Logic (IMLL), cf. [3]. The calculus $\mathbf{L(A)}$ is defined as follows:

Axioms:

$A \to A \; (1_A) \quad \to I \; (unit)$

Structutral Rules:

$$\frac{\Gamma \to A \quad A, \Delta \to B}{\Gamma, \Delta \to B} \text{(cut)} \quad \frac{\Delta \to I \quad \Sigma \to A}{\Delta, \Sigma \to A} \text{(wkn)} \quad \frac{\Gamma \to A}{\Gamma' \to A} \text{(perm)}$$

Logical Rules:

$$\frac{\Gamma \to A \quad \Delta \to B}{\Gamma, \Delta \to A \otimes B} (\to \otimes) \quad \frac{A, B, \Gamma \to C}{A \otimes B, \Gamma \to C} (\otimes \to)$$

$$\frac{A, \Gamma \to B}{\Gamma \to A \multimap B} (\to \multimap) \quad \frac{\Gamma \to A \quad B, \Delta \to C}{\Gamma, A \multimap B, \Delta \to C} (\multimap \to)$$

Here Γ, Δ, Σ are lists of formulas. A list of formulas $\Gamma = A_1, ..., A_n$ may be seen as an abbreviation of $\overline{\Gamma} = (...(A_1 \otimes ...) \otimes A_n) \otimes I$. The transformation $\mathbf{C}$ of L-derivations into F-derivations and $\mathbf{D}$ of F-derivations into L-derivations are described in detail in [14, 10]. Let us present here just one case as an example:

$$\mathbf{C}(\frac{\Gamma \xrightarrow{\psi} A \quad B, \Delta \xrightarrow{\varphi} C}{\Gamma, A \multimap B, \Delta \to C}) =$$

$$= (\mathbf{C}(\varphi) \circ ((e_{AB} \circ (\mathbf{C}(\psi) \otimes 1_{A \multimap B})) \otimes 1_{\overline{\Delta}})) \circ \zeta \; : \; \overline{\Gamma, A \multimap B, \Delta} \to C,$$

with $\zeta : \overline{\Gamma, A \multimap B, \Delta} \to (\overline{\Gamma} \otimes (A \multimap B)) \otimes \overline{\Delta}$ a central isomorphism (unique up to $\equiv$).

Via $\mathbf{C}$ we define the equivalence $\equiv$ on L-derivations as induced by $\equiv$ on F-derivations. The relation $\equiv$ on $\mathbf{L(A)}$-derivations is a congruence w.r.t. the

---

[5]G.E.Mints considered systems similar to $\mathbf{F(A)}$ (he called them "Hilbert-type systems") and the systems of natural deduction for CC, SCC, MCC, SMCC and CCC categories. Future study has shown that the systems of natural deduction are better for the development of deciding algorithms but sequential calculi are more flexible when the transformations of derivations and diagrams are studied.

application of rules. It is substitutive as well [6]. If we consider another equivalence relation $\sim$ on morphisms of $\mathbf{F}(\mathbf{A})$ it is also transferred to $\mathbf{L}(\mathbf{A})$ via $\mathbf{C}$. (Similarly, for every valuation $v : \mathbf{A} \to K$ the interpretation $|-|_v$ on derivations is defined via $\mathbf{C}$: $|d|_v = |\mathbf{C}(d)|_v$.)

The notion of sign and balanced sequent is generalized naturally to $\mathbf{L}(\mathbf{A})$. If $\Gamma = A_1, ..., A_n$ then the signs of occurrences of atoms in $\Gamma \to A$ are the same as in $A_1 \multimap (A_2 \multimap ...(A_n \multimap A)...)$. $\Gamma \to A$ is balanced if every atom occurs there exactly twice with opposite signs.

One of the most common transformations of derivations is cut-elimination.

THEOREM 5.

For every derivation $d : \Gamma \to A$ in $\mathbf{L}(\mathbf{A})$ there exists a cut-free derivation $d' \Gamma \to A$ such that $d' \equiv d$. (This also implies $d' \sim d$ for any $\sim$.) If $\Gamma \to A$ is balanced then all the sequents in its cut-free derivation are balanced [7].

Cut-elimination here is a standard algorithm used in proof-theory. For comparison, a cut-elimination procedure described in the algebraic notation in [6] is much more heavy and difficult to use. A similar theorem is true for all the logical calculi corresponding to the closed categories considered above.

Another useful transformation is the reduction of formula's depth described in detail in [14].

DEFINITION 6. The sequent $\Gamma \to A$ is called 2-sequent if A contains no more than one connective and each member of Γ no more than two connectives.

Some formulas may be replaced by isomorphic ones (reducing further the number of possibilities).

DEFINITION 7. $\Gamma \to A$ is called a pure 2-sequent if A has one of the forms $x, a \otimes b, a \multimap x$ and each member of Γ has one of the forms $x, a \multimap x, a \multimap (b \otimes c), (a \otimes b) \multimap x, (a \multimap x) \multimap y$. Here x, y stand for I or atoms, a, b are atoms.

Every derivation can be transformed into a derivation of some 2-sequent

---

[6] A sequent calculus for CCC may be described in a similar way. It is enough to modify it as follows:
(a) To replace the axiom $\to I$ by $\Delta \to I$ (Δ being an arbitrary list of formulas);
(b) To add the structural rule of contraction

$$\frac{A, A, \Gamma \to B}{A, \Gamma \to B}.$$

The resulting calculus represents exactly the Intuitionistic Propositional Logic with I as constant "true", $\otimes$ as conjunction and $\multimap$ as implication. The transformations $\mathbf{C}$ and $\mathbf{D}$ are modified accordingly, and the equivalence relation on derivations of $\mathbf{L}(\mathbf{A})$ is induced by equivalence in $\mathbf{F}(\mathbf{A})$ via $\mathbf{C}$. (The sequent calculi for MC, SC and Non-Symmetric Closed Categories and the transformations $\mathbf{C}$ and $\mathbf{D}$ are built in an analogous way.)

[7] In fact, not only *cut* but also all the trivial applications of *wkn* (with $\to I$ as left premise) can be eliminated, and several successive applications of *perm* can be replaced by one application. By abuse of terminology we shall assume below that in all *cut*-free derivations these simplifications are done as well. For all the logical calculi mentioned above (except for CCC) and any given sequent there exists only a finite number of *cut*-free derivations in this sense.

using two operations (followed by isomorphisms to obtain a *pure* 2-sequent):

$$\Gamma \xrightarrow{d} B \mapsto \frac{\Gamma \xrightarrow{d} B \quad p \xrightarrow{id} p}{\Gamma, B \multimap p \to p} (p\, fresh)$$

and *cut* with left premises of the form $p \multimap C, A[p] \to A[C]$ or $C \multimap p, A[p] \to A[C]$ (p fresh) [8]. There also exists the inverse transformation using substitutions $[C/p]$ and *cuts* with $\to C \multimap C$. (Due to the previous theorem *cut* can always be eliminated.)

THEOREM 8. *(Reduction to 2-sequents.) Let d_1, d_2 be two derivations of the same (balanced) sequent S. Then there exist two derivations d_1', d_2' of the same (balanced) pure 2-sequent S' such that for any relation $\sim$ (including $\equiv$ itself) d_1, d_2 are $\sim$-equivalent iff d_1', d_2' are $\sim$-equivalent.*

Another useful property is faithfullness. Using this property we may reduce the problem of equivalence of derivations that have identical inferences in the end to the equivalence of derivations of their premises.

DEFINITION 9. *(Cf. [10]) An equivalence $\sim$ is faithful w.r.t. the rule R if, for any two derivations φ, φ' of the same sequent ending by the inference of R having as premises some derivations of the same sequents, $\varphi \sim \varphi'$ iff the derivations of the premises are $\sim$-equivalent.*

In the case of SMCC faithfulness is easy to prove for the rules $\to\multimap$, $\to \otimes$, $\otimes \to$, *wkn*. Far from obvious, but true (see [10], theorem 4.15), it holds also for $\multimap\to$. It may not hold for other systems.

3 Commutative and Non-Commutative Diagrams in the Free SMCC.

Below we shall call diagrams not only pairs $f, g : A \to B$ in $\mathbf{F(A)}$ but also pairs of $\mathbf{L(A)}$-derivations of the same sequent.

A sequent S is called *proper* iff it does not contain occurrences of subformulas of the form $A \multimap B$ where B is constant (contains only I) and A is not constant.

THEOREM 10. *(The Kelly-Mac Lane coherence theorem reformulated for $\mathbf{L(A)}$, cf. [5].) Let $f, g : \Gamma \to A$ and the sequent $\Gamma \to A$ be proper. If f and g have the same graph[9] then $f \equiv g$.*

---

[8] Here a single occurrence of C is replaced by p. The form depends on the variance (sign) of this occurrence of C in A. One takes a standard derivation of these sequents, which always exists in "symmetric" calculi, i.e., logical systems for CCC, SMCC, SCC categories. The theorem that follows holds in each of these systems.

[9] In particular, if the sequent is balanced.

EXAMPLE 11. If the sequent is not proper, f may be non-equivalent to g.
The following diagram (called "triple-dual" diagram) is not commutative

(1)
$$\begin{array}{ccc} ((a \multimap I) \multimap I) \multimap I & \xrightarrow{\quad 1 \quad} & ((a \multimap I) \multimap I) \multimap I \\ & \searrow{\scriptstyle k_a \multimap 1} \quad \nearrow{\scriptstyle k_{a \multimap I}} & \\ & a \multimap I & \end{array}$$

where a is a variable and $k_a = (1 \multimap e_{aI}) \circ d_{a(a \multimap I)} : a \to (a \multimap I) \multimap I$ is the standard "embedding of a into its second dual".

Non-commutativity of this diagram may be checked formally in $\mathbf{F(A)}$ (the equivalence relation $\equiv$ is decidable). It is non-commutative also in certain models such as the SMCC of vector spaces or the SMCC of modules over a commutative ring with unit. One may note that it is always commutative in the full subcategory of vector spaces of finite dimension. On the contrary, it is not always commutative for finitely generated modules.

It will be useful to consider another diagram which is commutative (with respect to any relation $\sim$) iff the triple-dual diagram is commutative:

(2) $f, g : (a \otimes b \multimap I), ((b \multimap I) \multimap I), ((a \multimap I) \multimap I) \vec{\to} I.$

The fact that (2) is commutative iff (1) is commutative is more easily checked in $\mathbf{L(A)}$. It is also a good illustration of application of proof-theoretical methods.

Let $f_0 = 1_{((a \multimap I) \multimap I) \multimap I}$, and let g_0 denote the derivation corresponding to $k_{a \multimap I} \circ (k_a \multimap 1_I)$. We perform a *cut*-elimination. It is easily checked that *cut*-free derivations will be equivalent to the derivations ending by $\to \multimap$. By faithfulness (in this case corresponding to adjunction) we pass from the pair

$$f_0, g_0 : ((a \multimap I) \multimap I) \multimap I \to ((a \multimap I) \multimap I) \multimap I$$

to the pair

$$f_0^-, g_0^- : (((a \multimap I) \multimap I) \multimap I), ((a \multimap I) \multimap I) \to I.$$

Afterwards we perform the reduction to a pure 2-sequent, first applying (simultaneously to f_0^-, g_0^-) *cut* with

$$h : b \multimap (a \multimap I), ((b \multimap I) \multimap I) \to ((a \multimap I) \multimap I) \multimap I,$$

(it is an easy exercice to find the derivation h) and then (again via *cut*) the isomorphism $i : a \otimes b \multimap I \to b \multimap (a \multimap I)$. All *cuts* can be eliminated afterwards. The result is the pair of derivations $f, g : a \otimes b \multimap I, (b \multimap I) \multimap I, (a \multimap I) \multimap I \to I$ (one has $(b \multimap I) \multimap I$, and another $(a \multimap I) \multimap I$ as the main formula of last application of $\multimap \to$). One may return to f_0, g_0 via substitution $[b \multimap I/a]$, *cuts* with isomorphisms and $\to (b \multimap I) \multimap (b \multimap I)$, and application of the rule $\to \multimap$. Since all the steps preserve $\sim$, $f_0 \sim g_0 \iff f \sim g$.

The pair of derivations f, g is also an example of so called critical pair. There is a full description of non-equivalent pairs of derivations in $\mathbf{L(A)}$ based on the notion of critical pairs. The idea of critical pair was suggested by Voreadou [16], but the proof of her main theorem used an erroneous lemma. Here we give the formulation of the corrected theorem that was proved in [14]. Without loss of generality (due to theorems 8, 3), and to simplify the formulations we shall consider only the case of balanced pure 2-sequents.

DEFINITION 12. *A pair of derivations of the same balanced pure 2-sequent S is critical if*

(1) $d_1 \equiv \dfrac{\Gamma, A' \multimap I \overset{d'_1}{\to} A \quad I \overset{1_I}{\to} I}{\Gamma, A' \multimap I, A \multimap I \to I} \multimap\to,\ d_2 \equiv \dfrac{\Gamma, A \multimap I \overset{d'_2}{\to} A' \quad I \overset{1_I}{\to} I}{\Gamma, A' \multimap I, A \multimap I \to I} \multimap\to$
, *perm*;

(2) *a cut-free derivation of S can end only by some application of $\multimap\to$;*

(3) *the derivations d'_1, d'_2 are not $\equiv$-equivalent to derivations ending by $\multimap\to$.*
The pair is minimal if Γ does not contain single atoms as its members.

Let α be some substitution of I for variables. In [14] the "substitutions with purification" were defined. Let $d : \Gamma \to A$ be a derivation of some 2-sequent. Then $\alpha * d$ is the derivation obtained from d by α and *cuts* with isomorphisms that will make its final sequent pure. The derivation $\alpha * d$ is defined up to $\equiv$, but its final sequent is defined without ambiguity.

THEOREM 13. *(Cf. [14].) Let d_1, d_2 be derivations of a balanced sequent $\Gamma \to A$ and d'_1, d'_2 the corresponding derivations of a balanced pure 2-sequent. Then $d_1 \equiv d_2$ iff there exists a substitution α of I for variables such that $\alpha * d'_1, \alpha * d'_2$ is a minimal critical pair*[10].

4 The "Triple-Dual" Conjecture.

CONJECTURE 14. *Commutativity of the triple-dual diagram implies commutativity of all the diagrams of canonical maps $f, g : A \to B$ with balanced $A \to B$. More precisely: let $\sim$ be the smallest equivalence relation that satisfies all axioms of SMCC, is substitutive and the triple-dual diagram is commutative w.r.t. $\sim$. Then for all $f, g : A \to B$ with balanced $A \to B$ in $\mathbf{F(A)}$ we have $f \sim g$.*

An argument in favor of this conjecture is that the following theorem holds.

THEOREM 15. *(Soloviev, 1990 [13].) If $\sim$ is the smallest equivalence relation that satisfies all the axioms of SMCC, is substitutive, the triple-dual diagram is commutative w.r.t. $\sim$, and for all f, g and atom a*
(*) $[a \multimap I/a]f \sim [a \multimap I/a]f \Rightarrow f \sim g$,
then $f \sim g$ for all $f, g : A \to B$ in $\mathbf{F(A)}$ with the same graph[11].

---

[10]The conditions on the left premises that require verification of equivalence are applied to the (finite number of) derivations with smaller final sequent. This theorem may be used recursively to obtain deciding algorithms for $\equiv$. In [15] an algorithm of low polynomial complexity was described.
[11]Equivalently: with balanced $A \to B$.

As recently checked Antoine El Khoury, the commutativity of the triple-dual diagram implies (without the assumption (*)) the commutativity of all diagrams $f, g : A \to B$ with balanced $A \to B$ containing no more than 3 variables.

5 Main Results

Commutativity of the diagrams considered below **does not imply** the commutativity of the triple-dual diagram, so the equivalence relation generated by these identities are between $\equiv$ (minimal relation) and the relation generated by commutativity of the triple-dual.

First non-trivial "intermediate" equation was obtained due to a suggestion of M. Spivakovsky, developed later by L. Mehats and S. Soloviev [10].

The diagram (3) studied in [10] was obtained from

$$(2) \quad f, g : (a \otimes b \multimap I), ((b \multimap I) \multimap I), ((a \multimap I) \multimap I) \overset{\to}{\to} I$$

by *cut* with (unique) $h : (((a \multimap I) \otimes (b \multimap I)) \multimap I) \multimap I \to (a \otimes b \multimap I)$.

Let k be a field, $k[x, y]$ the related polynomial ring in two variables, $I = k[x, y]/(x^2, xy, y)$ and $M(k, I)$ the SMCC generated by I and k (as an I-module). It was shown in [10] that in $M(k, I)$ (3) is commutative while (2) and (1) are not (lemma 5.8). So, if we add to the axioms of SMCC the equation corresponding to (3), the diagrams (2) and (1) will remain non-commutative.

In this paper we describe certain sequence $D_2, ..., D_k, ..., D_m, ...$ of diagrams and certain models K_k such that in K_k the diagrams $D_2, ..., D_k$ are not commutative and there exists $m > k$ such that $D_m, ...$ are commutative (we don't know whether $D_k, ..., D_{m-1}$ are commutative).

Below we shall write A^* instead of $A \multimap I$. Let A^n denote the n-th "tensor power" of an object A, $A^n = (A \otimes ...) \otimes A$, and f^n the n-th "tensor power" of a morphism f, $f^n = (f \otimes ...) \otimes f$. For example, $e_{aI}^n : (a^* \otimes a)^n \to I^n$.

Let b_I^n be defined by $b_I^1 = b_I : I \otimes I \to I,$, ..., $b_I^n = b \circ (b_I^{n-1} \otimes 1_I) : I^{n+1} \to I$.

To obtain the diagrams $D_2, ..., D_k, ...$ we notice that there exists

$$h_k : (((a \multimap I)^k \multimap I) \multimap I) \to (a^k \multimap I).$$

In **F(A)** $h_k = \pi_{((a^*)^k)^{**} a^k I}(1_{((a^*)^k)^{**}} \otimes \pi_{a^k(a^*)^k I}(b_I^{k-1} \circ (e_{aI}^k \circ \xi)))$ Here ξ is an appropriate central isomorphism[12].

---

[12] In **L(A)** to h_k corresponds the following derivation:

$$\frac{\dfrac{a \to a \quad I \to I \qquad a \to a \quad I \to I}{a, a \multimap I \to I \qquad a, a \multimap I \to I}}{a, a \multimap I, a, a \multimap I \to I}$$

$$\frac{...}{\dfrac{a, a \multimap I, ..., a, a \multimap I \to I}{a, ..., a, (a \multimap I), ..., (a \multimap I) \to I}}$$

$$...$$

$$\dfrac{\dfrac{a^k, (a \multimap I)^k \to I}{(a \multimap I)^k, a^k \to I}}{\dfrac{a^k \to ((a \multimap I)^k \multimap I) \quad I \to I}{\dfrac{a^k, ((a \multimap I)^k \multimap I) \multimap I \to I}{((a \multimap I)^k \multimap I) \multimap I \to (a^k \multimap I)}}}$$

The diagram

$$(D_2^0)\ f_2^0, g_2^0 : (a^2)^*, a^{**}, a^{**} \overset{\rightarrow}{\rightarrow} I$$

is obtained from the diagram (2) by substitution of a for b (in other words, by identification of variables a and b). The diagram D_2

$$(D_2)\ f_2, g_2 : ((a^*)^2)^{**}, a^{**}, a^{**} \rightarrow I$$

is obtained from D_2^0 by *cut* with (the derivation corrsponding to) h_2.

We define[13] the diagram D_m^0, $m \geq 2$, as the result of substitution of a^{m-1} for b into diagram (2). The morphisms obtained from f, g by this substitution are denoted f_m^0, g_m^0.

The diagram D_m is obtained from D_m^0 by *cut* with h_m (f_m, g_m are resulting derivations):

$$(D_m)\ f_m, g_m : ((a^*)^m)^{**}, a^{**}, (a^{m-1})^{**} \overset{\rightarrow}{\rightarrow} I.$$

In order to obtain the models K_k we shall consider certain SMCCs of commutative semimodules over commutative semirings.

For all basic definitions concerning semirings and semimodules see [4]. Below we shall denote the "addition" of the semiring I by $+$ and "multiplication" by $*$. In case of a semimodule M we shall denote by $+_M$ its additive operation and $*_M$ the action of I on M; the index M will often be omitted.

PROPOSITION 16. *I-semimodiles over a commutative semiring I and their homomorphisms form an SMCC with tensor product $\otimes$ and internal hom-functor $\multimap$ defined in usual way.*

We consider the categories of semimodules over the semiring $I_n = \{0, ..., n\}$ with max as addition and with "bounded multiplication" $*$ as multiplication: $p * q = p \cdot q$ if $p \cdot q < n$ and $p * q = n$ otherwise.

Obviously I_n is a commutative semiring. When n is irrelevant or clear from the context it will be omitted.

Notice that in this category $b_M : M \otimes I \rightarrow M$ is defined by $b_M(x \otimes p) = p * x$, in partucular if $M = I$ then $b_I(p_1 \otimes p_2) = p_1 * p_2$. For an element $p_1 \otimes ... \otimes p_n \in I \otimes ... \otimes I$, $b_I^{n-1}(p_1 \otimes ... \otimes p_n) = p_1 * ... * p_n$.

We shall consider the semimodules M over I that have some additional properties.

(Top) There is a "top" element $T_M \in M, T_M \neq 0_M$ such that for all $x \in M$, $x + T_M = T_M + x = T_M$, if $x \in M, x \neq 0_M$ then $n * x = T_M$ and if $0 \neq k \in I$ then $k * T_M = T_M$.

Obviously, I itself does satisfy these conditions if we take $T_I = n$. For I_s considered as a semimodule over I s must be not greater than n.

LEMMA 17. *Let M_1, M_2 be two semimodules over I with top elements T_1 and T_2 respectively. Let $f : M_1 \rightarrow M_2$ be a homomorphism of semimodules, different from constant 0. Then $f(T_1) = T_2$ and for $x \in M_1$, $x \neq 0$, $f(x) \neq 0$. As a consequence, two morphisms $f, g : M_1 \rightarrow M_2$ always coincide at least on 0 and T_1.*

---

[13] The index m will correspond to the number of factors in tensor products.

DEFINITION 18. Let us call an I-semimodule r-reducible for some $r \in I, 1 < r \leq n$ if for every $x \in M$ $r * x = T_M$.

EXAMPLE 19. Let $M = I_2 = \{0,1,2\}$ considered as a semi-module over the semi-ring $I_4 = \{0,1,2,3,4\}$ (with the ordinary multiplication "bounded by 2" as the action). It satisfies (**Top**) with $T_M = 2$ and is 2-reducible. Of course M of this example is also 3- and 4-reducible.

We shall consider the semimodules M such that

(**r-red**) M is r-reducible for some $r \in I, 1 < r < n$.

LEMMA 20. Let M_1 and M_2 be an semi-modules over I satisfying (**top**) furthemore M_1 is r-reducible then :

(r_1) $M_1 \multimap M_2$ is r-reducible.

(r_2) $M_1 \otimes M_2$ is r-reducible.

(r_3) $M_1 \multimap I$ is r-reducible.

THEOREM 21. The category generated by $\otimes$ and $\multimap$ from I and a set of some given r-reducible semimodules is a SMCC.
And all the objects of this category have a **Top** element and are r-reducible for example $I = \{0,1,2,3,4\}$ and $M = \{0,1,2\}$, exept the semimodules isomorphic to I.

This theorem will permit us to consider the SMCC generated by $\otimes$ and $\multimap$ from I and some given semimodule, for example $I = \{0,1,2,3,4\}$ and $M = \{0,1,2\}$ and be sure that all the objects of this category will have a top element and be r-reducible[14].

LEMMA 22. Let M be an r-reducible semimodule over I and $f : M \to I$ (we may say also that $f \in M \multimap I$). Then for all $x \in M$ $f(x) \geq n/k$.

Let h_m^- denote $a^m \otimes (a^*)^m \xrightarrow{\xi} (a^* \otimes a)^m \xrightarrow{e_{aI}^m} I^m \xrightarrow{b_I^{m-1}} I$ Consider the SMCC K of semimodules satisfying (**top**) and (**r-red**) over I. Now we can easily prove the following lemma.

LEMMA 23. Let n,r be as above, m such that $(n/r)^m \geq n$ and v an interpretation defined by $v(a) = M \in Ob(K)$. Then the morphism $|h_m^-|$: $(M^m) \otimes (M^*)^m \to I$ takes the value 0 if its argument is 0 and $T_I = n$ otherwise.

COROLLARY 24. Under the same conditions, the morphism $|h_m|$ takes only two values: 0 when its argument is 0 and $T_{(M^m)^*}$ otherwise (for every $M \in Ob(K)$).

LEMMA 25. For all n,r,m as in lemma 23 and every interpretation v in the SMCC K of I-semimodules satisying **top** and **r-red**, the diagram $|D_m|_v$ is commutative. So it is commutative with respect to the relation $\sim_K$.

---

[14]These properties are easier to verify and use than, for example, the property of being a semilattice. Notice that the structure of the objects of the SMCC generated by I and M is not necessarily simple. For example, the semimodule $M \multimap I$ will be generated (non freely) by $[2] : 1 \mapsto 2, [3] : 1 \mapsto 3, (M \multimap I) \otimes (M \multimap I)$ will have four generators, etc.

Let $2 \leq k$, $n = 3^k + 1, l = n/2$. Let $I = I_n$, $M = \{0, 1, ..., l\}$. Notice that M is $n/2$-reducible. Consider the SMCC K_k of all semimodules satisfying **top** and **l-red** generated by I and M.

LEMMA 26. *Let the interpretation v be defined by $v(a) = M \in Ob(K_k)$. The diagrams $|D_2|_v, ..., |D_k|_v$ are non-commutative.*

To prove this lemma we verify that the image $Im(|h_j|_v)$ contains a certain element p different from 0 and $T_{(M^j)^*}$, and there exist elements $\psi \in M^{**}, \varphi \in (M^{j-1})^{**}$ such that the two arrows of the diagram D_j^0 take different values on the argument (p, ψ, φ) ($2 \leq j \leq k$).

THEOREM 27. *There exist infinitely many different varieties of SMCC. Each of these varieties is defined by taking a (single) diagram D_m of the sequence above as a new axiom.*

To prove this theorem we use the fact that for any $k \geq 2$ there exists m (it is enough to take $m \geq log_2(3^k + 1)$) such that D_k cannot belong to the smallest equivalence relation generated by D_m. In other words, the commutativity of D_m does not imply the commutativity of D_k (by lemmas 25, 26).

6 Conclusion

To verify the commutativity of a diagram in a model can be very difficult. Instead of verifying commutativity of diagrams case by case one may hope that if one diagram is commutative then the commutativity of another will follow.

By theorem 27 there exist infinitely many distinct equivalence relations $\sim_K$ on derivations of **F(A)** (or **L(A)**). This fact shows the importance of the study of dependency of diagrams in SMCC and closed categories with weaker structure.

Taking into account the existence of efficient deciding algorithms for commutativity of diagrams in free closed categories, the first step would be to verify whether a diagram is commutative in the free case. If it is not commutative, the study of dependency may follow. In particular, one may find some "key" diagrams whose commutativity will imply the commutativity of others (cf. the axiomatization of equivalence relations by critical pairs considered in [10]).

We believe that this direction of research will provide a new and promising application of proof theory to categorical algebra.

We would like to express our thanks to Kosta Dosen and Zoran Petric, who helped clarify many points and improve considerably the presentation of our work, to Nikolai Vasilyev for fruitful discussions concerning its algebraic aspects and to Grigory Mints who initiated in 1970es Sergei Soloviev to this direction of research.

BIBLIOGRAPHY

[1] K. Dosen and Z. Petric. The maximality of the typed lambda calculus and of cartesian closed catgeories. Belgrade, *Publications de l'Institut Mathématique*, Nouvelle Série, tome 68(82) (2000),pp.1-19.

[2] S. Eilenberg and G. M. Kelly. A generalization of the functorial calculus.-*J.of Algebra*, 1966.

[3] G.-Y. Girard, Y. Lafont. Linear logic and lazy computation. In: Proc.TAPSOFT 87 (Pisa), v.2, p.52-66, LNCS v.250 , 1987.

[4] J. Golan. Semirings and their applications. Kluwer Acad. Publishers, Dordrecht, 1999.
[5] G.M. Kelly and S. Mac Lane. Coherence in Closed Categories. *Journal of Pure and Applied Algebra*, 1(1):97–140, 1971.
[6] G.M.Kelly. A cut-elimination theorem. *Lecture Notes in Mathematics*, 281 (1972), pp 196-213.
[7] J. Lambek. Deductive Systems and Categories. I. *Math. Systems Theory*, 2, 287-318, 1968.
[8] J. Lambek. Deductive Systems and Categories II. Lect. Notes in Math., v.86, Springer, 1969, pp. 76-122.
[9] J. Lambek. Deductive Systems and Categories III. Lect. Notes in Math., v.274, Springer, 1972, pp. 57-82.
[10] L. Mehats, S. Soloviev. Coherence in SMCCs and equivalences on derivations in IMLL with unit. *Annals of Pure and Applied Logic*,
[11] G.E. Mints. Closed categories and Proof Theory. *Journal of Soviet Mathematics*, 15, 45–62, 1981.
[12] G. E. Mints. Category theory and proof theory (in Russian), in: *Aktualnye voprosy logiki i metodologii nauki*, Naukova Dumka, Kiev, 1980, 252-278. (English translation, with permuted title, in: G.E. Mints. Selected Papers in Proof Theory, Bibliopolis, Naples, 1992.)
[13] S.V. Soloviev. On the conditions of full coherence in closed categories. *Journal of Pure and Applied Algebra*, 69:301-329, 1990.
[14] S. Soloviev. Proof of a conjecture of S. Mac Lane. *Annals of Pure and Applied Logic*, 90 (1997), pp.101-162.
[15] S. Soloviev, V. Orevkov. On categorical equivalence of Gentzen-style derivations in IMLL. *Theoretical Comp. Science*, 303 (2003), pp. 245-260.
[16] R. Voreadou. Coherence and non-commutative diagrams in closed categories. *Memoirs of the AMS*, v. 9, issue 1, N 182, Jan. 1977.

The Substitution Method Revisited
W.W. TAIT

It is a pleasure for me to contribute a paper in honor of Grisha Mints. In view of his interesting work involving the epsilon-substitution method, I am returning to some work I did on that topic in 1960-1. I was primarily trying to understand the *concept* behind Ackermann's consistency proof for first-order number theory (1940), which for me was too heavy on syntax and too light on ideas. I found a satisfactory treatment the basic idea of which, I think, could have been behind Ackermann's approach.[1] But, aside from presenting my work on this topic in a lecture at the Eighth Logic Colloquium held at Oxford in 1963, I put it aside in favor of trying to extend it to full second-order number theory, a project that came to an abrupt halt in the winter of 1961-2. The work on the epsilon-calculus for first-order systems of arithmetic and predicate logic finally appeared in 1965 in two somewhat bloated papers (Tait 1965a, Tait 1965b). I am happy to have this occasion to present a leaner and cleaner exposition of that work. I will end with a brief discussion of why I believed that the method I have applied for first-order number theory would not extend to second-order number theory.

1 Preliminaries

Let
$$x, y_1, y_2, \ldots$$
be a fixed list of distinct individual variables.

Definition A *matrix* and in particular an *n-matrix* is a quantifier-free formula $A(x, y_1, \ldots, y_n)$ in a first-order language, with $n \geq 0$, whose free variables are $x, y_1, \ldots, y_n$ and such that for each i

- Every term in A other than $y_1, \ldots, y_n$ contains x.
- y_i has exactly one occurrence in A.
- The occurrence of y_i is to the left of y_{i+1} for $(i < n)$. □

Every quantifier-free formula $B(x)$ of a first-order theory is uniquely of the form $B(x) = A(x, \mathbf{t})$, where $A(x, \mathbf{y})$ is a matrix, $\mathbf{y} = y_1, \ldots, y_n$, and the $\mathbf{t} = t_1, \ldots, t_n$ do not contain x.

---
[1] On the other hand, Mints himself has lectured on an approach to eliminating epsilon-terms in PA which he seems to have extended to the theory of one elementary inductive definition. This method has some claim to be the development of the intuition behind Hilbert's original belief that epsilon-terms could be eliminated.

Let $\mathcal{T}$ be classical first-order number theory. For simplicity, we take the non-logical axioms of $\mathcal{T}$ to be the axioms of primitive recursive arithmetic PRA. These include the axioms for the predecessor function $pred$

$$pred\ 0 = 0 \qquad pred(x+1) = x$$

The logical axioms are those for the propositional connectives and existential quantifier introduction

$$A(t) \implies \exists x A(x)$$

The logical rules of inference are modus ponens and the rule of existential quantifier elimination

$$\frac{B(x) \implies C}{\exists x B(x) \implies C}$$

where x is not in C. Finally, there is the rule of mathematical induction:

$$\frac{B(0) \quad B(x) \implies B(x+1)}{B(s)}$$

Definition. For each $k < \omega$, the theory $\mathcal{T}^k$ is defined by induction.

$$\mathcal{T}^0 = \mathcal{T}.$$

$\mathcal{T}^{k+1}$ results from adding for each n-matrix $A = A(x, \mathbf{y})$ of $\mathcal{T}^k$ ($0 \le n$) but not of $\mathcal{T}^m$ for any $m < k$

- To the language of $\mathcal{T}^k$ a distinct new n-ary function constant f_A. f_A is called the *Skolem function constant* for A. (0-ary function constants are just individual constants.)

- To the axioms of $\mathcal{T}^k$ all substitution instances of

$$A(x, \mathbf{y}) \implies A(f_A \mathbf{y}, \mathbf{y})$$

called the *axioms of the first kind* for f_A.

- In the case of arithmetic, we also add all substitution instances of

$$A(x, \mathbf{y}) \implies f_A \mathbf{y} \ne x + 1$$

called the *axioms of the second kind* for f_A.

- If $A(x)$ is an axiom of $\mathcal{T}^k$ and s is a term of $\mathcal{T}^{k+1}$, then $A(s)$ is an axiom of $\mathcal{T}^{k+1}$. $\square$

Set

$$\mathcal{T}^* = \bigcup_k \mathcal{T}^k.$$

Note that the axioms of each $\mathcal{T}^k$ and hence those of $\mathcal{T}^*$ are closed under substitutions. $\mathcal{T}^*$ is sometimes called the "ϵ-calculus" and the substitution method the "ϵ-substitution method" because Hilbert wrote $\epsilon x B(x)$ for $f_A \mathbf{t}$ when A is a matrix and $B(x) = A(x, \mathbf{t})$.

Definition. The *rank* of a Skolem function constant f_A in $\mathcal{T}^*$ is the least n such that f_A is in $\mathcal{T}^{n+1}$.

Thus, if f_A is of rank n, then all of the Skolem function constants in the matrix A are of rank $< n$. If, for every matrix $A(x, \mathbf{y})$ in T^*, we abbreviate

$$\exists x A(x, \mathbf{t}) := A(f_A \mathbf{t}, \mathbf{t})$$

then every axiom

$$B(s) \implies \exists x B(x)$$

transforms into an axiom of the first kind

$$B(s) \implies B(f_A \mathbf{t})$$

for a suitable Skolem function f_A and each inference

$$\frac{B(x) \implies C}{\exists x B(x) \implies C}$$

where x is not in C, becomes the substitution

$$\frac{B(x) \implies C}{B(f_A(\mathbf{t})) \implies C}$$

Thus, we may consider $\mathcal{T}^*$ to be a quantifier-free system.

Consider now instances of the rule of mathematical induction

$$\frac{B(0) \qquad B(x) \implies B(x+1)}{B(s)}$$

From $\neg B(s)$ we can infer $\neg B(f_A \mathbf{t})$, where $\neg B(x) = A(x, \mathbf{t})$, by an axiom of the first kind. So it follows from the first premise of the induction that $f_A \mathbf{t} \neq 0$ and so $f_A \mathbf{t} = pred(f_A \mathbf{t}) + 1$. By the second premise of the induction then, $\neg B(pred(f_A \mathbf{t}))$, which contradicts an axiom of the second kind for f_A. Thus we deduce $B(s)$.

In this way, given a deduction in $\mathcal{T}$ of a formula C, we obtain a deduction of C in the quantifier-free system $\mathcal{T}^*$ which contains, along with the logical and non-logical axioms $\mathcal{T}$, closed under substitution of the new terms of $\mathcal{T}^*$, only the new axioms of the first and second kind for the Skolem functions. The only rule of inference is *modus ponens*.

2 Eliminating Skolem Functions in First-Order Number Theory

Let $\mathcal{D}$ be a deduction in $\mathcal{T}^*$ of a formula C. We can assume that all of the free variables in $\mathcal{D}$ are in C, since all others can be replaced throughout by 0. Let **z** be a list of the free variables in C. All of the axioms in $\mathcal{D}$ other than the axioms for Skolem functions are obtained from axioms of $\mathcal{T}$ by substituting terms of the form $f_A \mathbf{t}$ for variables. Let

$$f_{A_1}, \ldots, f_{A_m}$$

be all of the distinct Skolem constants occurring in $\mathcal{D}$, listed in order of non-decreasing rank, so that f_{A_j} occurs in the matrix $A_i(x, \mathbf{y})$ only if $i > j$. The axioms occurring in $\mathcal{D}$ for Skolem functions are of course finite in number. We shall call these axioms the *critical formulas* of $\mathcal{D}$.

We will show that, for a certain extension $\mathcal{T}^+$ of the quantifier-free part PRA of $\mathcal{T}$, for each finite set of axioms for Skolem functions

$$f_{A_1}, \ldots, f_{A_m}$$

we can define numerical-valued functions

$$\phi_1, \ldots, \phi_m$$

of **z** such that the result of replacing each term $f_i \mathbf{t}_i$ by $\phi_i \mathbf{z} \mathbf{t}_i$ for $i = 1, \ldots, m$ transforms each of the given axioms for Skolem functions into a theorem of $\mathcal{T}^+$. It will follow that the result of this substitution in C is a theorem of $\mathcal{T}^+$.

$\mathcal{T}^+$ is the quantifier-free system $PRA^2_{\epsilon_0}$ of second-order primitive recursive arithmetic with definition by recursion on each ordinal $\alpha < \epsilon_0$. We give a brief description of this system; but our construction below of the required ϕ_i will not include a detailed formalization in $\mathcal{T}^+$.

$\mathcal{T}^+$ contains variables of types $\omega^n \Longrightarrow \omega$ for $n \geq 0$. (When $n = 0$, these are the numerical variables.) Functions of one or more variables ranging over these types, whose values are of one of these types, are introduced by explicit definition, primitive recursion and by recursion on some $\alpha < \epsilon_0$. The order type ϵ_0 is represented in some standard way by a primitive recursive ordering $\prec$ of ω with least element 0. I will use lower case Greek letters to denote 'ordinals', i.e. natural numbers in their role as ordinals $< \epsilon_0$. Corresponding to the addition of the ordinal numbers represented by α and β, there is the primitive recursive function $\alpha \oplus \beta$ and, corresponding to raising the ordinal represented by α to the power 2 is the primitive recursive function E^α. Define

$$[x, y] = \begin{cases} x & \text{if } x \prec y \\ 0 & \text{otherwise} \end{cases}$$

Then definition by recursion on a limit ordinal $\alpha < \epsilon_0$ has the form

$$\Phi \mathbf{g} 0 = \Psi \mathbf{g}$$

and for $0 \prec x \prec \alpha$

$$\Phi \mathbf{g} x = \Xi \mathbf{g} x (\Phi \mathbf{g} [\Theta \mathbf{g} x, x])$$

and for $\alpha \preceq x$

$$\Phi \mathbf{g} x = 0_\tau.$$

Here $\mathbf{g}$ is a list of distinct variables of arbitrary types. $\Phi \mathbf{g} x, \Psi \mathbf{g}, \Xi \mathbf{g} x u$ are all of the same type $\tau = \omega^k \Longrightarrow \omega$ ($k \geq 0$) when u is a variable of type τ. $\Theta \mathbf{g} x$ is of type ω and 0_τ is some standard object of type τ, say the function with constant value 0. The formulas of $\mathcal{T}^+$ are built up from equations between terms of the same type by means of the propositional connectives. The axioms of $\mathcal{T}^+$ are those of identity, zero and successor, the defining equations of the function constants of each type and the axioms of propositional logic. The rules of inference are modus ponens, the rule of mathematical induction, and the rule of substitution

$$\frac{A(s) \qquad s\mathbf{x} = t\mathbf{x}}{A(t)}$$

where s and t are terms of some type $\omega^n \Longrightarrow \omega$ with $n > 0$ and $\mathbf{x}$ is a list of n distinct numerical variables that occur in neither s nor t.

$PRA^1_{\epsilon_0}$ is the result of restricting the variables to numerical variables and the function constants to types $\omega^n \Longrightarrow \omega$. The system $PRA^2_{\alpha+1}$ is a conservative extension of $PRA^1_{2^\alpha+1}$ and, in particular, $PRA^2_{\epsilon_0}$ is conservative over $PRA^1_{\epsilon_0}$. (Tait 1965a, Theorem 4)

Remark Our construction below would remain valid if we took the original system $\mathcal{T}$ to be, not the result of adding quantification to PRA, but the result of adding quantification to $PRA^1_{\epsilon_0}$. But then, from many points of view, that is the natural system of first-order number theory. □

$f = f_{A_m}$ has the highest rank of Skolem function constants in $\mathcal{D}$. The Skolem function constants in $A(x, \mathbf{y}) = A_m(x, \mathbf{y})$ are among $f_{A_1}, \ldots, f_{A_{m-1}}$. Let $\mathbf{g} = \mathbf{z}, f_{A_1}, \ldots, f_{A_{m-1}}$. The critical formulas in $\mathcal{D}$ for f then are substitution instances of formulas of the form

$$A(s_i(f, \mathbf{g}), \mathbf{y}) \Longrightarrow A(f\mathbf{y}, \mathbf{y})$$

or

$$A(s_i(f, \mathbf{g}), \mathbf{y}) \Longrightarrow f\mathbf{y} \neq s_i(f, \mathbf{g}) + 1$$

for $i = 0, \ldots p$ for some $p < \omega$. Set

$$S f \mathbf{g} = Max_{i \leq p} s_i(f, \mathbf{g}).$$

Then it suffices to find f as a function of the $\mathbf{g}$ satisfying

$$f\mathbf{y} \simeq \mu x \leq S f \mathbf{g}.A(x, \mathbf{y})$$

where $\simeq$ means that the two terms are equal if the right hand side is defined, i.e. if there is an $x \leq S f \mathbf{g}$ such that $A(x, \mathbf{y})$. Call this the *principal semi-equation*.

At the first step, where f is the Skolem function constant of highest rank, the $s_i(f, \mathbf{g})$, and so $Sf\mathbf{g}$, are just terms of $\mathcal{T}^*$. But in order to proceed by induction, we need to be able to solve the principal semi-equation in a more general setting in which $S f \mathbf{g}$ is a functional of f and $\mathbf{g}$ in $PRA^2_{\epsilon_0}$.

We do that by first defining successive approximations $\theta_n = \Theta_n \mathbf{g}$ of a solution for f as follows:

$$\theta_0 \mathbf{y} = 0$$

$$\theta_{n+1} \mathbf{y} = \mu x \leq Max\{\theta_n \mathbf{y}, S\theta_n \mathbf{g}\} A(x, \mathbf{y}).$$

where the latter is understood to be 0 if no such x exists. If, as a function of $\mathbf{g}$, we can determine an n such that

$$S\theta_n \mathbf{g} = S\theta_{n+1} \mathbf{g}$$

then $f = \theta_n$ is a solution of the principal semi-equation. Call this the *principal equation*.

Remark The principal equation arises in another proof-theoretic context, namely, in deriving the so-called *no-counterexample interpretation* of B from the witnesses of the no-counterexample interpretation of A and $A \implies B$, when A and B are arithmetic formulas. (Gödel 1938a) refers to a solution of it using "Souslin's schema," meaning bar recursion, but does not give details. (Kohlenbach 1999) actually carries out the derivation using an extensional form of bar recursion. This is discussed in (Tait 2005). I will sketch here the derivation given in (Tait 1965a) where bar recursion is avoided using induction non ordinals $< \epsilon_0$.

In 1962 Paul Cohen showed me a handwritten manuscript in which he presented the present procedure for eliminating the Skolem functions in first-order number theory, intuitively applying bar recursion to solve the principal equation. He had at that time no prior knowledge of the work in Hilbert's school on the ϵ-calculus. □

Suppose that f is a numerical constant, so that $\mathbf{y}$ is null. In that case, it is immediate that $\theta_2 = \theta_1$ and so $f = \theta_1$ solves the principal equation. So we may assume that f is a function constant of some type $\omega^{m+1} \implies \omega$ and, by contracting the arguments, we can assume that $m = 0$. The nth *sequence number*

$$\bar{h} n$$

of h is defined to be h if h is a number and it is the usual sequence number $\langle h0, \ldots, h(n-1) \rangle^\#$ of $\langle h0, \ldots, h(n-1) \rangle$ of h if h is a numerical function of one variable. If it is a numerical function of m variables with $m > 1$, then $\bar{h} n = \bar{h}' n$, where h' is the function of one variable with $h' \langle x_1, \ldots, x_m \rangle^+ = h x_1 \cdots, x_m$ and $\langle x_1, \ldots, x_m \rangle^+$ denotes the standard bijection from ω^m onto ω. For sequences of numbers and numerical functions, set

$$\overline{h_0, \ldots, h_p} n = \langle \bar{h}_0 n, \ldots, \bar{h}_p n \rangle^+.$$

A numerical-valued function S in $\mathcal{T}^+$ of $f, \mathbf{g}$ can be associated with a triple consisting of:

- An ordinal $\alpha_S < \epsilon_0$
- A non-decreasing function $\Phi_S : \omega \Longrightarrow \omega$ in $\mathcal{T}^+$ and
- A function $\Psi_S : \omega \Longrightarrow \alpha_S$ in $\mathcal{T}^+$

such that the following are theorems of $\mathcal{T}^+$:

$$\Phi_S(\overline{f,g}n) = 0 \Rightarrow \Psi_S(\overline{f,g}(n+1)) \prec \Psi_S(\overline{f,g}n)$$

and

$$\Phi_S(\overline{f,g}n) > 0 \Rightarrow Sf\mathbf{g} = \Phi_S(\overline{f,g}(n+m)) - 1$$

for all m. In fact when S is defined in $PRA^2_{\epsilon_0}$ using only recursions on β, then $\alpha_S < \omega^\beta$. (Tait 1965a, §5)

Let our given S be so represented by α_S, Φ_S and Ψ_S. We show how to obtain a solution

$$n = N\mathbf{g}$$

of the principal equation in $\mathcal{T}^+$. Since

$$\Phi_S(\overline{\theta_n, \mathbf{g}}k) = 0 \Rightarrow \Psi_S(\overline{\theta_n, \mathbf{g}}(k+1)) < \Psi_S(\overline{\theta_n, \mathbf{g}}k) < \alpha_S$$

it follows that

$$r_n = \mu x[\Phi_S(\overline{\theta_n, \mathbf{g}}x) > 0]$$

is definable by recursion on α_S. For each $m \geq 0$

(1) $\quad S\theta_m\mathbf{g} = \Phi_S(\overline{\theta_m\mathbf{g}}r_m) - 1$

Let

$$m_{n,1} < \ldots < m_{n,p_n}$$

be all the $x < r_n$ such that $\theta_n x = 0$. Set

$$\gamma_{n,i} = \Psi_S(\overline{\theta_n, \mathbf{g}}(m_{n,i} + 1))$$

Thus

$$\gamma_{n,1} \succ \ldots \succ \gamma_{n,p_n}$$

Now set

$$\gamma_n = E^{\gamma_{n,1}} \oplus \cdots \oplus E^{\gamma_{n,p_n}} \prec E^{\alpha_S}.$$

Now assume that $S\theta_n\mathbf{g} \neq S\theta_{n+1}\mathbf{g}$. Then $\Phi_S(\overline{\theta_n, \mathbf{g}}r_n) \neq \Phi_S(\overline{\theta_{n+1}, \mathbf{g}}r_{n+1})$. So $\overline{\theta_n\mathbf{g}}r_n \neq \overline{\theta_{n+1}\mathbf{g}}r_n$, since otherwise $r_{n+1} = r_n$ and so, by (1), $S\theta_n\mathbf{g} = S\theta_{n+1}\mathbf{g}$. Hence $\overline{\theta_n}r_n \neq \overline{\theta_{n+1}}r_n$. Let x be the least number such that $\theta_n x \neq \theta_{n+1}x$. Then $x < r_n$ and $\theta_n x = 0$. I.e. $x = m_{n,i}$ for some $i = 1, \ldots, p_n$. Thus, for $j < i, m_{n+1\,j} = m_{n,j}$ and so $\gamma_{n+1,j} = \gamma_{n,j}$. If there is no $m_{n+1,i}$, then clearly $\gamma_{n+1} \prec \gamma_n$. If $m_{n+1,i}$ exists, then it is $> m_{n,i}$ and so the sequence

$$\langle \overline{\theta_n\mathbf{g}}0, \ldots, \overline{\theta_n\mathbf{g}}m_{n,i} \rangle = \langle \overline{\theta_{n+1}\mathbf{g}}0, \ldots, \overline{\theta_{n+1}\mathbf{g}}m_{n,i} \rangle$$

is a proper initial subsequence of $\langle \overline{\theta_{n+1}\mathbf{g}}0, \ldots, \overline{\theta_{n+1}\mathbf{g}}m_{n+1,i}\rangle$. Hence $\gamma_{n+1,i} \prec \gamma_{n,i}$ and therefore $\gamma_{n+1} \prec \gamma_n$. We have proved that

$$S\theta_n \mathbf{g} \neq S\theta_{n+1}\mathbf{g} \Rightarrow \gamma_{n+1} \prec \gamma_n.$$

Hence

$$N\mathbf{g} = \mu x S\theta_x \mathbf{g} = S\theta_{x+1}\mathbf{g}$$

can be defined by recursion on E^{α_S}. Thus our solution of the principal semi-equation is

$$\Theta_{N\mathbf{g}}\mathbf{g}.$$

Substituting this for f in the critical formulas, we have reduced the number of Skolem function constants by one.

3 Eliminating Skolem Functions in the Case of Predicate Logic

Now we consider the case in which $\mathcal{T}$ is a first-order theory whose axioms are quantifier-free and closed under substitution. $\mathcal{T}^*$ is defined exactly as above, except that there are no axioms of the second kind for Skolem functions.

Again, let $\mathcal{D}$ be a deduction of C in $\mathcal{T}^*$ which we can assume to contain no variables other than those in C, since the others can be replaced by some individual constant. (If there are none, add one, carry out the following elimination procedure and then replace the constant by a variable.) Let $\mathbf{z}$ be a list of the distinct variables in C. All of the axioms in $\mathcal{D}$ other than the axioms for Skolem functions are obtained from valid numerical formulas by substituting terms containing Skolem function constants for variables and let

$$f_{A_1}, \ldots, f_{A_m}$$

be all of the distinct Skolem function constants occurring in $\mathcal{D}$, listed in order of non-decreasing rank. Only axioms for these constants of the first kind occur in $\mathcal{D}$.

Again we define the functions

$$\phi_1, \ldots, \phi_m$$

such that replacing each term $f_i \mathbf{t_i}$ by $\phi_i \mathbf{z} \mathbf{t_i}$ for $i = 1, \ldots, m$ transforms the axioms for the f_{A_i} into theorems of a suitable quantifier-free extension $\mathcal{T}^+$ of the quantifier-free part of $\mathcal{T}$. The result of this substitution in C will then be a theorem of $\mathcal{T}^+$.

$\mathcal{T}^+$ contains variables over individuals and over individual-valued functions of n individuals ($n > 0$) and it contains constants for functions of these variables whose values may be individuals or functions from n individuals to individuals. These constants are introduced by explicit definition or by definition by cases. The remaining axioms are the quantifier-free axioms of $\mathcal{T}$, closed under substitution for individual terms. Definition by cases

$$\Phi \mathbf{z} = \begin{cases} \Psi \mathbf{z} & \text{if } A(\mathbf{z}) \\ \Xi \mathbf{z} & \text{if } \neg A(\mathbf{z}) \end{cases}$$

is obviously expressed by quantifier-free axioms. Since $B(\Phi \mathbf{z})$ is equivalent in s to

$$[A(\mathbf{z}) \implies B(\Psi \mathbf{z})] \wedge [\neg A(\mathbf{z}) \implies B(\Xi \mathbf{z})]$$

$\mathcal{T}^+$ is conservative over the quantifier-free part of $\mathcal{T}$.

Again, $f = f_{A_m}$ has the highest rank of Skolem function constants in $\mathcal{D}$. Let $\mathbf{g} = \mathbf{z}, f_{A_1}, \ldots, f_{A_{m-1}}$.

$$C(f, \mathbf{g}, \mathbf{y}) = \bigvee_{i=1}^{p} A_m(S_i(f, \mathbf{g}), \mathbf{y}) \implies A_m(f\mathbf{y}, \mathbf{y})$$

expresses the conjunction of axioms

$$A_m(S_i(f, \mathbf{g}), \mathbf{y}) \implies A_m(f\mathbf{y}, \mathbf{y})$$

for f. Every axiom for f in $\mathcal{D}$ results by substitution for the $\mathbf{y}$ in one of these axioms. Let θ be the constant function defined by

$$\theta \mathbf{y} = c$$

where c is some individual constant of $\mathcal{T}$. By repeated use of definition by cases, we define $\Phi \mathbf{g}$ to be $\theta \mathbf{y} = c$, if $C(\theta, \mathbf{g}, \mathbf{y})$ and, if $\neg C(\theta, \mathbf{g}, \mathbf{y})$, to be $S_i(\theta, \mathbf{g})$ for the least $i \leq p$ such that $A_m(S_i(\theta, \mathbf{g}), \mathbf{y})$. It follows that

$$C(\Phi \mathbf{g} \mathbf{y}, \mathbf{y})$$

is a theorem of $\mathcal{T}^+$. Substitute $\Phi \mathbf{g}$ for f in the axioms for the remaining Skolem functions $f_{A_1}, \ldots, f_{A_{m-1}}$ and now solve for this shorter list of Skolem functions.

Let $\mathcal{D}$ be a deduction in $\mathcal{T}$ of

$$C = \exists x_1 \forall y_1 \cdots \exists x_n \forall y_n A(x_i, y_j)$$

We obtain from $\mathcal{D}$ a deduction of

$$\exists x_1 \cdots x_n A(x_i, g_j x_1 \cdots x_j)$$

where the g_j are distinct new function variables, or in T^*

$$A(s_i, g_j s_1 \cdots s_j)$$

where the s_i are terms containing Skolem function constants. In $\mathcal{T}^+$ we then obtain

$$A(t_i, g_j t_1 \cdots t_j)$$

where the t_i are defined by multiple cases. Eliminating them we obtain a deduction in $\mathcal{T}$ of

$$\bigvee_{k_1 < r_1} \cdots \bigvee_{k_n < r_n} A(u_{k_i}, f_j u_{k_1} \cdots u_{k_n})$$

This is the *First ϵ-Theorem* of (Hilbert & Bernays 1939). (Tait 1965b, §6.3)

4 Failure of Continuity of Third-order Computable Functions.

We note that there is a third order function
$$F : [[\omega \Longrightarrow \omega] \Longrightarrow \omega] \Longrightarrow \omega$$
and a computable sequence $\langle \phi_n \mid n < \omega \rangle$ such that
$$\phi_n : [\omega \Longrightarrow \omega] \Longrightarrow \omega$$
and
$$\phi_n g > 0 \Longrightarrow \phi_{n+1} g = \phi_n g$$
but with
$$F\phi_n \neq F\phi_{n+1}$$
for all n. This casts some doubt on whether the method used above for eliminating ϵ-terms in first-order number theory will extend to second-order number theory. Of course, to be a counterexample, we would have to show that such 'discontinuous' third-order functions actually arise in solving for Skolem functions in second-order number theory. As far as I know, that question is open. (See also (Tait 1965b, the Remark in §5).)

Define
$$f : \omega \times \omega \Longrightarrow \omega$$
by
$$fxy = f_x y = \begin{cases} 1 & \text{if } x \geq y \\ 0 & \text{if } x < y \end{cases}$$

Let $\phi_n g$ be defined for each $n < \omega$ and $g : \omega \Longrightarrow \omega$ by
$$\phi_n g = \begin{cases} \mu y \leq n(gy = 0) & \text{if } \exists y \leq n(gy = 0) \\ 0 & \text{if otherwise} \end{cases}$$

Then
$$\phi_n(g) > 0 \Longrightarrow \phi_{n+1} g = \phi_n g.$$

But
$$\phi_n f_x = \begin{cases} x+1 & \text{if } x < n \\ 0 & \text{if otherwise} \end{cases}$$

Hence
$$\phi_n(\lambda x \phi_n f_x) = n.$$

So define the third-order function
$$F : [\omega \Longrightarrow \omega] \Longrightarrow \omega$$
by
$$F(\psi) = \psi(\lambda x \psi f_x).$$

Then for all n
$$F\phi_n = n < F\phi_{n+1}.$$

BIBLIOGRAPHY

Ackermann, W. (1940). Zur Widerspruchsfreiheit der Zahlentheorie, *Mathematische Annalen* **117**: 162–194.

Gödel, K. (1938a). Lecture at Zilsel's, *Collected Works, Vol. III*, Oxford: Oxford University Press, pp. 87–113.

Hilbert, D. & Bernays, P. (1939). *Grundlagen der Mathematik II*, Berlin: Springer-Verlag. A second edition was published in 1970.

Kohlenbach, U. (1999). On the no-counterexample interpretation, *The Journal of Symbolic Logic* **64**: 1491–1511.

Tait, W. (1965a). Functionals defined by transfinite recursion, *The Journal of Symbolic Logic* **30**: 155–174.

Tait, W. (1965b). The substitution method, *The Journal of Symbolic Logic* **30**: 175–192.

Tait, W. (2005). Gödel's reformulation of Gentzen's first consistency proof for arithmetic, *Bulletin of Symbolic Logic* **11**(2): 225–238.

Isomorphisms and strong finite projective classes of commutative semigroups

MICHAEL A. TAITSLIN

ABSTRACT. [1] In "Sverdlovsk notebook" (Sverdlovsk, 1969), I proposed a question: Are any too first-order equivalent finitely generated commutative semigroups isomorphic? In 1970, B.I.Zilber answered the question negatively. A question arises: In what language, any equivalent over the language finitely generated commutative semigroups are isomorphic? In the note, we propose such a language. Moreover, we prove that there is an algorithm which for a given finite set of generators, a given finite set of defining relations of a commutative semigroup for the generators, and a closed formula of the language decides whether the formula holds in the semigroup.

1 Introduction

In [4] and "Sverdlovsk notebook" (Ural State University, Sverdlovsk, 1969) [2], I proposed a question: Are any first-order equivalent finitely generated commutative semigroups isomorphic? In [10], B.I.Zilber answered the question to construct two elementary equivalent commutative semigroups with 4 generators and 7 defining relations each. Zilber proved that the semigroups are not isomorphic.

The Zilber's example shows that the signature $\langle + \rangle$ is not relevant to describe a finite generated commutative semigroup up to isomorphism in first-order logic.

In the paper we consider an expanded signature to add unary relation G_a and constant symbol a for each generator a of the investigated finitely generated commutative semigroup and for a finite set of linear combinations of the generators. Having finite sets of generators and defining relations for the generators, we construct effectively a closed first-order formula of the expanded signature such that the formula holds in a finite generated commutative semigroup iff the new semigroup is isomorphic to the investigated semigroup. The truth of the formula in a commutative semigroup means that we can define the relations G_a and the constants a by such a way that the formula is true.

The result was formulated in [4] but the presented proof was not presented in detail. We propose a detailed proof and a new corollary.

---

[1] To Grigory Mints' 70-th Anniversary
[2] "Sverdlovsk notebook" is a collection of problems in semigroup theory

2 Definitions

We use terminology from [5], §3. Let us recall some results and definitions. A reason is to make the paper self-sufficient.

The set $\omega = \{0,1,2,\dots\}$ is called the set of natural numbers. By $L(\mathfrak{A})$ we denote the free semigroup with the finite set $\mathfrak{A}$ of free generators in the class of commutative semigroups with zero. For $\mathfrak{A} = \{a_1, \dots, a_k\}$, we consider $L(\mathfrak{A})$ as the set of all linear forms on the letters $a_1, \dots, a_k$ with natural numbers as coefficients.

For $a, b \in L(\mathfrak{A})$, the pair (a,b) is called a defining relation for $\mathfrak{A}$. A semigroup $L(\mathfrak{A}, \mathfrak{B})$, given in the class of commutative semigroups with zero, by finite sets of generators $\mathfrak{A}$ and defining relations $\mathfrak{B}$, is considered as a factor semigroup of the semigroup $L(\mathfrak{A})$. For $x \in L(\mathfrak{A})$, by $\overline{x}_{\mathfrak{B}}$ we denote the image of x under the canonical mapping $L(\mathfrak{A}) \longrightarrow L(\mathfrak{A}, \mathfrak{B})$. If it is understandable what $\mathfrak{B}$ is meant, we write $\overline{x}$ instead of $\overline{x}_{\mathfrak{B}}$.

By $M(\mathfrak{A}, \mathfrak{B})$ we denote the semigroup given in the class of commutative semigroups with cancellation and with zero, by finite sets of generators $\mathfrak{A}$ and defining relations $\mathfrak{B}$. We also treat $M(\mathfrak{A}, \mathfrak{B})$ as a factor semigroup of the semigroup $L(\mathfrak{A})$. For $x \in L(\mathfrak{A})$, by $[x]_{\mathfrak{B}}$ we denote the image of x under the canonical mapping $L(\mathfrak{A}) \longrightarrow M(\mathfrak{A}, \mathfrak{B})$. If it is understandable what $\mathfrak{B}$ is meant, we write $[x]$ instead of $[x]_{\mathfrak{B}}$.

For $a \in \mathfrak{A}$ and $f \in L(\mathfrak{A})$, we denote by $(f)_a$ the coefficient of the letter a in the form f. For $f, g \in L(\mathfrak{A})$, we write $f \leqslant g$ iff $(f)_a \leqslant (g)_a$ for all $a \in \mathfrak{A}$. If $f \leqslant g$, we say that f is *less* than g if f and g are different. We say that a form $f \in S$ is *minimal* in a set $S \subseteq L(\mathfrak{A})$ of forms iff g is not less than f for any $g \in S$. We say that a form $f \in S$ is *maximal* in a set $S \subseteq L(\mathfrak{A})$ of forms iff f is not less than g for any $g \in S$.

We write $\mathfrak{A}_1 \subset \mathfrak{A}$ iff $\mathfrak{A}_1$ is a subset of $\mathfrak{A}$ and $\mathfrak{A}_1 \neq \mathfrak{A}$.

$\emptyset$ denotes the empty set. $L(\emptyset)$ denotes the zero semigroup.

If $\mathfrak{A}_1 \subset \mathfrak{A}$, we consider the semigroup $L(\mathfrak{A}_1)$ as a sub-semigroup of the semigroup $L(\mathfrak{A})$. We also suppose that $\mathfrak{A}$ is a subset of $L(\mathfrak{A})$. θ denotes zero of the semigroup $L(\mathfrak{A})$.

For $\mathfrak{A}_1 \subset \mathfrak{A}$ and $f \in L(\mathfrak{A})$, we denote by $\mathfrak{A}_1(f)$ the form in $L(\mathfrak{A} \setminus \mathfrak{A}_1)$ such that $(\mathfrak{A}_1(f))_a = (f)_a$ for any $a \in \mathfrak{A} \setminus \mathfrak{A}_1$.

Let $q(\mathfrak{B})$ be the number of relations in $\mathfrak{B}$. Let $\lambda(\mathfrak{B})$ be the largest coefficient in these relations. We always assume that $\lambda(\mathfrak{B}) > 0$. By $h(\mathfrak{A}, \mathfrak{B})$ we denote the form

$$\sum_{a \in \mathfrak{A}} \lambda(\mathfrak{B})(q(\mathfrak{B}) + 1)a$$

in $L(\mathfrak{A})$.

For any $\mathfrak{A}_1 \subset \mathfrak{A}$, we denote by $\mathfrak{N}(\mathfrak{A}_1, \mathfrak{B})$ the set of all $x \in L(\mathfrak{A} \setminus \mathfrak{A}_1)$ such that the inequality

$$\overline{x + y} \neq \overline{h(\mathfrak{A}, \mathfrak{B}) + z}$$

holds in $L(\mathfrak{A}, \mathfrak{B})$ for all $y \in L(\mathfrak{A}_1)$, $z \in L(\mathfrak{A})$. By $\mathfrak{M}(\mathfrak{A}_1, \mathfrak{B})$ we denote the set of maximal elements of $\mathfrak{N}(\mathfrak{A}_1, \mathfrak{B})$. Finally, by $\mathfrak{M}_1(\mathfrak{A}_1, \mathfrak{B})$ we denote the set

$$\{x \in L(\mathfrak{A} \setminus \mathfrak{A}_1) \mid (\exists y)(y \in \mathfrak{M}(\mathfrak{A}_1, \mathfrak{B}) \& x \leqslant y)\}.$$

It follows from the theorem of Dickson ([1], p.129) that the sets $\mathfrak{M}(\mathfrak{A}_1, \mathfrak{B})$ and $\mathfrak{M}_1(\mathfrak{A}_1, \mathfrak{B})$ are finite.

We divide the set $\mathfrak{M}_1(\mathfrak{A}_1, \mathfrak{B})$ into reduction classes, assigning $x, y \in \mathfrak{M}_1(\mathfrak{A}_1, \mathfrak{B})$ to the same class iff there exists $z, u \in L(\mathfrak{A}_1)$ such that $\overline{x+z} = \overline{y+u}$ in $L(\mathfrak{A}, \mathfrak{B})$. By $i(\mathfrak{A}_1, \mathfrak{B})$ we denote the number of reduction classes, and by $\mathfrak{M}_{2,1}(\mathfrak{A}_1, \mathfrak{B}), \ldots, \mathfrak{M}_{2,i(\mathfrak{A}_1, \mathfrak{B})}(\mathfrak{A}_1, \mathfrak{B})$ all the different classes.

For $j \in \{1, \ldots, i(\mathfrak{A}_1, \mathfrak{B})\}$, consider $x, y \in \mathfrak{M}_{2,j}(\mathfrak{A}_1, \mathfrak{B})$ and $\mathfrak{A}_3 \subset \mathfrak{A}_1$. We denote by $\mathfrak{N}(\mathfrak{A}_1, \mathfrak{A}_3, x, y, \mathfrak{B})$ the set of all those $z \in L(\mathfrak{A}_1 \setminus \mathfrak{A}_3)$ such that the inequality $\overline{x+z+u} \neq \overline{y+v}$ holds in $L(\mathfrak{A}, \mathfrak{B})$ for all $u \in L(\mathfrak{A}_3)$ and $v \in L(\mathfrak{A}_1)$. We denote by $\mathfrak{M}(\mathfrak{A}_1, \mathfrak{A}_3, x, y, \mathfrak{B})$ the set of maximal elements of $\mathfrak{N}(\mathfrak{A}_1, \mathfrak{A}_3, x, y, \mathfrak{B})$. By $\mathfrak{M}_1(\mathfrak{A}_1, \mathfrak{A}_3, x, y, \mathfrak{B})$ we denote the set

$$\{u \in L(\mathfrak{A}_1 \setminus \mathfrak{A}_3) \mid (\exists v)(v \in \mathfrak{M}(\mathfrak{A}_1, \mathfrak{A}_3, x, y, \mathfrak{B}) \& u \leqslant v)\}.$$

The sets $\mathfrak{M}(\mathfrak{A}_1, \mathfrak{A}_3, x, y, \mathfrak{B})$ and $\mathfrak{M}_1(\mathfrak{A}_1, \mathfrak{A}_3, x, y, \mathfrak{B})$ are finite (the theorem of Dickson, [1], p.129).

It was proved in [2] that

(1) the sets $\mathfrak{M}(\mathfrak{A}_1, \mathfrak{B})$, $\mathfrak{M}_1(\mathfrak{A}_1, \mathfrak{B})$, $\mathfrak{M}(\mathfrak{A}_1, \mathfrak{A}_3, x, y, \mathfrak{B})$ are effectively constructible from $\mathfrak{A}$, $\mathfrak{B}$, $\mathfrak{A}_1$, $\mathfrak{A}_3$, x, y;

(2) the set $\mathfrak{M}_1(\mathfrak{A}_1, \mathfrak{B})$ is effectively divisible into reduction classes with respect to $\mathfrak{A}$, $\mathfrak{B}$, $\mathfrak{A}_1$;

(3) there exists an effective procedure which, for $\mathfrak{A}$, $\mathfrak{B}$, $\mathfrak{A}_1$, x such that $\mathfrak{A}_1 \subset \mathfrak{A}$ and $x \in L(\mathfrak{A} \setminus \mathfrak{A}_1)$, constructs a finite set $\mathfrak{B}(\mathfrak{A}_1, x)$ of defining relations for $\mathfrak{A}_1$ such that, for any $u, v \in L(\mathfrak{A}_1)$, equality $\overline{u} = \overline{v}$ holds in $L(\mathfrak{A}_1, \mathfrak{B}(\mathfrak{A}_1, x))$ iff $\overline{x+u} = \overline{x+v}$ holds in $L(\mathfrak{A}, \mathfrak{B})$;

(4) for any $x, y \in L(\mathfrak{A})$, $\overline{h(\mathfrak{A}, \mathfrak{B}) + x} = \overline{h(\mathfrak{A}, \mathfrak{B}) + y}$ holds in $L(\mathfrak{A}, \mathfrak{B})$ iff $[x] = [y]$ in $M(\mathfrak{A}, \mathfrak{B})$ (Lemma of Ceitin);

(5) for any $x \in L(\mathfrak{A})$, either there exists $z \in L(\mathfrak{A})$ such that $\overline{x} = \overline{h(\mathfrak{A}, \mathfrak{B}) + z}$ in $L(\mathfrak{A}, \mathfrak{B})$, or there exists $\mathfrak{A}_1 \subset \mathfrak{A}$ such that $\mathfrak{A}_1(x) \in \mathfrak{M}_1(\mathfrak{A}_1, \mathfrak{B})$;

(6) for $\mathfrak{A}_1 \subset \mathfrak{A}$, $j \in \{1, \ldots, i(\mathfrak{A}_1, \mathfrak{B})\}$, $x, y \in \mathfrak{M}_{2,j}(\mathfrak{A}_1, \mathfrak{B})$, and $u \in L(\mathfrak{A}_1)$, either there exists $v \in L(\mathfrak{A}_1)$ such that $\overline{x+u} = \overline{y+v}$ in $L(\mathfrak{A}, \mathfrak{B})$, or there exist $\mathfrak{A}_3 \subset \mathfrak{A}_1$ and $v \in \mathfrak{M}(\mathfrak{A}_1, \mathfrak{A}_3, x, y, \mathfrak{B})$ such that $\mathfrak{A}_3(u) \leqslant v$.

3   A formula which claims that G_a distinguishes the sub-semigroup generated by a

LEMMA 1. *In any finite generated Abelian group, for any element a of the group, any sequence*

$$a_0, a_1, \ldots, a_n, \ldots$$

such that in the group,

$$a_i = 2a_{i+1} \vee a_i = 2a_{i+1} + a,$$

contains only finite number of different elements.

Proof. Any finite generated Abelian group is a finite direct sum of infinite cyclic groups and a finite group. Each a_i can be present as a sum of a linear combination of the generators of the infinite cyclic groups and an element of the finite group. It is enough to prove that the sequence of the linear combinations contains only finite number of different elements. So it is enough to prove for each coordinate, that there exists a natural number such that the sequence of the coordinates of the linear combinations contains only the number of different elements. But if the coordinate of a_0 is n, and the coordinate of a is k, then the module of the coordinate of a_i does not exceed $\max(|n|, |k|)$. ∎

LEMMA 2. *In any finite generated commutative semigroup with cancellation, for any element a of the semigroup, any sequence*

$$a_0, a_1, \ldots, a_n, \ldots$$

such that in the semigroup,

$$a_i = 2a_{i+1} \lor a_i = 2a_{i+1} + a,$$

contains only finite number of different elements.

Proof. Any finite generated commutative semigroup with cancellation is isomorphically embedded into a finite generated Abelian group. ∎

Having an element $\bar{a}$ of a finite presented commutative semigroup $L(\mathfrak{A}, \mathfrak{B})$ where $a \in L(\mathfrak{A})$, we denote by $\mathfrak{A}(a)_\mathfrak{B}$ the set of all $b \in \mathfrak{A}$ such that the equality $n\bar{a} = \bar{b} + \bar{z}$ holds in $L(\mathfrak{A}, \mathfrak{B})$ for some $z \in L(\mathfrak{A})$ and natural number n. The set $\mathfrak{A}(a)_\mathfrak{B}$ is closed in $L(\mathfrak{A}, \mathfrak{B})$. It means that for any $x \in L(\mathfrak{A}(a)_\mathfrak{B})$, and any $y, z \in L(\mathfrak{A})$, if $\bar{x} = \bar{y} + \bar{z}$ in $L(\mathfrak{A}, \mathfrak{B})$, then $y \in L(\mathfrak{A}(a)_\mathfrak{B})$. Thus, in this case, $\mathfrak{B}(\mathfrak{A}(a)_\mathfrak{B}, \theta)$ is the set of all relations from $\mathfrak{B}$ which are defining relations for $L(\mathfrak{A}(a)_\mathfrak{B})$.

We denote by γ_a the number $h(\mathfrak{A}(a)_\mathfrak{B}, \mathfrak{B}(\mathfrak{A}(a)_\mathfrak{B}, \theta))$.

Now we construct a formula $\kappa(a)$ of signature $\langle +, a, G_a \rangle$. We write $3a$ instead of $(a + a + a)$. The similar meaning has ia for any natural number i.

The formula $\kappa(a)$ is the conjunction of the following formulas:

(7) $(G_a(a) \& (\forall x)(\forall y)((G_a(x) \& G_a(y)) \to G_a(x+y)) \&$

$$(\forall x)((\forall y) x + y = y \to G_a(x)))$$

(it means that G_a is a sub-semigroup containing a and θ);

(8) $(\forall x)(G_a(x) \to \bigvee_{i=0}^{m-1} x = ia)$

if the sub-semigroup with zero generated by $\bar{a}$ in $L(\mathfrak{A}, \mathfrak{B})$ is finite and contains m elements;

$((\forall x)((G_a(x) \& (\exists y)(G_a(y) \& x + y = \gamma_a a)) \to \bigvee_{i=0}^{\gamma_a} x = ia) \&$

$(\forall x)(\forall y_1)(\forall y_2)(\forall y_3)(\forall t_1)(\forall t_2)(\forall t_3)(\forall z_1)(\forall z_2)(\forall z_3)$

$((G_a(x) \& G_a(y_1) \& G_a(y_2) \& G_a(y_3) \& (\exists y)(G_a(y) \& x = \gamma_a a + y) \&$

$$y_1 = t_1 + z_1 \& y_2 = t_2 + z_2 \& y_3 = t_3 + z_3 \&$$
$$x + t_1 + t_2 = x + t_1 + t_3) \to x + t_2 = x + t_3))$$

if the sub-semigroup with zero generated by $\bar{a}$ in $L(\mathfrak{A}, \mathfrak{B})$ is infinite (it means that for infinite G_a and any natural number $\alpha \geqslant \gamma_a$, in $L(\mathfrak{A}, \mathfrak{B})$, the cancellation rule

$$((\overline{\alpha a + x + y} = \overline{\alpha a + x + z}) \to (\overline{\alpha a + y} = \overline{\alpha a + z}))$$

holds for any $x, y, z \in L(\mathfrak{A}(a)_{\mathfrak{B}}))$;

(9) $(\forall x)(G_a(x) \to (\exists u)(G_a(u) \& (x = 2u + a \vee x = 2u)))$;

(10) $(\forall x)(G_a(x) \to (\exists u)(G_a(u) \& (x = u + \gamma_a a \vee \gamma_a a = x + u)))$.

The truth of $\kappa(a)$ in a finite defined commutative semigroup $L(\mathfrak{A}', \mathfrak{B}')$ means that the semigroup can be expanded to the structure of signature $\langle +, a, G_a \rangle$ such that $\kappa(a)$ holds in the expanded structure. It is obvious that $G_a(ia)$ is true for any natural number i in any expansion of $L(\mathfrak{A}', \mathfrak{B}')$ such that $\kappa(a)$ holds in the expanded structure (see (7)).

LEMMA 3. *For any $a \in L(\mathfrak{A})$, $\kappa(a)$ holds in the expansion of $L(\mathfrak{A}, \mathfrak{B})$, in which truth of $G_a(x)$ means that x is equal to $i\bar{a}$ for some natural number i and the interpretation of a is $\bar{a}$.*

Proof. (7), (9), and (10) are true in the expansion by the definition.

If the sub-semigroup with zero generated by $\bar{a}$ in $L(\mathfrak{A}, \mathfrak{B})$ is finite, the truth of (8) in the expansion is obvious.

If the sub-semigroup with zero generated by $\bar{a}$ in $L(\mathfrak{A}, \mathfrak{B})$ is infinite, the truth of (8) in the expansion is followed from the definitions of $\mathfrak{A}(a)_{\mathfrak{B}}$ and γ_a and Ceitin Lemma (see (4)). ∎

LEMMA 4. *If $\kappa(a)$ holds in an expansion of $L(\mathfrak{A}', \mathfrak{B}')$, then in the expansion, truth of $G_a(x)$ implies that x is equal to ia for a natural number i.*

Proof. By contradiction. Let $G_a(x)$ holds in the expansion and x is different from ia for any natural number i. By (9), take x as u_0 and an element u_1 such that $(G_a(u_1) \& (x = 2u_1 + a \vee x = 2u_1))$. If u_j has constructed, take an element u_{j+1} such that $(G_a(u_{j+1}) \& (u_j = 2u_{j+1} + a \vee u_j = 2u_{j+1}))$. It is obvious that for any natural number j, u_j is different from ia for any natural number i. It follows from (8) and (10) that $(\exists u)(G_a(u) \& (u_j = u + \gamma_a a))$ holds in the expansion. Now it follows from (8) that $(u_j = 2u_{j+1} + a \vee u_j = 2u_{j+1})$ holds in $M(\mathfrak{A}', \mathfrak{B}')$ for some $u_0, \ldots, u_j, \ldots$ But it contradicts to Lemma 2. ∎

4 A formula for equality in a finite defined commutative semigroup with cancellation

First of all, we write a first-order formula

$$\Phi_{11}(\mathfrak{A}, \mathfrak{B}, y_1, \ldots, y_k, z_1, \ldots, z_k)$$

of signature $\langle +, a, G_a \mid a \in \mathfrak{A}\rangle$ such that for any $d_1,\ldots, d_k, b_1,\ldots, b_k$ satisfying $G_{a_1}(d_1)$, $G_{a_1}(b_1)$, $\ldots$, $G_{a_k}(d_k)$, $G_{a_k}(b_k)$, $\Phi_{11}(\mathfrak{A}, \mathfrak{B}, d_1,\ldots, d_k, b_1,\ldots, b_k)$ implies that sums $d_1 + \cdots + d_k$ and $b_1 + \cdots + b_k$ are equal in $M(\mathfrak{A}, \mathfrak{B})$. If any generator from $\mathfrak{A}$ generates a finite sub-semigroup in $M(\mathfrak{A}, \mathfrak{B})$, $M(\mathfrak{A}, \mathfrak{B})$ is finite and $\Phi_{11}(\mathfrak{A}, \mathfrak{B}, y_1,\ldots, y_k, z_1,\ldots, z_k)$ can be taken as a first-order formula of signature $\langle +, a \mid a \in \mathfrak{A}\rangle$. The same holds if $k = 1$. Suppose $k > 1$ and $[a_1]$ generates an infinite sub-semigroup in $M(\mathfrak{A}, \mathfrak{B})$. To simplify notation, suppose that $c_i \in \mathfrak{A}$ and $(c_i, a_1 + a_i) \in \mathfrak{B}$ for $i = 2,\ldots, k$.

$\Phi_{11}(\mathfrak{A}, \mathfrak{B}, y_1,\ldots, y_k, z_1,\ldots, z_k)$ is

$$(\exists t_{1,1})\ldots(\exists t_{1,k})\ldots(\exists t_{q,1})\ldots(\exists t_{q,k})$$

$$((\bigwedge_{j=1}^{q}\bigwedge_{i=1}^{k} F_{a_1,a_i}(t_{j,1}, t_{j,i})) \& \bigwedge_{i=1}^{k}[y_i + \sum_{j=1}^{q} t_{j,i}(\mathfrak{A}_{j,i} - \mathfrak{B}_{j,i}) = z_i])$$

where $\mathfrak{B}$ is $\{(\mathfrak{A}_j, \mathfrak{B}_j) \mid j = 1,\ldots, q\}$; for any $j \in \{1,\ldots, q\}$, $\mathfrak{A}_j = \sum_{i=1}^{k} \mathfrak{A}_{j,i} a_i$, and $\mathfrak{B}_j = \sum_{i=1}^{k} \mathfrak{B}_{j,i} a_i$. As usual, for natural number i, $x - iy = z$ means that $x = z + iy$. We need only explain what is F_{a_1,a_i}.

$F_{a_1,a_i}(x, y)$ tells us that there exists a natural number j such that $\overline{x}$ is equal to $j\overline{a}_1$ and $\overline{y}$ is equal to $j\overline{a}_i$ in $L(\mathfrak{A}, \mathfrak{B})$.

If $[a_i]$ generates a finite sub-semigroup in $M(\mathfrak{A}, \mathfrak{B})$, the construction of $F_{a_1,a_i}(x, y)$ is obvious.

If $[a_1]$ and $[a_i]$ are linear independent in $M(\mathfrak{A}, \mathfrak{B})$, we can use $G_{c_i}(x + y)$ instead of $F_{a_1,a_i}(x, y)$.

In other cases, there exists a natural number m such that in $M(\mathfrak{A}, \mathfrak{B})$, any sum $x = x_1 + \cdots + x_k$ where $G_{a_j}(x_j)$ for $j = 1,\ldots, k$ can be effectively presented as $x + z_1 + \cdots + z_k = y_1 + \cdots + y_k$ where $G_{a_j}(y_j)$ and $G_{a_j}(z_j)$ for $j = 1,\ldots, k$, where z_i is zero, and where $\bigvee_{j=0}^{m} y_i = j[a_i]$ (see [3]). Moreover, two such presentations $x_1 + z_{1,1} + \cdots + z_{1,k} = y_{1,1} + \cdots + y_{1,k}$ and $x_2 + z_{2,1} + \cdots + z_{2,k} = y_{2,1} + \cdots + y_{2,k}$ represent equal in $M(\mathfrak{A}, \mathfrak{B})$ elements x_1 and x_2 iff $y_{1,i}$ coincides with $y_{2,i}$ and

$$y_{1,1} + \cdots + y_{1,k} + z_{2,1} + \cdots + z_{2,k} = y_{2,1} + \cdots + y_{2,k} + z_{1,1} + \cdots + z_{1,k}.$$

It means that if $[a_i]$ generates an infinite sub-semigroup in $M(\mathfrak{A}, \mathfrak{B})$ and $[a_1]$ and $[a_i]$ are linear dependent in $M(\mathfrak{A}, \mathfrak{B})$, we do not need to use $F_{a_1,a_i}(x, y)$.

5 Main formula

Consider a finite defined commutative semigroup $L(\mathfrak{A}, \mathfrak{B})$. Now we are going to construct a closed first-order formula $\Phi(\mathfrak{A}, \mathfrak{B})$ of signature $\langle +, a, G_a \mid a \in \mathfrak{A}\rangle$ such that truth of $\Phi(\mathfrak{A}, \mathfrak{B})$ in a finite generated commutative semigroup $L(\mathfrak{A}', \mathfrak{B}')$ means that $L(\mathfrak{A}, \mathfrak{B})$ and $L(\mathfrak{A}', \mathfrak{B}')$ are isomorphic. Suppose $\mathfrak{A} = \{a_1,\ldots, a_k\}$. If any generator from $\mathfrak{A}$ generates a finite sub-semigroup in $L(\mathfrak{A}, \mathfrak{B})$, $L(\mathfrak{A}, \mathfrak{B})$ is finite and $\Phi(\mathfrak{A}, \mathfrak{B})$ can be taken as a closed first-order formula of signature $\langle + \rangle$. The same holds if $k = 1$. Suppose $k > 1$ and $\overline{a}_1$ generates an infinite sub-semigroup in $L(\mathfrak{A}, \mathfrak{B})$.

$\Phi(\mathfrak{A}, \mathfrak{B})$ is the conjunction of the formulas $\bigwedge_{a \in \mathfrak{A}} \kappa(a)$,

$$(\forall x)(\exists y_1)\ldots(\exists y_k)((\bigwedge_{i=1}^{k} G_{a_i}(y_i)) \& x = y_1 + \cdots + y_k),$$

and a closed first-order formula $\Phi_1(\mathfrak{A}, \mathfrak{B})$ of signature $\langle +, a, G_a \mid a \in \mathfrak{A} \rangle$.
Let us construct $\Phi_1(\mathfrak{A}, \mathfrak{B})$.
$\Phi_1(\mathfrak{A}, \mathfrak{B})$ is

$$(\forall y_1)\ldots(\forall y_k)(\forall z_1)\ldots(\forall z_k)((\bigwedge_{i=1}^{k} G_{a_i}(y_i) \& \bigwedge_{i=1}^{k} G_{a_i}(z_i) \&$$

$$z_1 + \cdots + z_k = y_1 + \cdots + y_k) \to \Phi_{10}(\mathfrak{A}, \mathfrak{B}, y_1, \ldots, y_k, z_1, \ldots, z_k)).$$

$\Phi_{10}(\mathfrak{A}, \mathfrak{B}, y_1, \ldots, y_k, z_1, \ldots, z_k)$ is constructed by induction on k and is

$$(((\bigwedge_{\mathfrak{A}_1 \subset \mathfrak{A}} (\sum_{a_i \in (\mathfrak{A} \setminus \mathfrak{A}_1)} y_i) \notin \mathfrak{M}_1(\mathfrak{A}_1, \mathfrak{B})) \& \Phi_{11}(\mathfrak{A}, \mathfrak{B}, y_1, \ldots, y_k, z_1, \ldots, z_k)) \vee$$

$$(\bigvee_{\mathfrak{A}_1 \subset \mathfrak{A}} ((\sum_{a_i \in (\mathfrak{A} \setminus \mathfrak{A}_1)} y_i) \in \mathfrak{M}_1(\mathfrak{A}_1, \mathfrak{B}) \& \Phi_{12}(\mathfrak{A}, \mathfrak{B}, \mathfrak{A}_1, y_1, \ldots, y_k, z_1, \ldots, z_k)))).$$

$\Phi_1(\mathfrak{A}, \mathfrak{B})$ tells us that either there is $u \in L(\mathfrak{A})$ such that $y_1 + \cdots + y_k$ is equal to $h(\mathfrak{A}, \mathfrak{B}) + u$ in $L(\mathfrak{A}, \mathfrak{B})$ and then $[y_1 + \cdots + y_k] = [z_1 + \cdots + z_k]$ in $M(\mathfrak{A}, \mathfrak{B})$, or there is $\mathfrak{A}_1 \subset \mathfrak{A}$ such that $(\sum_{a_i \in (\mathfrak{A} \setminus \mathfrak{A}_1)} y_i) \in \mathfrak{M}_1(\mathfrak{A}_1, \mathfrak{B})$ and $\Phi_{12}(\mathfrak{A}, \mathfrak{B}, \mathfrak{A}_1, y_1, \ldots, y_k, z_1, \ldots, z_k)$. The last formula will be described below.

The correctness of $\Phi_1(\mathfrak{A}, \mathfrak{B})$ is followed from (5) and

LEMMA 5. *If $x, y \in L(\mathfrak{A})$ and $\overline{x} = \overline{y}$, then for any $\mathfrak{A}_1 \subset \mathfrak{A}$, $\mathfrak{A}_1(x) \in \mathfrak{M}_1(\mathfrak{A}_1, \mathfrak{B})$ iff $\mathfrak{A}_1(y) \in \mathfrak{M}_1(\mathfrak{A}_1, \mathfrak{B})$.*

Proof. By contradiction. Suppose $\mathfrak{A}_1(x) \in \mathfrak{M}(\mathfrak{A}_1, \mathfrak{B})$. If $v \in L(\mathfrak{A}_1)$, $w \in L(\mathfrak{A})$, and $\overline{\mathfrak{A}_1(y) + v} = \overline{h(\mathfrak{A}, \mathfrak{B}) + w}$, then

$$\overline{\mathfrak{A}_1(x) + v + (\mathfrak{A} \setminus \mathfrak{A}_1)(x)} =$$

$$\overline{\mathfrak{A}_1(y) + v + (\mathfrak{A} \setminus \mathfrak{A}_1)(y)} = \overline{h(\mathfrak{A}, \mathfrak{B}) + w + (\mathfrak{A} \setminus \mathfrak{A}_1)(y)}.$$

It means that $\mathfrak{A}_1(x) \notin \mathfrak{M}(\mathfrak{A}_1, \mathfrak{B})$. A contradiction. If $u \in L(\mathfrak{A} \setminus \mathfrak{A}_1)$ and $(\mathfrak{A}_1(y) + u) \in \mathfrak{M}(\mathfrak{A}_1, \mathfrak{B})$, then $(\mathfrak{A}_1(x) + u) \in \mathfrak{M}(\mathfrak{A}_1, \mathfrak{B})$. A contradiction. If $\mathfrak{A}_1(x) \in \mathfrak{M}_1(\mathfrak{A}_1, \mathfrak{B})$, then there exists $u \in L(\mathfrak{A} \setminus \mathfrak{A}_1)$ such that $(\mathfrak{A}_1(x) + u) \in \mathfrak{M}(\mathfrak{A}_1, \mathfrak{B})$. By the previous proof, $(\mathfrak{A}_1(y) + u) \in \mathfrak{M}(\mathfrak{A}_1, \mathfrak{B})$. Thus $\mathfrak{A}_1(y) \in \mathfrak{M}_1(\mathfrak{A}_1, \mathfrak{B})$. ∎

6 A construction of $\Phi_{12}(\mathfrak{A}, \mathfrak{B}, \mathfrak{A}_1, y_1, \ldots, y_k, z_1, \ldots, z_k)$

Let $\mathfrak{A}_1 \subset \mathfrak{A}$, $(\sum_{a_i \in (\mathfrak{A} \setminus \mathfrak{A}_1)} y_i) \in \mathfrak{M}_1(\mathfrak{A}_1, \mathfrak{B})$, $\bigwedge_{i=1}^{k} G_{a_i}(y_i)$, $\bigwedge_{i=1}^{k} G_{a_i}(z_i)$, and $z_1 + \cdots + z_k = y_1 + \cdots + y_k$. Then $(\sum_{a_i \in (\mathfrak{A} \setminus \mathfrak{A}_1)} z_i) \in \mathfrak{M}_1(\mathfrak{A}_1, \mathfrak{B})$ (see Lemma 5). Denote $z_1 + \cdots + z_k$ by z and $y_1 + \cdots + y_k$ by y. By definition, there exists $j \in \{1, \ldots, i(\mathfrak{A}_1, \mathfrak{B})\}$ such that $\mathfrak{A}_1(y), \mathfrak{A}_1(z) \in \mathfrak{M}_{2,j}(\mathfrak{A}_1, \mathfrak{B})$. By definition, for any $\mathfrak{A}_3 \subset \mathfrak{A}_1$, $\mathfrak{A}_3((\mathfrak{A} \setminus \mathfrak{A}_1)(y)) \notin \mathfrak{M}_1(\mathfrak{A}_1, \mathfrak{A}_3, \mathfrak{A}_1(y), \mathfrak{A}_1(z), \mathfrak{B})$.

Denote by $\mathfrak{R}(\mathfrak{A}_1(y), \mathfrak{A}_1(z), \mathfrak{B})$ the set of all $u \in L(\mathfrak{A}_1)$ such that $\mathfrak{A}_3(u) \notin \mathfrak{M}_1(\mathfrak{A}_1, \mathfrak{A}_3, \mathfrak{A}_1(y), \mathfrak{A}_1(z), \mathfrak{B})$ for any $\mathfrak{A}_3 \subset \mathfrak{A}_1$. Denote by $\mathfrak{R}_1(\mathfrak{A}_1(y), \mathfrak{A}_1(z), \mathfrak{B})$ the set of all minimal elements from $\mathfrak{R}(\mathfrak{A}_1(y), \mathfrak{A}_1(z), \mathfrak{B})$. The set $\mathfrak{R}_1(\mathfrak{A}_1(y), \mathfrak{A}_1(z), \mathfrak{B})$ is finite (the theorem of Dickson, [1], p.129).

It followed from (6) that for any $u \in \mathfrak{R}_1(\mathfrak{A}_1(y), \mathfrak{A}_1(z), \mathfrak{B})$, there exists $v(u) \in L(\mathfrak{A}_1)$ such that $\overline{\mathfrak{A}_1(y) + u} = \overline{\mathfrak{A}_1(z) + v(u)}$ in $L(\mathfrak{A}, \mathfrak{B})$.

By definition, there exists $u \in \mathfrak{R}_1(\mathfrak{A}_1(y), \mathfrak{A}_1(z), \mathfrak{B})$ such that $u \leqslant (\mathfrak{A} \backslash \mathfrak{A}_1)(y)$.

It demonstrates that $\Phi_{12}(\mathfrak{A}, \mathfrak{B}, \mathfrak{A}_1, y_1, \ldots, y_k, z_1, \ldots, z_k)$ can be taken as

$$\bigvee_{u \in \mathfrak{R}_1(\mathfrak{A}_1(y), \mathfrak{A}_1(z), \mathfrak{B})} (\exists w)((\mathfrak{A} \backslash \mathfrak{A}_1)(y) = u + w \ \&$$

$$\Phi_{10}(\mathfrak{A}_1, \mathfrak{B}(\mathfrak{A}_1, \sum_{a_i \in (\mathfrak{A} \backslash \mathfrak{A}_1)} z_i), (\mathfrak{A} \backslash \mathfrak{A}_1)(z) + w + v(u), (\mathfrak{A} \backslash \mathfrak{A}_1)(z))).$$

Here we use the following shorting: we write d instead of $n_1 d_1, \ldots, n_s d_s$ for natural numbers $n_1, \ldots, n_s$ if the form d on the free generators $d_1, \ldots, d_s$ is $n_1 d_1 + \cdots + n_s d_s$.

7 Conclusion

There exists an algorithm that for any finite defining commutative semigroup $L(\mathfrak{A}', \mathfrak{B}')$, determines whether or not $\Phi(\mathfrak{A}, \mathfrak{B})$ is true in $L(\mathfrak{A}', \mathfrak{B}')$. Indeed, $\Phi(\mathfrak{A}, \mathfrak{B})$ is true in $L(\mathfrak{A}', \mathfrak{B}')$ iff $L(\mathfrak{A}', \mathfrak{B}')$ and $L(\mathfrak{A}, \mathfrak{B})$ are isomorphic. There exists an algorithm to decide whether or not $L(\mathfrak{A}', \mathfrak{B}')$ and $L(\mathfrak{A}, \mathfrak{B})$ are isomorphic (see [9]).

BIBLIOGRAPHY

[1] A.H. Clifford and G.B. Preston. *The algebraic theory of semigroups*, volume 2. American Math.Soc., Providence, R.I., 1967.
[2] M.A. Taitslin. On elementary theories of commutative semigroups. *Algebra i logika*, 5(4):55–89, 1966. In Russian.
[3] M.A. Taitslin. On elementary theories of commutative semigroups with cancelation. *Algebra i logika*, 5(1):51–69, 1966. In Russian.
[4] M.A. Taitslin. Two remarks on commutative semigroups isomorphism. *Algebra i logika*, 6(1):95–116, 1967. In Russian.
[5] M.A. Taitslin. Algorithmic problem for commutative semigroups. *Sov.Math.Dokl.*, 9(1):201–204, 1968.
[6] M.A. Taitslin. On the isomorphism problem for commutative semigroups. *Sibirsky Matematichesky journal*, 9(2):375–401, 1968. In Russian.
[7] M.A. Taitslin. Equivalence of automata with respect to a commutative semigroup. *Algebra i logika*, 8(5):553–600, 1969. In Russian.
[8] M.A. Taitslin. On the isomorphism problem for commutative semigroups. *Math. USSR Sbornik*, 22(1):104–128, 1974.
[9] M.A. Taitslin. The isomorphism problem for the commutative semigroups has the positive solution. In *Model theory and its applications*, pages 75–81. Kazakh State University, Alma-Ata, 1980. In Russian.
[10] B.I. Zilber. On isomorphism and elementary equivalence of commutative semigroups. *Algebra and logic*, 9(6):667–671, 1970. In Russian.

Priority Arguments and Epsilon Substitution

HENRY TOWSNER

1 Introduction

While the priority argument has been one of the main techniques of recursion theory, it has seen only a few applications to other areas of mathematics [13, 16]. One possibility for another such application was pointed out by Kreisel: Hilbert's ϵ-substitution method, a technique for proving the 1-consistency of theories. Kreisel's observation was that the proof that the method works [1] bears a striking resemblance to the structure of a traditional finite injury priority argument.

Such a connection might have benefits for both fields. The ϵ-substitution method has powerful extensions [3, 4, 5] which might provide new tools for solving difficult recursion theoretic problems. In the other direction, the most popular proof theoretic technique for proving 1-consistency results, cut-elimination, has bogged down in technical details, and new ideas are need to make ordinal analytic results more accessible.

Unfortunately, Kreisel's observation has been difficult to turn into a concrete argument. After Yang [17], the reason is clear: the success of all finite injury priority arguments is exactly enough to prove the 1-consistency of the weak theory $I\Sigma_1$, and therefore finite injury arguments cannot be sufficient to prove the consistency of stronger theories. Using a general framework for priority arguments developed by Lerman and Lempp [10, 11, 12], Yang goes on to show that arguments on the n-th level of their hierarchy of priority arguments are equivalent to the 1-consistency of $I\Sigma_n$, and so it requires the full ω levels of that hierarchy to give 1-consistency for all of first-order arithmetic.

The better known infinite injury and monster injury priority arguments belong to the second and third levels of this hierarchy, and, as the name "monster" suggests, going to higher levels becomes impractical without some kind of general framework. The Lerman-Lempp framework is one of several that have been proposed [6, 7, 8, 9]. One technique, usually described using "workers on many levels," originally developed by Harrington, has been extended to hyperarithmetic levels.

We show in this paper that, if one is prepared to use one of these frameworks to describe the necessary priority argument, that the ϵ-substitution method can be proven to work using a priority argument. We follow Yang in using the Lerman-Lempp framework, although we know of no reason that other frameworks would not work just as well.

Currently, those few priority arguments that have been extended to hyperarithmetic levels have a fixed ordinal height α. This paper and Yang's suggest that this corresponds to the 1-consistency of Peano Arithmetic plus transfinite induction up to a particular ordinal. The ϵ-substitution has difficult but reasonably well-understood extensions to systems like ID_1 [2, 14], a system which adds a least fixed point to arithmetic, and (less well-understood) extensions to even stronger systems [3, 5]. We hope that these results can also be translated into the priority argument environment, giving a stronger recursion theoretic technique which might be capable of answering unsolved questions.

In the hopes of making the proof more accessible, we abandon the standard terminology of the ϵ-substitution method for more conventional terminology. We work in a quantifier-free language with Skolem functions: function symbols of the form $c^{\vec{y}}_{\exists x.\phi[x,\vec{y}]}$ where ϕ is quantifier-free and $\vec{y}$ is a sequence of variables. A term $c^{\vec{y}}_{\exists x.\phi[x,\vec{y}]}(\vec{t})$ is intended to represent a value n such that $\phi[n,\vec{t}]$ holds, if there is such an n. Note that we allow nesting, to represent Σ_n formulas for arbitrary n; for instance terms like $c^{\vec{y}}_{\exists x.\phi[c^{\vec{z}}_{\exists w.\psi[w,\vec{z}]},\vec{y}]}$ are allowed.

An ϵ-substitution is just a partial model for this language, providing an interpretation for the value of some Skolem functions when evaluated at some points. Such partial models may not satisfy all axioms, but we will be interested in satisfying only finitely many axioms at a time. When an ϵ-substitution fails to satisfy some axiom, it will always be possible to repair this in a canonical way by extended the substitution. The act of doing so, however, may force us to remove some other elements, since changing the value of one term may alter the interpretation of others.

The Lerman-Lempp framework for priority arguments uses a tower of trees, where the n-th tree controls conditions guiding Σ_n properties. We choose branches in the tree in stages, with each stage corresponding to a step in our construction. The important idea is that the branches we choose stabilize enough to give a well-formed construction. For instance, at the bottom level are Σ_1 properties; in our case, these form a tree where we can only change once: when we first reach a node with one of these conditions, unless our construction already witnesses the Σ_1 case, we assume a Π_1 outcome. If at a later stage we discover a witness, we backtrack and choose a different branch, abandoning some of our progress through the tree. But, having been witnessed, the Σ_1 outcome cannot change, so eventually we achieve a path through this tree. The next tree controls Σ_2 properties, which can change back and forth repeatedly; the key to the proof will be that, relative to the first tree, the second tree has controlled backtracking: that is, except when we backtrack in the first tree, the second tree behaves like the first tree. But since we can control the backtracking in the first tree, this gives an indirect control on the path we construct in the second tree. This process is then repeated to give enough control on all the trees to prove that the construction we want is well-behaved.

Unlike a typical priority argument, our setting is finitary. While this changes the phrasing of some arguments, the underlying concerns are the same: in a usual priority argument, we must arrange infinitely many conditions so that they have order type ω, while in this case, we must arrange finitely many (where some appear multiple times) so that they eventually run out. This proof could be modified to work with countably many conditions—for instance, all possible conditions—and therefore to prove that there is a recursively enumerable ϵ-substitution assigning correct values to all rank 1 Skolem functions (that is, all Skolem functions for Σ_1 formulas).

Rather than literally following the H-process, we prove termination of a modified process derived from our construction. The primary difference is that in certain situations we add additional information to our ϵ-substitution whose correctness is witnessed even if there is no axiom compelling us to do so. This turns out to better match our construction since it means we can decide locally, by examining only the ϵ-substitution, whether that information is present, rather than having to know what happened at previous stages to figure out whether it might have been added at some point.

The author would like to express particular gratitude to Grisha Mints for introducing him to the ϵ-substitution process—and, indeed, to proof theory itself—and whose suggestions were essential at every stage of the development of this paper. We would also like to thank the anonymous referee for suggesting many helpful clarifications.

2 Skolem Functions and ϵ-Substitutions

In this section, we present a simplified version of the ϵ-substitution method. For the standard presentation, as well as those lemmas whose proofs we have omitted, see [15].

We work in a Skolemized version of first-order arithmetic.

DEFINITION 1. Let $\mathcal{L}_0$ be the ordinary language of first-order arithmetic. In particular, it contains predicate symbols for each primitive recursively definable relation, and the function symbols 0 and **S** (and no others).

Given a language $\mathcal{L}$ define the Skolemization $\mathcal{L}'$ by adding, for each Σ_1 formula $\exists x.\phi[x, y_1, \ldots, y_k]$ such that $\phi[x, 0, \ldots, 0]$ contains no closed Skolem terms and $y_1, \ldots, y_k$ includes all free variables besides x in ϕ, add a k-ary *Skolem function* $c_{\exists x.\phi[x]}^{y_1,\ldots,y_k}$. A *Skolem term* is a term of the form $c_{\exists x.\phi[x]}^{y_1,\ldots,y_k}(t_1, \ldots, t_k)$.

Let $\mathcal{L}_{n+1} := \mathcal{L}'_n$, and let $\mathcal{L}_\omega := \bigcup \mathcal{L}_n$. Let $\mathcal{L}\epsilon$ be the quantifier-free part of $\mathcal{L}_\omega$.

A formula or term e has *rank* n, written $rk(e) = n$, if it belongs to $\mathcal{L}_n$ but no $\mathcal{L}_m$ for $m < n$.

Note that $\exists x.\phi[x, y_1, \ldots, y_n]$ may contain Skolem functions which depend on x. For example,
$$c_{\exists x.0=c_{\exists y.y=z}^{z}(x)}$$
is a permissible Skolem function.

DEFINITION 2. Within $\mathcal{L}\epsilon$, when $\phi[x, \vec{y}]$ is a formula with all free variables displayed such that $\phi[x, 0, \ldots, 0]$ contains no closed Skolem terms, we take $\exists x.\phi[x, \vec{t}]$ to be an abbreviation for $\phi[c_{\exists x.\phi}^{\vec{y}}(\vec{t}), \vec{t}]$.

Note that, with some effort, we may find for any formula $\psi[x]$ a formula $\phi[x, \vec{t}]$ meeting the conditions above. We could also, by analogy, abbreviate $\neg\phi[c_{\exists x.\neg\phi}^{\vec{y}}(\vec{t}), \vec{t}]$ by $\forall x.\phi[x, \vec{t}]$.

DEFINITION 3. The only rule of $PA\epsilon$ is modus ponens. The axioms are:

1. All propositional tautologies of the language $\mathcal{L}\epsilon$

2. All substitution instances of the defining axioms for predicate constants

3. Equality axioms $t = t$ and $s = t \rightarrow \phi[s] \rightarrow \phi[t]$

4. Peano axioms $\neg \mathbf{S}t = 0$ and $\mathbf{S}s = \mathbf{S}t \rightarrow s = t$

5. Critical formulas:

- $\phi[t, \vec{u}] \to \exists x.\phi[x, \vec{u}]$
- $\phi[0, \vec{u}] \wedge \neg\phi[t, \vec{u}] \to \exists x.(\phi[x, \vec{u}] \wedge \neg\phi[\mathbf{S}x, \vec{u}])$
- $\neg s = 0 \to \exists x.s = \mathbf{S}x$

This is a standard axiomization of Peano Arithmetic, except that $\exists x.\phi[x, \vec{t}]$ is an abbreviation for a statement about Skolem terms. Note that all critical formulas have a general form $\phi \to \psi[c]$ for a distinguished Skolem term c; when we refer to an arbitrary Skolem formula as being in this form, we of course mean ϕ, ψ, and c to be the particular subformulas listed explicitly above, rather than some other decomposition into this form which might occur in particular cases.

DEFINITION 4. *A particular collection of terms, the* numerals, *denoted $\mathbb{N}$, are defined inductively: 0 is a numeral, and if t is a numeral then $\mathbf{S}t$ is a numeral.*

Clearly the numerals have a canonical bijection with the natural numbers. The variables u and v will by used to denote numerals; we will sometimes abuse notation to use u or v as both a numeral and the corresponding number.

THEOREM 5. *If there is a proof of a closed formula ϕ in PA then there is a proof of ϕ in $PA\epsilon$ (where quantifiers are interpreted as abbreviations) containing only closed formulas.*

From here on, we assume that all formulas are closed (since in the Skolemized language there is no need for free variables).

2.1 ϵ-Substitutions

We will be interested in particular partial models of formulas in $\mathcal{L}\epsilon$ assigning values to finitely many values of the Skolem functions. We will only assign values to predicates of the form $c^{\vec{y}}_{\exists x.\phi[x,\vec{y}]}(\vec{t})$ where each t_i is a natural number, and will assign either a natural number u (asserting that $\phi[u, \vec{t}]$ holds) or a default value ? (leaving open the possibility that $\forall x.\neg\phi[x, \vec{t}]$).

DEFINITION 6. *A* canonical *term is a term of the form $c(\vec{t})$ where c is a Skolem function and each t_i is a numeral.*

To keep some continuity with other work in the area, we call these models ϵ-substitutions:

DEFINITION 7. *An ϵ-substitution is a function S such that:*

- *The domain of S is a set of canonical terms*

- *If $e \in \mathrm{dom}(S)$ then $S(e)$ is either a numeral or the symbol ?*

An ϵ-substitution is *total* if its domain is the set of all canonical terms.

We will frequently have a non-total ϵ-substitution which we wish to take to be "complete": that is, we wish to assign the default value to every canonical term not specifically assigned some other value.

DEFINITION 8.

The standard extension $\overline{S}$ of an ϵ-substitution S is given by

$$\overline{S} := S \cup \{(e, ?) \mid e \notin \text{dom}(S)\}$$

DEFINITION 9. We modify the function S to define $\hat{S}(t)$ on arbitrary terms by induction on t, and also to sequences of terms:

- If $\vec{s}$ is the sequence $s_1, \ldots, s_k$, set $\hat{S}(\vec{s}) := \hat{S}(s_1), \ldots, \hat{S}(s_k)$
- $\hat{S}(0) := 0$
- $\hat{S}(\mathbf{S}t) := \mathbf{S}\hat{S}(t)$
- If t is a canonical Skolem term in the domain of S and $S(t) = ?$ then $\hat{S}(t) := 0$
- If t is a canonical Skolem term in the domain of S and $S(t) \in \mathbb{N}$ then $\hat{S}(t) := S(t)$
- If t is a canonical Skolem term not in the domain of S then $\hat{S}(t) := t$
- If t is a non-canonical Skolem term of the form $c(\vec{s})$ and for some i, $\hat{S}(s_i)$ is not a numeral then $\hat{S}(t) := c(\hat{S}(\vec{s}))$
- If t is a non-canonical Skolem term of the form $c(\vec{s})$ and for every i, $\hat{S}(s_i)$ is a numeral then $\hat{S}(t) := \hat{S}(c(\hat{S}(\vec{s})))$

Note that the final clause is well-defined since, if $\hat{S}(s_i)$ is a numeral for each i then $\hat{S}(c(\hat{S}(\vec{s})))$ is defined by the other clauses.

DEFINITION 10.

- If ϕ is an atomic formula $Rt_1 \cdots t_n$ then $S \models Rt_1 \cdots t_n$ iff $\hat{S}(t_i)$ is a numeral for each i and $R\hat{S}(t_1) \cdots \hat{S}(t_n)$ holds in the standard model
- If ϕ is a negated atomic formula $\neg Rt_1 \cdots t_n$ then $S \models \neg Rt_1 \cdots t_n$ iff $\hat{S}(t_i)$ is a numeral for each i and $\neg R\hat{S}(t_1) \cdots \hat{S}(t_n)$ holds in the standard model
- $S \models \phi \wedge \psi$ iff $S \models \phi$ and $S \models \psi$
- $S \models \neg(\phi \wedge \psi)$ iff $S \models \neg\phi$ or $S \models \neg\psi$
- $S \models \phi \vee \psi$ iff $S \models \phi$ or $S \models \psi$
- $S \models \neg(\phi \vee \psi)$ iff $S \models \neg\phi$ and $S \models \neg\psi$

The unusual handling of negation is necessary because if S is not total, some formulas may be indeterminate.

DEFINITION 11. S *decides* ϕ if $S \models \phi$ or $S \models \neg\phi$.

$unev(\phi, S)$, the set of terms in ϕ not evaluated by S, consists of terms of the form $c(\vec{s})$ such that $\hat{S}(s_i)$ is a numeral for each i, but $\hat{S}(c(\vec{s}))$ is not a numeral.

LEMMA 12.

- If $S \models \phi$ then $S \not\models \neg\phi$.

- If S is total then S decides all closed formulas.
- If S does not decide a closed formula ϕ then $unev(\phi, S)$ is non-empty

DEFINITION 13.
$$S_{\leq r} := \{(e, u) \in S \mid rk(e) \leq r\}$$

LEMMA 14. *If S and S' have the same domain and same values for Skolem functions of rank $\leq r$ (that is, $S_{\leq r} = S'_{\leq r}$) and ϕ contains only Skolem functions of rank $\leq r$ then $S \vDash \phi$ iff $S' \vDash \phi$.*

The purpose of ϵ-substitutions is the following theorem:

THEOREM 15. *Suppose that for every proof of a formula ϕ in $PA\epsilon$, there is an ϵ-substitution S such that $S \vDash \phi$. Then Peano Arithmetic is 1-consistent.*

Proof. If $PA \vdash \exists x.\phi[x, \vec{t}]$ then $PA\epsilon \vdash \phi[c^{\vec{y}}_{\exists x.\phi}(\vec{t}), \vec{t}]$ where the terms in $\vec{t}$ are numerals. By assumption, there is an S such that $S \vDash \phi[c^{\vec{y}}_{\exists x.\phi}(\vec{t}), \vec{t}]$, and therefore $\phi[S(c^{\vec{y}}_{\exists x.\phi}(\vec{t})), \vec{t}]$ is a true quantifier-free formula. ∎

Importantly, this theorem is provable in PRA: we will give a computable procedure for finding such a substitution. First, we find simpler conditions under which $S \vDash \phi$ holds.

LEMMA 16.
- If ϕ is an axiom other than a critical formula then $S \vDash \phi$
- If $S \vDash \phi$ and $S \vDash \phi \rightarrow \psi$ then $S \vDash \psi$

Therefore to show that S satisfies the conclusion of a proof, it suffices to show that S satisfies each critical formula appearing in the proof.

DEFINITION 17.
- $\phi[[v, \vec{t}]] := \phi[v, \vec{t}] \wedge \bigwedge_{u < v} \neg \phi[u, \vec{t}]$
- $\mathcal{F}(S) := \{\phi[[v, \vec{t}]] \mid (c^{\vec{y}}_{\exists x.\phi(x)}(\vec{t}), v) \in S \wedge v \neq ?\}$
- S is *correct* if for any $\phi \in \mathcal{F}(S)$, $\overline{S} \vDash \phi$

$\phi[[v]]$ just states that v is the smallest value where $\phi[x]$ holds. A correct ϵ-substitution ensures that whenever it assigns a numeral to some canonical term, it is assigning the minimal correct witness to the Skolem function.

From here on, let $Cr = \{Cr_0, \ldots, Cr_N\}$ be a fixed sequence of closed critical formulas.

DEFINITION 18. We say S is *solving* if for each $I \leq N$, $\overline{S} \vDash Cr_I$.

Let S be a finite, correct, nonsolving ϵ-substitution. We will consider the critical formulas made false by $\overline{S}$ and select the first one of minimal rank to be fixed. For $I = 0, \ldots, N$, we define parameters needed for the H-process. Informally, what we want is to fix a representation of a critical formula Cr in the form

$$\psi \rightarrow \phi[e(Cr), t(Cr)]$$

where
$$e(Cr) = c_{\exists x.\phi}^{\vec{y}}(t(Cr)).$$

Accordingly, we want to define a reduction $red(Cr, S)$ which has the property of applying $\hat{S}$ to every Skolem term in Cr except for the main term $c_{\exists x.\phi}^{\vec{y}}$. That is, $red(Cr, S)$ applies S as much as possible while making sure $red(Cr, S)$ is still a critical formula.

DEFINITION 19. If Cr is a critical formula, we define the *key term*, $e(Cr)$, the *parameters* $t(Cr)$, and the *reduced form* $red(Cr, S)$, by:

- If Cr has the form $\phi[s, \vec{t}] \to \phi[c_{\exists x.\phi}^{\vec{y}}(\vec{t}), \vec{t}]$ then $t(Cr) := \vec{t}$, $e(Cr) := c_{\exists x.\phi}^{\vec{y}}(t(Cr))$, and

$$red(Cr, S) := \phi[\hat{S}(s), \hat{S}(\vec{t})] \to \phi[c_{\exists x.\phi}^{\vec{y}}(\hat{S}(\vec{t})), \hat{S}(\vec{t})]$$

- If Cr has the form

$$\phi[0, \vec{t}] \wedge \neg\phi[t', \vec{t}] \to \phi[c_{\exists x.\phi[x] \wedge \neg\phi[\mathbf{S}x]}^{\vec{y}}(\vec{t}), \vec{t}] \wedge \neg\phi[\mathbf{S}c_{\exists x.\phi[x] \wedge \neg\phi[\mathbf{S}x]}^{\vec{y}}(\vec{t}), \vec{t}]$$

then $t(Cr) := \vec{t}$, $e(Cr) := c_{\exists x.\phi[x] \wedge \neg\phi[\mathbf{S}x]}^{\vec{y}}(t(Cr))$, and

$$red(Cr, S) := \phi[0, \hat{S}(\vec{t})] \wedge \neg\phi[\hat{S}(t'), \hat{S}(\vec{t})]] \to$$
$$\phi[c_{\exists x.\phi(x) \wedge \neg\phi(\mathbf{S}x)}^{\vec{y}}(\hat{S}(\vec{t})), \hat{S}(\vec{t})] \wedge \neg\phi[\mathbf{S}c_{\exists x.\phi(x) \wedge \neg\phi(\mathbf{S}x)}^{\vec{y}}(\hat{S}(\vec{t})), \hat{S}(\vec{t})]$$

- If Cr has the form $\neg s = 0 \to s = \mathbf{S}(c_{\exists x.y=\mathbf{S}x}^{y}(s))$ then $t(Cr) := s$, $e(Cr) := c_{\exists x.y=\mathbf{S}x}^{y}(t(Cr))$, and

$$red(Cr, S) := \neg \hat{S}(s) = 0 \to \hat{S}(s) = \mathbf{S}(c_{\exists x.y=\mathbf{S}x}^{y}(\hat{S}(s)))$$

We define $e(Cr, S) := e(Cr)(\hat{S}(t(Cr)))$.

Note that if $S \vDash \neg Cr$ where Cr has the form $\phi \to \psi[c]$ then there is a fixed u such that for any correct $S' \supseteq S$ deciding each $\psi[v]$ for $v \leq u$, there is some $v \leq u$ such that $S' \vDash \psi[[v]]$.

3 Finite Injury Relationships

We present the key idea behind the construction we will later introduce, the finite injury relationship between two trees. We are interested in a map λ from a tree T_1 to a tree T_2 with the property that well-foundedness of T_2 will guarantee well-foundedness of T_1. A particularly simple way to do this would be a "zero injury" relationship: if $x \subsetneq y$ in T_1 then $\lambda(x) \subsetneq \lambda(y)$ in T_2. The finite injury relationship is more flexible; in addition to allowing $\lambda(x)$ to extend $\lambda(y)$, $\lambda(x)$ might "correct" some choice of branch in $\lambda(y)$, but in such a way that the choice made at each node may only be "corrected" finitely many times. This ensures that eventually, the choice at each node stabilizes, so an infinite branch in T_1 would give rise to an infinite branch in T_2.

For our purposes, we use a simplified form, where branches are labeled with elements from $\mathbb{N} \cup \{?\}$ and the only possible correction is from ? to a value in $\mathbb{N}$.

DEFINITION 20. Let T_1, T_2 be trees such that the branches of T_2 are labeled by $\mathbb{N} \cup \{?\}$, and let $\lambda : T_1 \to T_2$ be given. We say λ is a *finite injury relationship* if whenever $x \subsetneq y$, either $\lambda(x) \subsetneq \lambda(y)$ or there is an $\alpha^\frown \langle ? \rangle \subseteq \lambda(x)$ such that $\alpha^\frown \langle u \rangle \subseteq \lambda(y)$ for some $u \in \mathbb{N}$.

Note that if we take the underlying set of T_2 to be partially ordered by $u < ?$ for all $u \neq ?$, this is the same as saying that λ is order-preserving from the extension ordering on T_1 to the Kleene-Brouwer ordering on T_2.

DEFINITION 21. We say $\lambda : T_1 \to T_2$ is *weakly finite injury* if whenever $x \subsetneq y$ either $\lambda(x) \subseteq \lambda(y)$ or there is an $\alpha^\frown \langle ? \rangle \subseteq \lambda(x)$ such that $\alpha^\frown \langle u \rangle \subseteq \lambda(y)$ for some $u \in \mathbb{N}$, and for any infinite sequence $x_0 \subsetneq x_1 \subsetneq \cdots \subsetneq x_n \subsetneq \cdots$, there is some n such that $\lambda(x_0) \neq \lambda(x_n)$.

This weakens the finite injury condition to allow finite runs where λ is constant.

LEMMA 22. *If T_2 is well-founded and $\lambda : T_1 \to T_2$ is weakly finite injury then T_1 is well-founded.*

Proof. Let $x_0 \subsetneq x_1 \subsetneq \cdots$ be an infinite branch in T_1. Then we may inductively construct an infinite branch μ in T_2 such that for each m, there is some i such that $\mu \upharpoonright m \subseteq \lambda(x_j)$ whenever $j \geq i$.

Suppose we have constructed $\gamma = \mu \upharpoonright m$, and let i be such that $j \geq i$ implies $\gamma \subseteq \lambda(x_j)$. Then, since $\gamma \subseteq \lambda(x_i)$ there is some $n > i$ such that $\gamma^\frown \langle u \rangle \subseteq \lambda(x_n)$ for some $u \in \mathbb{N} \cup \{?\}$. If $u \neq ?$ then it must be that $\gamma^\frown \langle u \rangle \subseteq \lambda(x_j)$ whenever $j \geq n$.

Otherwise, there are two possibilities. Either $\gamma^\frown \langle ? \rangle \subseteq \lambda(x_j)$ for all $j \geq n$, in which case we are done, or there is some $j > n$ and some $u \in \mathbb{N}$ such that $\gamma^\frown \langle u \rangle \subseteq \lambda(x_j)$, and therefore $\gamma^\frown \langle u \rangle \subseteq \lambda(x_k)$ whenever $k \geq j$. In either case, we have constructed $\mu \upharpoonright m+1$. ∎

In Section 6 we will show that when λ is finite injury, an ordinal bound on the height of T_2 can be converted to a bound on the height of T_1. (A similar argument could be made for λ weakly finite injury, but would be made significantly more complicated by the need to handle runs where λ is constant.)

4 Examples of Priority Trees

First, we describe the general motivation behind our priority construction. We are attempting a computation that depends on various parameters whose "ideal" value is non-recursive (specifically, the true values of Skolem functions). Fortunately, we don't need to know the true value of these parameters, only values which suffice to satisfy certain conditions, the critical formulas appearing in the proof.

Since the critical formulas contain parameters, which can themselves change in the course of our construction, a single critical formula may give rise to multiple conditions, as the values assigned to its parameters are changed. The first step of the construction will be the process of unwinding critical formula with parameters to a tree of formulas without parameters.

It is convenient to arrange conditions in a tree, where the nodes represent conditions and the branches representing the possible values that can be assigned to that condition. In our case, the core conditions will turn out to be canonical Skolem terms, only some of which will be the key terms of a critical formula. The others will be parameters needed to compute the correct way to satisfy critical formulas.

Having built such a tree, we will proceed in stages. At each stage we will proceed up from a node, choosing appropriate branches, until we reach a node associated with a critical formula. At this point, we will stop and consider how to satisfy that critical formula. We will then choose a branch, adding a Skolem term to our ϵ-substitution. This may invalidate previous choices, so we may have to backtrack; we will ensure that when we do so, we always backtrack to a node where we had chosen ?, and instead choose an integer. This ensures that our process is finite injury.

When dealing with higher rank critical formulas—that is, questions whose ideal solution is Σ_N for some potentially large N—we will have to use a tower of N trees. Roughly, the $n+1$-st tree will behave like a finite injury argument relative to the n-th tree: that is, as long as the n-th tree is simply accounting for information from the $n+1$-st, the $n+1$-st will behave in a finite injury way. When the $n+1$-st tree reaches a level n-condition, the n-th tree may force us to throw out some information from the $n+1$-st tree and start that process over, causing the $n+1$-st tree to exhibit more complicated behavior. So while it is difficult to describe the behavior of the $n+1$-st tree relative to the $n-1$-st directly (it is roughly that of an infinite injury argument), and essentially impossible to describe its behavior relative to the first level, we can describe each level's behavior as being finite injury relative to the previous level.

We first exhibit the simplified proof for the case where all Skolem terms have rank 1, which substantially simplifies the process of computing a solving substitution from our construction.

4.1 The Case of Rank 1

Suppose we have a set $Cr_0, \ldots, Cr_{K-1}$ of critical formulas such that $rk(e(Cr_I)) = 1$ for each $I < K$. We first produce a tree T_2 consisting of sequences from $\mathbb{N} \cup \{?\}$ with length at most K, and assign to each node α of length $I < K$ the formula $form_2(\alpha) := Cr_I$.

We fix an ω-ordering $\prec$ of T_2 so that when $\alpha \subsetneq \beta$ then $\alpha \prec \beta$. Next we construct another tree, T_1, also branching over $\mathbb{N} \cup \{?\}$. We will assign to each node in T_1 other than the leaves either a critical formula (whose key term is canonicl) or a canonical Skolem term. Formally, for each non-leaf $\alpha \in T_1$, exactly one of $e_1(\alpha)$ and $form_1(\alpha)$ will be defined. We will denote undefined values by $\bot$.

DEFINITION 23. We say S settles a node $\beta \in T_2$ if $red(form_2(\beta), S)$ contains no canonical terms other than the key term and $e(form_2(\beta), S)$ belongs to the domain of S.

If S settles every node of T_2, $e_1(S) := form_1(S) := \bot$. Otherwise let β be the $\prec$-least node of T_2 such that S does not settle β. If $red(form_2(\beta), S)$ contains canonical terms besides the key term, define $e_1(S)$ to be a canonical term contained in $red(form_2(\beta), S)$ and $form_1(S) := \bot$. Otherwise, we define $form_1(S) := red(form_2(\beta), S)$ and $e_1(S) := \bot$.

Finally, set $e_1(\bot) := form_1(\bot) := \bot$.

Now we may define by simultaneous recursion, for each node α, an ϵ-substitution $S(\alpha)$ and either $e_1(\alpha)$ or $form_1(\alpha)$:

DEFINITION 24.

- $e_1(\alpha) := e_1(S(\alpha))$

- $form_1(\alpha) := form_1(S(\alpha))$

- $S(\langle\rangle) := \emptyset$

- $S(\alpha^\frown\langle u\rangle) := \begin{cases} S(\alpha) \cup \{e(form_1(\alpha)), u)\} & \text{if } form_1(\alpha) \neq \bot \\ S(\alpha) \cup \{(e_1(\alpha), u)\} & \text{if } e_1(\alpha) \neq \bot \\ \bot & \text{otherwise} \end{cases}$

Set
$$T_1 := \{\alpha \mid S(\alpha) \neq \bot\}.$$

Here and below, we take a *path* through a tree to be either a node in that tree or an infinite sequence of nodes $x_0 \subsetneq x_1 \subsetneq \ldots$ all in that tree. Given any node $\alpha \in T_1$, we define a path through T_2:

DEFINITION 25.

- $\langle\rangle \subseteq \lambda(\alpha)$

- If $\beta \subseteq \lambda(\alpha)$ and $\widehat{S(\alpha)}(e(form_2(\beta)))$ is an integer n then $\beta^\frown\langle n\rangle \subseteq \lambda(\alpha)$

Note that when $\beta \subseteq \alpha$, $\lambda(\beta) \subseteq \lambda(\alpha)$. Furthermore, λ is weakly finite injury: it is easy to see that if β is $\prec$-least such that $S(\alpha)$ does not settle β then in fact there is a fixed n such that whenever α' is an extension of α with $lh(\alpha') \geq lh(\alpha) + n$ then $S(\alpha')$ settles β (because there are a fixed number of Skolem terms in $form_2(\beta)$ which might need to be assigned values). So if $\alpha_0 \subsetneq \alpha_1 \subsetneq \cdots \subsetneq \alpha_n \subsetneq \cdots$ then any $\beta \preceq \lambda(\alpha_0)$ must eventually be settled by $S(\alpha_n)$ for large enough n, and in particular $\lambda(\alpha_0) \subsetneq \lambda(\alpha_n)$ for large enough n.

Select a sequence of nodes through T_1' as follows:

- $\alpha_0 = \langle\rangle$

- If $e_1(\alpha_n)$ is defined, set $\alpha_{n+1} := \alpha_n^\frown \langle\overline{S(\alpha_n)}(e_1(\alpha_n))\rangle$

- If $form_1(\alpha_n)$ is defined to be $\phi \to \psi[c]$ and $S(\alpha_n) \vDash \phi \to \psi[0]$ then $\alpha_{n+1} := \alpha_n^\frown\langle\overline{S(\alpha_n)}(e(form_1(\alpha_n)))\rangle$

- If $form_1(\alpha_n)$ is defined to be $\phi \to \psi[c]$ and $S(\alpha_n) \vDash \neg(\phi \to \psi[0])$ then there is some n such that $S(\alpha_n) \vDash \psi[[n]]$. If there is some $\gamma \subseteq \alpha_n$ such that $e_1(\gamma) = e(form_1(\alpha_n))$ then set $\beta := \gamma$, otherwise set $\beta := \alpha_n$. Then set $\alpha_{n+1} := \beta^\frown\langle n\rangle$

This is a finite injury process from the integers to T_1', and therefore terminates at some node α. (Indeed, we could observe more directly that this process terminates because it is decreasing in the Kleene-Brouwer ordering on T_1'.) Observe that $S(\alpha)$ is correct and satisfies every critical formula along the path up to $\lambda(\alpha)$, and is therefore a solving substitution.

5 The Main Construction

5.1 Trees

Let $Cr_0, \ldots, Cr_{K-1}$ be a fixed sequence of critical formulas. The tree T_{N+1} will consist of sequences of length K; assign to each node α other than the leaves the critical formula Cr_I where I is the length of α; denote this by $form_{N+1}(\alpha)$.

To each node other than leaves in the trees T_i, $i \leq N$, we will assign either a canonical Skolem term of rank $\leq N$, which we will denote $e_i(\alpha)$, or a critical formula with canonical key term, which we will denote $form_i(\alpha)$.

Assuming we have made these assignments for T_{i+1}, fix a constructive ω-ordering $\prec$ of T_{i+1} with the property that if $\alpha \subsetneq \beta$ then $\alpha \prec \beta$.

DEFINITION 26. We say S *settles* a node β in T_{i+1} if one of the following holds:

- $e_{i+1}(\beta) \neq \bot$, $rk(e_{i+1}(\beta)) \leq i$, and $e_{i+1}(\beta) \in \text{dom}(S)$
- $e_{i+1}(\beta) \neq \bot$ and $rk(e_{i+1}(\beta)) > i$
- $form_{i+1}(\beta) \neq \bot$, $rk(form_{i+1}(\beta)) \leq i$, and $e(red(form_{i+1}(\beta), S)) \in \text{dom}(S)$
- $form_{i+1}(\beta) \neq \bot$, $rk(form_{i+1}(\beta)) > i$, $form_{i+1}(\beta)$ has the form $\phi \rightarrow \psi[c]$, and $S \vDash \phi \rightarrow \psi[0]$
- $form_{i+1}(\beta) \neq \bot$, $rk(form_{i+1}(\beta)) > i$, $form_{i+1}(\beta)$ has the form $\phi \rightarrow \psi[c]$, $S \vDash \neg(\phi \rightarrow \psi[0])$, and $S \vDash \psi[[n]]$ for some n
- $e_{i+1}(\beta) = form_{i+1}(\beta) = \bot$

If S settles every node of T_{i+1}, $e_i(S) := form_i(S) := \bot$. Otherwise let β be the $\prec$-least node of T_{i+1} such that S does not settle β. We define e_{i+1} or $form_{i+1}$ by the cases above:

- If $e_{i+1}(\beta) \neq \bot$, $rk(e_{i+1}(\beta)) \leq i$, but $e_{i+1}(\beta) \notin \text{dom}(S)$ then $e_i(S) := e_{i+1}(\beta)$ and $form_i(S) := \bot$
- It cannot be the case that $e_{i+1}(\beta) \neq \bot$ and $rk(e_{i+1}(\beta)) > i$
- If $form_{i+1}(\beta) \neq \bot$ and $rk(form_{i+1}(\beta)) \leq i$, but $e(red(form_{i+1}(\beta), S)) \in \text{dom}(S)$ then $form_i(S) := red(form_i(\beta), S)$ and $e_i(S) := \bot$
- If $form_{i+1}(\beta) \neq \bot$, $rk(form_{i+1}(\beta)) > i$, $form_{i+1}(\beta)$ has the form $\phi \rightarrow \psi[c]$, but S does not decide $\phi \rightarrow \psi[0]$ then let $e_i(S)$ be an element of $unev(\phi \rightarrow \psi[0])$ and $form_i(S) := \bot$
- If $form_{i+1}(\beta) \neq \bot$, $rk(form_{i+1}(\beta)) > i$, $form_{i+1}(\beta)$ has the form $\phi \rightarrow \psi[c]$, $S \vDash \neg(\phi \rightarrow \psi[0])$, but there is no n such that $S \vDash \psi[[n]]$ then let n be least such that S does not decide $\psi[n]$ and let $e_i(S)$ be an element of $unev(\phi[n], S)$ and $form_i(S) := \bot$

This definition is made clearer by imagining the following computation in T_{i+1}: given S, we begin traversing the tree by attempting, at each node β, to choose a correct value for $e_{i+1}(\beta)$ or $e(red(form_{i+1}(\beta), S))$. When the rank of this term is $\leq i$, we simply query S, and either choose the path given by S, or give up if $dom(S)$ does not contain this term.

When $e_{i+1}(\beta)$ is defined and has rank $> i$, we again query S, but if $dom(S)$ does not contain the term, we assume the default. (This is consistent with our view that S is the sole authority on terms of rank $\leq i$, but that computations in our tree may uncover new terms at rank $> i$.) Finally, when we have a critical formula of rank $> i$, we attempt to choose the correct value: if $S \vDash \phi \to \psi[0]$, we go up the path corresponding to 0, and if $S \vDash \psi[[n]]$ we go up the path corresponding to n; if S does not contain enough terms to settle the truth of this formulas, again we give up.

The tree T_{i+1} can be seen as the tree of all possible such computations. Then the definitions $e_i(S)$ and $form_i(S)$ are precisely the first query to S that fails in any branch of the tree. (Note, however, that we worry about ensuring that S resolves queries in all branches of the tree, not only those which would seem to be computed given S. This is because adding terms to S can sometimes alter earlier steps of the computation.)

Before proceeding, we have to note that the first tree, T_{N+1}, has slightly unusual behavior, since $form_{N+1}(\beta)$ may contain ϵ-terms other than the key term when $\beta \in T_{N+1}$; no other tree will allow this, so when $i < N$ and $\beta \in T_{i+1}$, we will have $e(red(form_{i+1}(\beta, S))) = e(form_{i+1}(\beta))$.

For $i < N$, this process then leads to a natural definition of a substitution from a node $\beta \in T_{i+1}$, so whenever $\alpha^\frown \langle u \rangle \subseteq \beta$, the substitution contains either $(e_{i+1}(\alpha), u)$ or $(e(form_{i+1}(\beta), S), u)$ as appropriate. We may then construct T_i from T_{i+1} by assigning to a node $\alpha \in T_i$ the task of resolving the first query T_{i+1} makes which is unanswered by the substitution corresponding to α. Formally, we define by simultaneous recursion, for each node $\alpha \in T_i$, an ϵ-substitution $S(\alpha)$ and either $e_i(\alpha)$ or $form_i(\alpha)$:

DEFINITION 27.

- $e_i(\alpha) := e_i(S(\alpha))$
- $form_i(\alpha) := form_i(S(\alpha))$
- $S(\langle\rangle) := \emptyset$
- $S(\alpha^\frown \langle u \rangle) := \begin{cases} S(\alpha) \cup \{e(form_i(\alpha)), u)\} & \text{if } form_i(\alpha) \neq \bot \\ S(\alpha) \cup \{(e_i(\alpha), u)\} & \text{if } e_i(\alpha) \neq \bot \\ \bot & \text{otherwise} \end{cases}$

Set
$$T_i := \{\alpha \mid S(\alpha) \neq \bot\}.$$

LEMMA 28. *If $\alpha_0 \subsetneq \alpha_n \subsetneq \cdots$ is an infinite sequence in T_i then for any $\beta \in T_{i+1}$, there is some n such that $S(\alpha_n)$ settles β.*

Proof. By induction along $\prec$. Suppose β is the $\prec$-least node in T_{i+1} such that no $S(\alpha_n)$ settles β. Then we may choose some m such that $S(\alpha_m)$ settles every $\gamma \prec \beta$. If $e_{i+1}(\beta) \neq \bot$ then $S(\alpha_{m+1})$ settles β. So $form_{i+1}(\beta)$ is defined. If $S(\alpha_m) \vDash \neg \phi \to \psi[0]$ then there is some n such that, if S extends $S(\alpha_m)$ and decides $\psi[i]$ for each $i \leq n$ then $S \vDash \psi[[i]]$ for some $i \leq n$; since $\bigcup_{i \leq n} unev(\psi[[i]], S(\alpha_m))$ is finite, it follows that $S(\alpha_k)$ resolves β for k sufficiently large.

In the remaining case, $S(\alpha_m)$ does not decide $\phi \to \psi[0]$; but then $unev(\phi \to \psi[0], S(\alpha_m))$ is finite, so there is a k such that $S(\alpha_k)$ decides $\phi \to \psi[0]$, and the previous argument applies. ∎

5.2 Building a Solving Substitution

Now we describe the actual construction of a particular solving substitution, using the trees $T_1, \ldots, T_{N+1}$. Analogous to the definition of $\lambda : T_1 \to T_2$ in example case, we define functions $\lambda_i : T_i \to T_{i+1}$. The definition is slightly more complicated because the natural definition of $\lambda_i(\alpha)$ is not given by a sequence of extensions; instead we will find $\lambda_i(\alpha)$ as the final element of a sequence α_+^n of elements from T_{i+1}.

Roughly speaking, the definition will proceed in the manner of the computation which was described above, with one additional complication. Recall that, when $e_{i+1}(\beta)$ is defined and has rank $> i$, we allowed the computation to proceed with a default value when S had not already assigned one. But it might be that later on in the computation, we encounter a node β' such that $e(form_{i+1}(\beta)) = e_{i+1}(\beta)$, and furthermore, that while $dom(S)$ does not contain $e_{i+1}(\beta)$, S does determine that the critical formula $form_{i+1}(\beta)$ should assign a value this term. Rather than try to proceed in the face of this conflict, we backtrack to β and proceed down a different branch. Note that when this happens, always takes some $\beta^\frown \langle ? \rangle$ and replaces it with $\beta^\frown \langle u \rangle$; in particular, it can only happen once at each node.

DEFINITION 29. We say S is *consistent* if for any $\phi \in \mathcal{F}(S)$, $\overline{S} \not\vdash \neg\phi$.

DEFINITION 30. For a node $\alpha \in T_i$ such that $S(\alpha)$ is consistent, we define a sequence of nodes α_+^n in T_{i+1} by recursion as follows. α_+^0 is $\langle \rangle$. If $S(\alpha)$ does not settle α_+^n or α_+^n is a leaf then the process terminates. Otherwise we split into cases.

If $e := e_{i+1}(\alpha_+^n)$ is defined and $rk(e) \leq i$ then $S(\alpha)(e)$ is defined and $\alpha_+^{n+1} := \alpha_+^n{}^\frown \langle S(\alpha)(e) \rangle$. If $rk(e) = i+1$ and there is some u such that $S(\alpha) \vDash \phi[[u, \vec{t}]]$ where $e(\alpha)$ is $c_{\exists x.\phi[x,\vec{y}]}^{\vec{y}}(\vec{t})$ then $\alpha_+^{n+1} := \alpha_+^n{}^\frown \langle u \rangle$, and if there is no such u, $\alpha_+^{n+1} := \alpha_+^n{}^\frown \langle ? \rangle$.

If $f := form_{i+1}(\alpha_+^n)$ is defined to be $\phi \to \psi[c]$ and $rk(f) \leq i$ then $S(\alpha)(e(f))$ is defined and $\alpha_+^{n+1} := \alpha_+^n{}^\frown \langle S(\alpha)(e(f)) \rangle$. Otherwise $rk(f) = i+1$, and either $S(\alpha) \vDash \phi \to \psi[c]$, in which case $\alpha_+^{n+1} := \alpha_+^n{}^\frown \langle ? \rangle$, or there is a u such that $S(\alpha) \vDash \psi[[u]]$. In this case, if there is some $\gamma \subseteq \alpha_+^n$ such that $e_{i+1}(\gamma) = e(form_{i+1}(\alpha_+^n), S(\alpha))$ then set $\beta := \gamma$, otherwise set $\beta := \alpha_+^n$. Set $\alpha_+^{n+1} := \beta^\frown \langle u \rangle$.

LEMMA 31. *For any $\alpha \in T_i$, if $S(\alpha)$ is consistent then for every n,*

- $S(\alpha_+^n)_{\leq i} \subseteq S(\alpha)$

- *If $\phi \in \mathcal{F}(S(\alpha_+^n))$ then $S(\alpha) \vDash \phi$*

- $S(\alpha_+^n)$ *is consistent*

Proof. By induction on n. The first two claims are immediate from the definition. To see the third, if $(\epsilon x.\phi[x], u) \in S(\alpha_+^{n+1})$ with $u \neq ?$ then either $(\epsilon x.\phi[x], u) \in S(\alpha_+^n)$ or $S(\alpha) \vDash \phi[[u]]$. Since $(S(\alpha_n))_{<rk(\epsilon x.\phi[x])} \subseteq S(\alpha_{n+1}) \cap S(\alpha)$ and both $S(\alpha_{n+1})$ and $S(\alpha)$ are consistent, it follows that $S(\alpha_{n+1}) \not\vdash \neg\phi[[u]]$. ∎

LEMMA 32. *For any $\alpha \in T_i$ such that $S(\alpha)$ is consistent, the function $n \mapsto \alpha_+^n$ is weakly finite injury.*

Proof. It is clear from the definition that either $\alpha_+^n \subsetneq \alpha_+^{n+1}$ or we are in the case where $form_{i+1}(\alpha_+^n)$ is defined to be $\phi \to \psi[c]$, $rk(f) > i$, $S(\alpha) \vDash \neg(\phi \to \psi[c])$, there is a u such that $S(\alpha) \vDash \psi[[u]]$ (since α settles α_+^n), and there is a $\beta^\frown \langle v \rangle \subseteq \alpha_+^n$ with $e_{i+1}(\beta) = e(form_{i+1}(\alpha_+^n, S(\alpha)))$. But it must be that $v = ?$, since otherwise the consistency of $S(\alpha)$ implies that $\overline{S(\alpha)} \nvDash \neg \phi[[v]]$; since $S(\alpha) \vDash \phi[[u]]$, $u = v$, which implies $S(\alpha) \vDash \phi \to \psi[c]$, which is a contradiction. ∎

Define $T'_{N+1} := T_{N+1}$, and having defined T'_{i+1}, define $\lambda_i : T_i \to T'_{i+1}$ by setting $\lambda_i(\alpha)$ to be the final α_+^n in this process (that is, to be α_+^n with n least such that either $S(\alpha)$ does not settle α_+^n or α_+^n is a leaf). Define T'_i to be those nodes $\alpha \in T_i$ such that $S(\alpha)$ is consistent and for no $\beta \subsetneq \alpha$ is $\lambda_i(\beta)$ a leaf—that is, leaves in T'_i are initial nodes α such that $\lambda_i(\alpha)$ is a leaf. We will show inductively that each λ_i is weakly finite injury on T'_i, and therefore that T'_i is well-founded, which will in turn ensure that λ_{i-1} is well-defined.

LEMMA 33. *λ_i is weakly finite injury on T'_i.*

Proof. Suppose $\alpha \subsetneq \beta$, and proceed by induction along α_+^n and β_+^n. If $\alpha_+^n = \beta_+^n$, it is clear that if the process terminates at β_+^n then it also terminates at α_+^n. If the process terminates at α_+^n but not β_+^n, the fact that $n \mapsto \alpha_+^n$ is weakly finite injury implies that either $\lambda_i(\alpha) \subseteq \lambda_i(\beta)$ or there is $\gamma^\frown \langle ? \rangle \subseteq \lambda_i(\alpha)$ such that $\gamma^\frown \langle u \rangle \subseteq \lambda_i(\beta)$ for some u. It is apparent from the definition and the consistency of $S(\alpha) \subseteq S(\beta)$ that these are the only possibilities: if α_+^{n+1} and β_+^{n+1} are both defined then they must be equal.

We must check that there are no infinite sequences $\alpha_0 \subsetneq \alpha_1 \subsetneq \cdots$ such that $\lambda(\alpha_0) = \lambda(\alpha_n)$ for all n. This follows immediately from Lemma 28 and the fact that $S(\alpha)$ never settles $\lambda(\alpha)$ except when α is a leaf. ∎

Now choose a path through T'_1 as follows:

- Define α_0 to be $\langle \rangle$
- If $e_1(\alpha_n)$ is defined, set $\alpha_{n+1} := \alpha_n^\frown \langle \overline{S(\alpha_n)}(e_1(\alpha_n)) \rangle$
- If $form_1(\alpha_n)$ is defined to be $\phi \to \psi[c]$ and $S(\alpha_n) \vDash \phi \to \psi[0]$ then $\alpha_{n+1} := \alpha_n^\frown \langle \overline{S(\alpha_n)}(e(form_1(\alpha_n))) \rangle$
- If $form_1(\alpha_n)$ is defined to be $\phi \to \psi[c]$ and $S(\alpha_n) \vDash \neg(\phi \to \psi[0])$ then there is some u such that $S(\alpha_n) \vDash \psi[[u]]$. If there is some $\gamma \subseteq \alpha_n$ such that $e_1(\gamma) = e(form_1(\alpha_n))$ then set $\beta := \gamma$, otherwise set $\beta := \alpha_n$. Then set $\alpha_{n+1} := \beta^\frown \langle n \rangle$.

This process is a weakly finite injury function from the natural numbers to T'_1, and therefore terminates after finitely many steps at some node α.

LEMMA 34. *$S(\alpha_n)$ is correct for each n.*

Proof. Note that, for rank 1 Skolem terms, consistency and correctness are identical. The argument is then the same as the argument that the sequence $n \mapsto \alpha_+^n$ preserves consistency. ∎

As in the arguments above, the process $n \mapsto \alpha_n$ is finite injury, so it terminates at some element γ_1. Let $\gamma_i := \lambda_{i-1}(\cdots \lambda_1(\gamma_1) \cdots)$ for each i. $\bigcup_{i \leq N} S(\gamma_i)$ is correct by Lemma 31. To check that it is solving, it suffices to see that it settles each leaf of T_{N+1}; but this follows from the fact that γ_1 is a leaf of T'_1, and therefore each γ_i is a leaf of T'_i, so $\lambda_N(\gamma_N)$ is a leaf of T_{N+1}.

This give a constructive proof of:

THEOREM 35. *If $Cr_0, \ldots, Cr_{K-1}$ are critical formulas then there is a correct ϵ-substitution S such that $S \vDash Cr_I$ for each $I < K$.*

Combining this with Theorem 5 gives:

THEOREM 36. *PA is 1-consistent.*

6 Ordinal Analysis

LEMMA 37. *Suppose $f : T_1 \to T_2$ is a finite injury relation, and T_2 has height α. Then there is a height function $o : T_1 \to \omega^\alpha$ such that $x \subsetneq y$ implies $o(y) < o(x)$.*

Proof. Let $h : T_2 \to \alpha$ be such that $s \subsetneq t$ implies $h(t) < h(s)$. Then we define $o : T_1 \to \omega^\alpha$ as follows:

$$o(x) = \left(\sum_{a \frown \langle ? \rangle \subseteq f(x)} \omega^{h(a)} \right) + \omega^{h(x)+1}$$

We must show that this is order-preserving. Let $y \subsetneq x$, and suppose $f(y) \subsetneq f(x)$. Then we have $f(y) \frown \langle u \rangle \subseteq f(x)$; if $u = ?$ then

$$o(x) = \left(\sum_{a \frown \langle ? \rangle \subseteq f(y)} \omega^{h(a)} \right) + \omega^{h(f(y))} + \left(\sum_{f(y) \frown \langle ? \rangle \subseteq a \frown \langle ? \rangle \subseteq f(x)} \omega^{h(a)} \right) + \omega^{h(f(x))+1}$$

So it suffices to show that

$$\omega^{h(f(y))} + \left(\sum_{f(y) \frown \langle ? \rangle \subsetneq a \frown \langle ? \rangle \subseteq f(x)} \omega^{h(a)} \right) + \omega^{h(f(x))+1} < \omega^{h(y)+1}$$

But this is clear, since $h(f(x)) + 1 \leq h(f(y)) < h(f(y)) + 1$ and $h(a) < h(f(y))$ whenever $f(y) \frown \langle ? \rangle \subseteq a$.

If $u \neq ?$ then this is even simpler, since the $\omega^{h(f(y))}$ term is omitted.

Now suppose that $b \frown \langle ? \rangle \subseteq f(y)$ and $a \frown \langle u \rangle \subseteq f(x)$. Then

$$o(y) = \beta + \omega^{h(f(y))} + \gamma$$

for suitable $\gamma < \omega^{h(f(y))} < \beta$, and

$$o(x) = \beta + \delta$$

where $\delta < \omega^{h(f(y))}$. Therefore $o(x) < o(y)$. ∎

BIBLIOGRAPHY

[1] Wilhelm Ackermann. Zur Widerspruchsfreiheit der Zahlentheorie. *Math. Ann.*, 117:162–194, 1940.
[2] Toshiyasu Arai. Epsilon substitution method for $ID_1(\Pi_1^0 \vee \Sigma_1^0)$. *Ann. Pure Appl. Logic*, 121(2-3):163–208, 2003.
[3] Toshiyasu Arai. Epsilon substitution method for $[\Pi_1^0, \Pi_1^0]$-FIX. *Arch. Math. Logic*, 44(8):1009–1043, 2005.
[4] Toshiyasu Arai. Ideas in the epsilon substitution method for Π_1^0-FIX. *Ann. Pure Appl. Logic*, 136(1-2):3–21, 2005.
[5] Toshiyasu Arai. Epsilon substitution method for Π_2^0-FIX. *J. Symbol Logic*, 71:1155–1188, 2006.
[6] C. J. Ash. Stability of recursive structures in arithmetical degrees. *Ann. Pure Appl. Logic*, 32(2):113–135, 1986.
[7] C. J. Ash. Labelling systems and r.e. structures. *Ann. Pure Appl. Logic*, 47(2):99–119, 1990.
[8] M. J. Groszek and T. A. Slaman. Foundations of the priority method, i: Finite and infinite injury. manuscript.
[9] J. F. Knight. A metatheorem for constructions by finitely many workers. *J. Symbolic Logic*, 55(2):787–804, 1990.
[10] S. Lempp and M. Lerman. Priority arguments using iterated trees of strategies. In *Recursion Theory Week, 1989*, number 1482 in Lecture Notes in Mathematics, pages 277–296, Berlin, Heidelberg, New York, 1990. Springer-Verlag.
[11] S. Lempp and M. Lerman. The existential theory of the poset of r.e. degrees with a predicate for single jump reducibility. *Jour. Symb. Logic*, 57:1120–1130, 1992.
[12] S. Lempp and M. Lerman. Iterated trees of strategies and priority arguments. *Arch. Math. Logic*, 36:297–312, 1997.
[13] D. A. Martin. Borel determinacy. *Ann. of Math.*, 2:363–371, 1975.
[14] Grigori Mints. Extension of epsilon substitution method to ID1, streamlined version. preprint, 2003.
[15] Grigori Mints, Sergei Tupailo, and Wilfried Buchholz. Epsilon substitution method for elementary analysis. *Archive for Mathematical Logic*, 35:103–130, 1996.
[16] R. M. Solovay. Degrees of models of true arithmetic. preliminary version, 1984.
[17] Yue Yang. Iterated trees and fragments of arithmetic. *Arch. Math. Logic*, 34(2):97–112, 1995.

Grigori Mints and Computer Science
ENN TYUGU

ABSTRACT. A survey is presented of works of Grigori Mints from the eighties of the last century where logic was applied to program synthesis and semantics of specification languages. It demonstrates examples of fruitful application of logic in computing. The main results of these works are rather practical: defining limits of application of the intuitionistic propositional calculus to program synthesis and introduction of a practical program synthesis method that is based on type inference.

1 Coming to Tallinn

We in Tallinn established contacts with Grigori Mints in 1980. He started participating in the summer- and winter-schools organized with the best Soviet computer scientists and mathematicians as lecturers. A few years before that, Grigori had participated in a successful large project of construction of a natural deduction theorem-prover together with N. Shanin, S. Maslov et al. [20] which required considerable amount of programming, but at that time, he was neither highly experienced nor interested in computer science and computer applications. It was a pleasant surprise that he became interested in the program synthesizer that we were developing with a strong practical orientation. The synthesizer called PRIZ [12] had been developed as a result of a long experimentation and worked on mainframes similar to IBM370. We were trying to explain the synthesis algorithm in terms of logic, but found little understanding from logicians. Grigori spent many hours questioning us about the synthesis algorithm and trying to fit it into logic.

Grigori lost his researcher position in St. Petersburg, because he was a quiet dissident – not actively outspoken, but enough to be persona non grata for the official Soviet science scene. For a shorter period he even worked in a software house in St. Petersburg as a programmer, writing code in the IBM assembler language. Hillar Aben – Director of the Institute of Cybernetics of Estonian Academy of Sciences – listened to my pleading and agreed to give Grigori a researcher position at the Institute of Cybernetics in Tallinn in the end of 1980. This was a stroke of good luck for the fields of computer science and logic in Estonia. Grigori spent ten years in Tallinn – until 1991, when he took a position as a professor at Stanford University. He started educating us in logic, published an easily readable preprint on logical foundation of program synthesis [8], and gave lectures. He was teaching logic, doing research, organizing scientific meetings (including a large meeting on relations between computing and logic COLOG-88 [5]) and communicating with scientists of different countries as much as the circumstances permitted it. Institute of Cybernetics hosted S. McLane, H. Barendregt, S. Maslov, P. Martin-Löf, J. McCarthy,

P. Suppes, J.-Y. Girard, V. Lifshits, A. Slisenko, E. Griffor and many other mathematicians and logicians as visitors invited by Grigori Mints.

2 Structural synthesis in the beginning

In the beginning of the eighties we started calling the program synthesis that was used in PRIZ "structural synthesis of programs" [21], because it relied on structural properties of computations, taking into the account only which variables were inputs and outputs of building blocks of a synthesized program – like it is generally accepted now in composition of web services, e.g. [19]. Actual relations between the values of inputs and outputs were ignored. However, functional inputs were permitted, and the values of these inputs had to be synthesized as well.

A goal of the synthesis (a synthesis problem) was presented by defining only input and output variables of the required program. Complete specification of a synthesis problem was presented as a set of all objects that are candidates for being computed (i.e. that may be involved in solving the problem) and a set of functions that can be used for computing these objects. The elements of the first set could be called variables, because they could get values computed by the elements of the second set, but these were variables with single assignment, hence not conventional program variables. Besides these two sets, a set of functional variables that are inputs of the functions was given. Such a variable had to be evaluated during the program synthesis process, if a function having it as an input was used in computations.

A specification for structural synthesis of a program could be easily represented as a graph with nodes as elements of the sets from above, and incidence relation of nodes representing data dependencies (i.e. inputs and outputs) of functions. In order to explain the synthesis algorithm, a symbolic notation was introduced for the specification as well. Each function was represented by a formula $x_1, \ldots, x_m \to y_1, \ldots, y_n\{f\}$ called a computability statement, where $x_1, \ldots, x_m$ were the inputs and $y_1, \ldots, y_n$ the outputs of the function f. There was a problem that some of the inputs were functional variables that had to be evaluated during the program synthesis, and not during the execution of the program. Inputs and outputs of a functional variable were given in a specification as well. Therefore such a variable itself could be described by a formula that gives its inputs and outputs: $u_1, \ldots, u_k \to v_1, \ldots, v_l$. The order of inputs and outputs of a function could be ignored at the derivation of an algorithm, hence a formula for a function took the form $a \to b$ with a and b sets of its inputs and outputs. Different versions of the synthesis algorithm were described by respective derivation rules for stepwise construction of an algorithm. As an example of a derivation rule here is a rule for a case when no functional variables were present

$$\frac{A \to B \cup C \quad (C \cup G \to F)}{A \cup G \to F},$$

where A, B, C, G, F are metavariables denoting sets of variables. Handling functional inputs of the building blocks required higher order notations, and this was not exactly defined for the structural synthesis of programs. This was how we saw the structural synthesis of programs (SSP) before Grigori became involved.

3 Completeness of structural synthesis

We were quite sure that the synthesis algorithm of PRIZ was complete in some sense (which we had not defined yet). But, I remember very well one morning, when Grigori just arrived by train from St. Petersburg, where he still was working part time as a programmer in a company called "Lengipromjasomolprom", and gave an example of a specification and a goal, asking whether the goal is solvable by the PRIZ system. It became soon clear that this goal was unsolvable by the synthesis algorithm implemented in PRIZ. This event started a discussion of exact representation of the *SSP* in logic and, naturally, attempts to improve the synthesis algorithm. As a result, the synthesis algorithm was changed, and the first paper on logic of *SSP* appeared [10].

The following is a brief summary of the results on completeness of SSP. Let us assume for simplicity that the output of a function is always only one (maybe structured) value. A precise logical description of a building block f, represented earlier by a computability statement

$$x_1, x_2, \ldots, x_k \to y\{f\}$$

is as follows:

$$\forall u_1 \forall u_2 \ldots \forall u_k (X_1(u_1) \wedge X_2(u_2) \wedge \ldots X_k(u_k) \to Y(f(u_1, u_2, \ldots, u_k))), \quad (*)$$

where $X_1, X_2, \ldots, X_k$ are unary predicate symbols denoting computability of a proper value of x_i, so that the formula $X_i(s_i)$ can be read as "the value s_i is a suitable value for the input (or output) x_i". In the case of m functional variables $\varphi_1, \varphi_2, \ldots, \varphi_m$ as additional inputs of a building block F, the logical formula describing the building block takes the form

$$\bigwedge_{1 \leq i \leq m} (\forall s_{i,1} \forall s_{i,2} \ldots \forall s_{i,k_i} (U_{i,1}(s_{i,1}) \wedge U_{i,2}(s_{i,2}) \wedge \ldots U_{i,k_i}(s_{i,k_i}) \to$$
$$V_i(\varphi_i(s_{i,1}, s_{i,2}, \ldots, s_{i,k_i})))) \to \quad (**)$$
$$\forall u_1 \forall u_2 \ldots \forall u_k (X_1(u_1) \wedge X_2(u_2) \wedge \ldots X_k(u_k) \to$$
$$Y(F(u_1, u_2, \ldots, u_k, \varphi_1, \varphi_2, \ldots, \varphi_m))).$$

This is the general form of formulas appearing in the logical language of structural synthesis. The nested implications on the left side and functional variables may be missing in many specifications of building blocks like in (*). To present the logical rules of structural synthesis of programs we will use the abbreviation (***) instead of the form (**)

$$\bigwedge_{1 \leq i \leq m} (\boldsymbol{U_i} \to V_i\{\varphi_i\}) \to (\boldsymbol{X} \to Y\{F\}), \quad (***)$$

where the structure of a formula is preserved, but all bound variables as well as inessential indices are omitted, $\boldsymbol{U_i}$ and $\boldsymbol{X}$ denote conjunctions of unary predicates. Let us note that this formula includes functional variables $\varphi_1, \varphi_2, \ldots, \varphi_m$

and a higher-order functional constant F. From now on it will be assumed that the formulas are closed – universally quantified with respect to the variables $\varphi_1, \varphi_2, \ldots, \varphi_m$ as well.

The structural synthesis rules SSR in a sequent notation were defined as follows:

$$\frac{\Rightarrow \bigwedge_{1 \leq i \leq m}(\boldsymbol{U_i} \to V_i\{\varphi_i\}) \to \quad \Gamma_i \Rightarrow \boldsymbol{U_i} \to V_i\{g_i\},}{\to (\boldsymbol{X} \to Y\{F\});\quad i=1,2,\ldots,m}{\Gamma_1, \Gamma_2, \ldots, \Gamma_m \Rightarrow \boldsymbol{X} \to Y\{F^*\}} \quad (\to --)$$

$$\frac{\Gamma \Rightarrow \boldsymbol{X} \to Y\{f\}; \quad \Sigma_i \Rightarrow X_i(t_i), i=1,2,\ldots,n}{\Gamma, \Sigma_1, \Sigma_2, \ldots, \Sigma_n \Rightarrow Y\{f(t_1, t_2, \ldots, t_n)\}} \quad (\to -)$$

$$\frac{\Gamma, \boldsymbol{X} \Rightarrow Y(t)}{\Gamma \Rightarrow \boldsymbol{X} \to Y\{\lambda \boldsymbol{x}.t\}} \quad (\to +)$$

Axioms $\Gamma, H \Rightarrow H$ for any formula H are also assumed. $\boldsymbol{X}$ on the left side of the sequent in the rule $(\to +)$ denotes a list of unary predicates. Σ, Γ denote lists of formulas. F^* denotes the result of substitution of g_i for $\varphi_i, i = 1,2,\ldots,m$ in F. We distinguish between terms for unary predicates and functional terms for implications by using different brackets: () and {} respectively.

Grigori proved the following two theorems in [10] that demonstrate the completeness of SSR.

THEOREM 1. *Let $C_1, \ldots, C_k$ be formulas of the form (***) and let C be a goal $\boldsymbol{U} \to V$. Then a sequent $C'_1, \ldots, C'_k \Rightarrow C'$ is derivable in the calculus of natural deduction if and only if $\Rightarrow C$ is derivable from $\Rightarrow C_1, \ldots, \Rightarrow C_k$ by the rules SSR, where $C'_1, \ldots, C'_k, C'$ are respective formulas of the form (**).*

THEOREM 2. *Let $C_1, \ldots, C_k$ be formulas of the form (***), $\boldsymbol{W}$ a list of unary predicates denoting computability, and let $\boldsymbol{K}$ be a conjunction of the goals $\boldsymbol{U_i} \to V_i$, $1 \leq i \leq m$. Then the sequent*

$$C'_1, \ldots, C'_k, \boldsymbol{W} \Rightarrow \boldsymbol{K}$$

is derivable by natural deduction if and only if all the sequents

$$\boldsymbol{W} \Rightarrow \boldsymbol{U_i} \to V_i, \ 1 \leq i \leq m$$

have normal deductions from $\Rightarrow C_1, \ldots, \Rightarrow C_k$ according to the rules of SSR. Here $C'_1, \ldots, C'_k$ denote respective formulas of the form (**).

Proof of the "if" part of these theorems is rather obvious – SSR are admissible rules of intuitionistic logic for the abbreviated form (***) of formulas. Proof of these theorems in the reverse direction is based on the observation that any natural deduction can be transformed into deduction in the long normal form, and deductions in the long normal form can be easily transformed into deductions with SSR [7, 10].

4 It can be a propositional logic

The formula (**) includes only unary predicates, and the quantifiers are used so that we can move quantifiers close to the predicates with respective bound variables and transform the quantified subformulas into the form $\exists u U(u)$. Indeed, instead of a conventional specification $\forall x(P(x) \to \exists y R(x,y))$ of a program (or its building block) where input x and output y are bound by a relation R, the SSP uses a specification $\forall x(P(x) \to \exists y R(y))$. It can be presented in the equivalent form $\exists x P(x) \to y R(y)$. Now the closed subformulas $\exists x P(x)$ and $\exists y R(y)$ can be considered as propositions, i.e. the specification becomes $P \to R$, where P denotes computability of the input and R denotes computability of the output of a program. Bearing in mind that the proofs in SSP are constructed in intuitionistic logic, one should not worry about the description of computations – the realizations of formulas will be programs, more precisely – lambda terms that can be easily converted to programs. This is how the authors reasoned when writing the paper [13].

The logical language of SSP becomes now an implicative fragment of the propositional language with restricted nestedness of implications. The general form of the formulas of a specification is (***), assuming that a special case without nested implications is also accepted. But now the formulas obtain precise meaning instead of being just abbreviations of quantified formulas. The symbols g_i and F denote now lambda terms that are realizations of implications $U_i \to V_i$ and $(X \to Y)$ respectively. They are constants for preprogrammed building blocks of programs, and more complex lambda-terms built for derived formulas. The inference rules for this language remain in essence the same as shown above. Changing slightly the rule $(\to --)$ gives us derivations where each step exactly corresponds to the application of a preprogrammed function – a computation step. The changed rule requires that all inputs of a function are computed, and the rule is as follows:

$$\frac{\Rightarrow \bigwedge_{1 \leq i \leq m}(U_i \to V_i\{\varphi_i\}) \to (X \to Y\{F\}); \quad \Gamma_i \Rightarrow U_i \to V_i\{g_i\},\ \Sigma_j \Rightarrow X_j(t_j),\quad i=1,2,\ldots,m;\quad j=1,2,\ldots,n}{\Gamma_1,\Gamma_2,\ldots,\Gamma_m,\Sigma_1,\Sigma_2,\ldots,\Sigma_n \Rightarrow Y\{F(g_1,\ldots,g_m,t_1,\ldots,t_n)\}} (\to --)$$

Now the rule $(\to -)$ becomes a special case of the rule $(\to --)$, hence it is not needed any more.

Grigori pointed out that the language of SSP is expressive and SSR rules are complete in the following sense. Given a list L of intuitionistic propositional formulas and a proposition B, the list L can be transformed into a list L' of formulas of the form (***) such that $L' \Rightarrow B$ is derivable in SSR if and only if $L \Rightarrow B$ is derivable in intiutionistic logic [10]. This transformation is performed by introducing new propositional variables for every subformula $A * B$ with a connective $*$ that spoils the form (***) and applying equivalent substitution theorem – substituting a new variable W instead of a subformula $A * B$, and adding the implications $W \to A * B$ and $A * B \to W$ to the list L so that the derivability is preserved. Elimination of $\bot$ as well as elimination of $\vee$ on the right side of an implication requires more effort and expands the list L' polynomially, because new implications have to be introduced for every existing propositional variable, see [10].) These transformations are suggested by the following two second order equivalences:

$$(p \vee q) \leftrightarrow \forall x((p \to x) \to ((q \to x) \to x)) \quad \text{and} \quad \bot \leftrightarrow \forall x x.$$

We made a theorem-proving experiment – proved all intuitionistic propositional theorems (more than one hundred theorems) from S. Kleene's "Introduction to metamathematics" [4], first, encoding the theorems in the input language of the PRIZ system and then using the *SSP* program synthesizer as a theorem prover [23]. This happened to be an interesting experiment, because the prover that had been initially designed for program synthesis gave some interesting proofs. An example is a derivation of the intuitionistic analog $((((A \to B) \to A) \to A) \to B) \to B$ of the valid classical formula $((A \to B) \to A) \to A$. Because of the deeper nestedness of implications, this formula had to be rewritten by introducing new propositional variables X, Y denoting respectively $A \to B$ and $X \to A$. This gave a new formula $((Y \to A) \to B) \to B$. Taking B as a goal, one had to derive the sequent $\Rightarrow B$ from the sequent

$$\Rightarrow (Y \to A) \to B,$$

using possibly also the axioms that were added according to the equivalent replacement condition:

$$\Rightarrow X \to (A \to B)$$
$$\Rightarrow (A \to B) \to X$$
$$\Rightarrow Y \to (X \to A)$$
$$\Rightarrow (X \to A) \to Y.$$

The proof was as follows:

$$\frac{\Rightarrow (Y \to A) \to B; \quad \dfrac{\Rightarrow Y \to (X \to A); \quad \dfrac{\Rightarrow (A \to B) \to X; \quad \dfrac{\Rightarrow (Y \to A) \to B; \quad Y, A \Rightarrow A}{A \Rightarrow B}(\to --)}{Y \Rightarrow Y; \quad \Rightarrow X}(\to --)}{Y \Rightarrow A}(\to --)}{\Rightarrow B}(\to --)$$

5 Specifications as types

In practical applications, specifications for program synthesis are written in a language different from the logical language used for synthesis. Looking from a practical side – program synthesis has become today a part of a compilation technique of declarative problem-oriented languages, and most of the logic is hidden in a compiler. Although this approach was not common in the eighties of the last century, the *SSP* was used already in combination with a user-friendly declarative language. A question of the precise semantics of this language arouse. In the paper [15], an attempt was made by G. Mints, J. Smith and E. Tyugu to define the semantics of a small specification language in three different ways: by translating it into logic and applying *SSP* (*logical semantics*); by considering a specification as a presentation of a type, and proving that the type of a goal is inhabited (*types semantics*); by interpreting the specification in a set theory (*sets semantics*). The equivalence of these three semantics was shown. Considering the idea of formulas as types, the relatedness of these semantics should not be a surprise.

Let us look at the semantics of the kernel language of specifications for *SSP* presented in [11] and [15]. A specification written in this language is a sequence of statements of the following form:

$$a : (x : s; \ldots; y : t),$$

where $a, x, \ldots, y$ are new names, a will denote an object and $x, \ldots, y$ will denote its components that are also objects having types given by the type specifiers $s, \ldots, t$. Names of the components $x, \ldots, y$ used outside of the object a are $a.x, \ldots, a.y$, i.e. the prefix a is added to the names. Longer compound names like $a.x.u$ may also appear, depending on the types specified by $s, \ldots, t$. A type specifier can be

- name of a primitive type
- name of a object specified earlier
- an expression of one of the following forms

$$u_1, \ldots, u_m \to u_{m+1}$$
$$u_1, \ldots, u_m \to u_{m+1}\{f\},$$

where $u_1, \ldots, u_m, u_{m+1}$ are names of components, f is a name of a predefined function.

Logical semantics of the language was defined by giving a method of constructing a program for a given specification and a solvable goal. This method consisted, first, of rules for translating a specification in the language of SSP, and second, applying the SSP. The translation was rather straightforward: unfolding the specifications and copying the implications expressing computability for all introduced objects and their components. Besides that, one had to introduce extra implications to represent a structure of objects. For each compound object a with components $x, \ldots, y$ the following new implications had to be introduced:

$$a \to x, \ldots, a \to y \text{ and } x, \ldots, y \to a.$$

These implications described the computability of components of a and the computability of a itself from its components.

It is easy to see that the kernel language describes objects of simple types, and it is easy today to apply type inference to find a proper term for a given goal. However, a more general approach with dependent types was used for types semantics in the work [15] done together with Jan Smith from the Gothenburg computer science group, where program synthesis in Martin-Löfs type theory was investigated. The types $(\Pi x : A)B$ and $(\Sigma x : A)B$ were used for representing $A \to B$ and $A \wedge B$ respectively. For a specification S with representation $\Theta(S)$ in the type language, and a solvable goal $a \to b$, the types semantics gave a term f such that $\Theta(S) \vdash f : (\Pi x : A)B$, where $\vdash$ denotes the derivability in Martin Löfs type theory as described in [18]. Here again, first the representation $\Theta(S)$ of the specification was found, and then the term f was built.

6 Induction in a propositional language

The program synthesis used in the PRIZ system permitted to synthesize recursive programs from axioms given in the language of SSP. In order to formulate an extension of derivation rules for synthesis of recursive programs, the notion of a sequent was extended so that inductive proofs became possible. Expressions of the form $[A]$ for formulas A were allowed to occur on the left side of a sequent for expressing the induction. The following rule for recursion was added:

$$\frac{\Gamma, [\boldsymbol{X} \to Y], \boldsymbol{X} \Rightarrow Y}{\Gamma, \boldsymbol{X} \Rightarrow Y} \quad (\text{Rec})$$

This rule was obtained from the usual transfinite induction rule by suppressing individual variables like we did in Section 3 for all bound variables. The rule $(\to - - R)$ had to be extended as well:

$$\frac{\Rightarrow \bigwedge_{1 \leq i \leq m}(U_i \to V_i) \to (\boldsymbol{X} \to Y); \quad \Gamma_i \Rightarrow U_i \to V_i, \ i = 1, 2, \ldots, l}{\Gamma_1, \Gamma_2, \ldots, \Gamma_l, [U_i \to V_i]_{l+1 \leq i \leq m} \Rightarrow \boldsymbol{X} \to Y} \quad (\to - - R)$$

Also the programming language had to be extended with a recursion functional. This approach was quite innovative in the eighties. Introducing induction without explicitly requiring well-foundedness meant that one could not ensure termination of a synthesized program. The proof of termination had to be done by other means (remained a responsibility of the user). A simple example of a recursive synthesis is the following synthesis of factorial:

$$\cfrac{\cfrac{\cfrac{\Rightarrow (N \to F) \to (N \to F)}{[N \to F] \Rightarrow N \to F} (\to - - R)}{N \Rightarrow F} (\text{Rec})}{\Rightarrow N \to F} (\to +)$$

7 Propositional logic programming and more synthesis

Although logic programming is defined broadly as "using logic for program construction", it is often thought of as predicate Horn clause programming, e.g. Prolog. But, more generally, logic programming means exploiting the basic truth that the structure of a program is similar to a constructive proof of the fact that the desired result of the program exists (can be computed). This enables one to use results obtained in logic for building correct program schemas or for determining schemas of computations and control the computations. We made an attempt to present program development in PRIZ, i.e. the structural synthesis of programs, as propositional logic programming, and to extend the scope of logic programming in this way [13, 17]. The control of computations in

Prolog is performed in runtime by unification, it may cause backtracking and unnecessary computations. Control in PRIZ is encoded in the preprogrammed higher order functions that call synthesized parts of programs. The search with backtracking is done before the actual computations, and unnecessary computations are avoided. Comparing the control in Prolog and PRIZ in a naïve way, we can say that a clause

$$P(X, Y) :- Q(\ldots), \ldots, R(\ldots)$$

means in Prolog "if $P(X,Y)$ may be useful, then try to use sequentially $Q(\ldots), \ldots, R(\ldots)$, if this is not successful, then try to find some other way." In PRIZ, the control of computations is preprogrammed in the implementations of formulas with nested implications. A formula

$$(u \to v) \wedge \ldots (s \to t) \to (x \to y)\{F\}$$

means in PRIZ "if F may be useful, then try to find functions for solving $u \to v, \ldots, s \to t$ and apply them as prescribed by F."

Grigori collected a number of papers from the Soviet logicians and computer scientists, and we put them in a special issue of the Journal of Logic Programming [2]. The papers by I. Babaev [1], by M. Kanovich [3], and by G. Mints and E. Tyugu [13] in this journal are on propositional logic programming. An interesting paper of V. Mikhailov and N. Zamov [6] is about the synthesis of technological algorithms that is also a program synthesis in some sense.

Grigori explained another synthesis algorithm used in PRIZ for so called independent subtasks in modal logic [17]. He also wrote a comprehensive and elegant paper about the complexity of proof search in different fragments of the intuitionistic propositonal logic [9], and I think he got the idea that this kind of a paper may be useful for computer scientists, when discussing several modifications of the PRIZ synthesizer, e.g. independent subtasks.

The independent subtasks are useful because of less complex search needed for synthesis. A subtask is called independent, if its solvability does not depend on the availability of inputs of other subtasks. In the case when all subtasks are independent, the order of testing the solvability of subtasks is inessential as long as no new variables are computed. Hence, for finding in the synthesis process a new applicable function, one has to make at most n solvability tests of subtasks where n is the number of subtasks. A subtask solvability test can be performed in linear time with respect of the number of occurrences of propositional variables in all formulas. As the result, the time complexity of search in this case is polynomial. (It is exponential in the general case of SSP as it is for the proof search in the intuitionistic propositional calculus.)

For the independent subtasks, one used another rule $(\to --)ind$ instead of the SSR rule $(\to --)$:

$$\frac{\Rightarrow \bigwedge_{1 \leq i \leq m}(U_i \to V_i\{\varphi_i\}) \quad U'_i \Rightarrow V_i\{g_i\}, \quad \Sigma_j \Rightarrow X_j(t_j),}{\to (X \to Y\{F\}); \quad i = 1, 2, \ldots, m; \quad j = 1, 2, \ldots, n} \quad (\to --)ind$$
$$\Rightarrow Y\{F(g_1, \ldots, g_m, t_1, \ldots, t_n)\}$$

where U'_i denoted a list of variables occurring in the conjunction U_i. One can see that the second premise of this rule did not contain formulas Γ_i that made the complexity of the search with the rule $(\to\ --)$ exponential, this is why the complexity of search with $(\to\ --)ind$ was simpler. The rule $(\to\ --)ind$ together with the SSR rules $(\to\ -)$ and $(\to\ +)$ were called $SSRind$.

Unfortunately for the PRIZ developers, using the rule $(\to\ --)ind$ did not fit in the logic as we understood it. To help us, Grigori translated the formulas and sequents of the synthesis with independent subtasks in the language of modal logic by rewriting implications $A \to B$ of the subtasks as strict implications $(A \supset B)$. He proved that a sequent in this language is derivable in the modal logic S4 if and only if the original sequent in the language of SSP is derivable in $SSRind$ [17].

8 Concluding remarks

Working together with computer scientists in Tallinn, Grigori became an advocate of using logic in computer science. It was his idea to organize together with P. Martin-Löf and P. Lorents an international conference where logicians and computer scientists could meet. The International Conference on Computer Logic COLOG-88 [5] was a big success and it indeed brought together researchers from computing and logic from many countries. He participated also several years in the Soviet New Generation Computer Project START, helping to develop the logic of an object-oriented software environment NUT [22] that supported program synthesis.

Speaking about Grigori's influence on computer science one must not ignore his influence on young scientists whose educator he has been. A number of well-known professors found their way to logic and computer science due to Grigori, for example, Mati Pentus (now professor of Moscow University), Tarmo Uustalu and Tanel Tammet (professors of Tallinn Technical University). I am not speaking about Grigori's role in teaching logic in general, but still wish to point out Sergei Tupailo as a special example. Sergei was Grigori's doctoral student, he defended one Ph.D. thesis at Stanford University and another at Tartu University. He is dedicated to foundations of mathematics, and is known by his works on NF.

I take the complete responsibility for all mistakes and for the form of the presentation of the results here, although most of the ideas presented in this paper belong to Grigori Mints, who has had invaluable influence on the computer scientists who were lucky to work with him.

BIBLIOGRAPHY

[1] I. Babajev. Problem Specification and Program Synthesis in the System SPORA. J. of Logic Programming, v. 9, No. 2&3 (1990) 141–157.
[2] Journal of Logic Programming, v. 9, No. 2&3 (1990)
[3] M. Kanovich. Efficient Program Synthesis in Computational Models. J. Logic Programming, v. 9, No. 2&3 (1990) 159–177.
[4] S. Kleene. Introduction to Metamatemathics. North-Holland (1952)
[5] P. Martin-Löf, G. Mints (editors): COLOG-88, International Conference on Computer Logic, Proceedings LNCS, v. 417, Springer (1988)
[6] V. Mikhailov, N. Zamov. Deductive Synthesis of Solutions for Technological Tasks. J. Logic Programming, v. 9, No. 2&3 (1990) 195–220.
[7] G. Mints, E. Tyugu. The completeness of structural synthesis rules. Soviet Math. Doklady. V. 25 (1982) p. 2334–2336.

[8] G. Mints. Logical foundation of program synthesis. Institute of Cybernetics (in Russian). Tallinn (1982).
[9] G. Mints. Complexity of Subclasses of the Intuitionistic Propositional Calculus. Bit, v. 32, No. 1 (1992) 64–69.
[10] G. Mints, E. Tyugu: Justifications of the Structural Synthesis of Programs. Sci. Comput. Program. 2(3): 215–240 (1982)
[11] G. Mints, E. Tyugu. Semantics of a declarative language. Information Processing leters 23, Elsevier (1986) p. 147–151.
[12] G. Mints, E. Tyugu: The Programming System PRIZ. J. Symb. Comput. 5(3): 359–375 (1988)
[13] G. Mints, E. Tyugu: Propositional Logic Programming and the Priz System. J. Log. Program. 9 (2&3): 179–193 (1990)
[14] G. Mints, E. Tyugu: The Programming System PRIZ. Baltic Computer Science LNCS, v. 502, Springer (1991) 1–17
[15] G. Mints, J. M. Smith, E. Tyugu: Type-theoretical Semantics of Some Declarative Languages. Baltic Computer Science LNCS, v. 502, Springer (1991) p. 18–32
[16] G. Mints, T. Tammet: Condensed Detachment is Complete for Relevance Logic: A Computer-Aided Proof. J. Autom. Reasoning 7(4): 587–596 (1991)
[17] G. Mints. Propositional Logic Programming. In. J. Hayes, D. Michie, E. Tyugu (eds.) Machine Intelligence v. 12 (1991) 17–37
[18] B. Nordström, K. Petersson, J. M. Smith. Programming in Martin-Löf's Type Theory. Chalmers (1989)
[19] J. Rao. "Semantic Web Service Composition via Logic-based Program Synthesis". PhD Thesis. Department of Computer and Information Science, Norwegian University of Science and Technology, December 10, 2004.
[20] N. Shanin, G. Davidov, S. Maslov, G. Minc, V. Orevkov. A. Slisenko. An algorithm for a machine search of a natural logical deduction in a propositional calculus. (in Russian) Academy of Sciencies of the USSR, Steklov Mat. Inst., Leningrad department, "Nauka", Moscow (1965).
[21] E. Tyugu: The structural synthesis of programs. Algorithms in Modern Mathematics and Computer Science LNCS, v. 122, Springer (1979) 290–303
[22] E. Tyugu: Three New-Generation Software Environments. Commun. ACM 34(6): 46–59 (1991)
[23] B. Volozh, M. Matskin, G. Mints, E. Tyugu. Theorem proving with the aid of a program synthesizer. Cybernetics, No. 6, (1982) 63–70.

Complexity and stable evolution of circuits

S. VAKULENKO AND D. GRIGORIEV

ABSTRACT. We consider the viability problem for random dynamical systems, in particular, for circuits. A system is viable only if the system state stays in a prescribed domain Π of a phase space. We assume that the circuit structure is coded by a code evolving in time. We introduce the notion of stable evolution of the code and the system: evolution is stable if there is a $\delta > 0$ such that the probability P_T to be in Π within time interval $[0,T]$ satisfies $P_T > \delta$ as $T \to \infty$.

We show that for certain large classes of systems, the stable evolution has the following fundamental property: the Kolmogorov complexity of the code cannot be bounded by a constant as time $t \to \infty$. For circuit models, we describe examples of stable evolution of complicated boolean networks for a difficult case when the domain Π is unknown.[1]

1 Introduction

One of the main characteristics of biological systems is that these systems support their own life functions. In particular, a biological system tries to keep the values of the main characteristics of each cell – such as temperature, pressure, pH (acidity measure), concentrations of different reagents – within a certain domain of values that makes the biological processes possible. These domains of values are called *viability domains*, and the process of supporting the life functions – by keeping the values inside viability domains – is called *homeostasis*. The concept of homeostatis was first developed by the French physiologist Claude Bernard; it is now one of the main concepts of biology; see, e.g., [5].

The homeostasis process is notoriously difficult to describe in precise mathematical terms. At first glance, homeostasis is similar to the well-known and well-studied notion of stability: in both cases, once a system deviates from the desirable domain, it is pushed back. However, a more detailed analysis shows that these notions are actually different:

- the usual mathematical descriptions of stability mean that a system will indefinitely remain in the desired state, for time $t \to \infty$, while

- a biological cell (and the whole living being) eventually dies.

This difference has been emphasized in a recent paper by M. Gromov and A. Carbone: "Homeostasis of an individual cell cannot be stable for a long

---

[1] We dedicate this paper to Professor Grisha Mints. We admire the breadth of his interests.

time as it would be destroyed by random fluctuations within and off cell" ([13], p. 40).

One might argue that while individuals die, their children survive and thus, species remain. However, it turns out that the biological species are unstable too. This conclusion was confirmed, e.g., by L. Van Valen based on his analysis of empirical data; see, e.g., [23, 30]. Moreover, he concluded that the species extinction rate is approximately constant for all the species, so this species change is not just a problem for unfit species.

The species extinction does not necessarily mean complete extinction, it usually means that a species evolves and a new mutated better-fit species replaces the original one. From this viewpoint, the evolution is "stable" – in the sense that it keep life on Earth effectively functioning. However, as M. Gromov and A. Carbone mention, it is very difficult to describe this "stability" in precise terms: "There is no adequate mathematical formalism to express the intuitively clear idea of replicative stability of dynamical systems" ([13], p. 40). The problem of describing this idea in precise terms is called the *viability problem*.

Specifically, we need to formalize two ideas:

- First, that biological systems are unstable (in particular, under random perturbations).

- Second, that these systems can be stabilized by replication (evolution).

In this paper, we show that an important progress in solving both aspects of the viability problem can be achieved if we use the notion of Kolmogorov complexity. In our formalizations, we will use the basic concepts and ideas proposed by M. Gromov and A. Carbone [13], L. Van Valen [30], and L. Valiant [31, 32].

2 Systems under consideration

Let us describe the models that we will use to describe biological systems. Let n denote the number of quantities that characterize the current state of a given system. This means that the state of the system can be described by a tuple $u = (u_1, \ldots, u_n)$ consisting of the values of all these characteristics. The set H of all possible states of a system is therefore equal to $H = \mathbb{R}^n$.

The state of a biological system evolves with time. In practice, even when we have measuring instruments that "continuously" monitor the state of a biological system, we can only perform a finite number measurements in each time interval. So, in effect, we only know the values of the corresponding characteristics at certain moments of time. Thus, to get a good description of the observed data, it makes sense to consider discrete-time models.

Usually, there is a certain frequency with which we perform measurements, so we get values measured at moments T_0, $T_1 = T_0 + \Delta T$, $T_2 = T_0 + 2\Delta T$, ..., $T_t = T_0 + t \cdot \Delta T$, ... It is therefore convenient to call the integer index t of the moment T_t the t-th moment of time, and talk about the state $u(t)$, $t = 0, 1, \ldots$, at the t-th moment of time.

The state $u(t)$ of a system at a given moment of time affects the state of the system $u(t+1)$ at the next moment of time. The next state of the system is determined not only by its previous state: biological systems operate in an environment in which unpredictable ("random") fluctuations occur all the time. Let m be the number of parameters that describe such fluctuations; then, the current state of these fluctuations can be described by an m-dimensional vector $\xi(t) = (\xi_1(t), \ldots, \xi_m(t))$.

Once we know the current state of the system $u(t)$ and the current state $\xi(t)$ of all the external parameters that affect this system, we should be able to determine the next state $u(t+1)$. In other words, we consider the following dynamics:

(1) $\quad u_i(t+1) = f_i(u(t), \xi(t)), \quad t = 0, 1, \ldots$

with initial conditions $u_i(0) = \varphi_i$.

To specify evolution, we must therefore describe the transition functions f_i and the random process $\xi(t)$. We will do this in the following two subsections.

2.1 Transition functions

The transition functions f_i describe physical processes and thus ultimately come from physics. Most equations of fundamental physics – equations of quantum mechanics, electrodynamics, gravity, etc. – are partial differential equations with polynomial right-hand sides. Other physical phenomena are described by partial differential equations that use fundamental fields – i.e., solutions to the fundamental physics equations – as solutions. The resulting dependencies can be again used in the right-hand sides of other physics equations, etc. The resulting functions are known as *Pfaffian* functions; these functions are formally defined as follows (see [15]):

Definition 1.

- *By a* Pfaffian chain, *we mean a sequence of real analytic functions*

$$f_1(x), f_2(x), \ldots, f_r(x)$$

defined on $\mathbb{R}^n$ *which, for every* $j = 1, \ldots, r$, *satisfy a system of partial differential equations*

$$\frac{\partial f_j}{\partial x_k} = g_{kj}(x, f_1(x), \ldots, f_j(x)), \quad j = 1, \ldots, n,$$

with polynomials g_{kj}.

- *For each Pfaffian chain, the integer r is called its* length, *and the largest of the degrees of polynomials g_{kj} is called its* degree.

- *A function $f(x)$ is called* Pfaffian *if it appears in a Pfaffian chain.*

It is known that Pffafian functions satisfy many important properties; in particular:

- the sum and the product of two Pfaffian functions f_1 and f_2 of lengths r_i and degrees d_i are again Pffafian functions, of length $r_1 + r_2$ and degree $d_1 + d_2$;

- superpositions of Pfaffian functions are also Pfaffian.

Results from the theory of Pfaffian functions and the powerful computational tools that are based on these results are described in [10, 12, 15].

So, in this paper, we consider dynamical systems (1) with Pfaffian functions f_i. The class of such systems will be denoted by **Kh**:

Class Kh. *This class consists of the systems (1) for which f_i are Pfaffian functions.*

We will also consider several subclasses of this class, subclasses which are known to be useful in applications. Two of these subclasses are related to the fact that when the fluctuations are small and/or the deviation of the state from a nominal state is small, we can expand the dependence f_i into Taylor series and keep only the first few terms (or even only the first term) in this expansion – because higher-order terms can be safely ignored. In this case, we end up with a polynomial (or even linear) dependence.

Usually, the deviation of the state of a biological system from its nominal state can be reasonably large, so terms which are quadratic in this dependence cannot be ignored; however, random fluctuations can be small. When the random fluctuations are so small that we can only keep terms which are linear in ξ, we get the following class which is well studied in control theory:

Class Kl. *This class consists of the systems (1) in which the transition functions f_i have the form $f_i(u, \xi) = g_{0i}(u) + \sum_{k=1}^{m} \xi_k g_{ki}(u)$, with polynomial g_{ki}.*

When the fluctuations are larger and their squares can no longer be ignored, we get a more general class of systems:

Class Kp. *This class consists of the systems (1) in which the transition functions f_i are polynomial in u and ξ.*

Comment. Here, l in **Kl** stands for *linear* (meaning linear dependence on ξ), while **p** in **Kp** stands for *polynomial*.

Another important class comes from the situation when a random fluctuation simply means selecting one of the finitely many options.

Class Kr. *Let us assume that we have a finite family of maps $u \to \tilde{f}^{(k)}(u) = (\tilde{f}_1^{(k)}(u), ..., \tilde{f}_n^{(k)}(u))$, $u \in \mathbf{R}^n$, where $k = 1, ..., m'$. Assume that $f_i = \tilde{f}_i^{(k(t))}(u) + \lambda \cdot g_i(u, \xi)$, where $\lambda > 0$ is a parameter, $\tilde{f}_i^{(k)}$ and g_i are Pfaffian, and $k(t)$ is a random index: at each moment t we make a random choice of k with probabilities $p_k \geq 0$, $p_1 + p_2 + ... + p_{m'} = 1$ (these choices at different moments of time are done independently).*

In the particular case when all the maps $u \to \tilde{f}^{(k)}(u)$ are contractions and $\lambda = 0$, we obtain so-called *iterated function systems*; see, e.g., [14].

It is important to mention that the class **Kh** contains many neural and genetic *circuit models*. Genetic circuits were proposed to take into account theoretical ideas and experimental information on gene interaction; see, e.g.,

[11, 18, 21, 26]; see [25] for a review. In this paper, we consider the following model

$$(2) \quad u_i(t' + \tau) = \sigma\left(\sum_{j=1}^{N} K_{ij}(t')u_j(t') + h_i - \xi_i(t')\right), \quad u_i(0) = x_i,$$

where $t' = 0, \tau, 2\tau, \ldots, d \cdot \tau$, $i = 1, 2, \ldots, N$, d and N are positive integers, $\tau > 0$ is a real parameter, and $x = (x_1, \ldots, x_N)$ is an initial condition. It is usually assumed that the function σ is a strictly monotone increasing function for which $\lim_{z \to -\infty} \sigma(z) = 0$ and $\lim_{z \to \infty} \sigma(z) = 1$. Such systems have interesting applications to biology, e.g., to the morphogenesis problem [29].

Circuits (2) can simulate all Turing machines [16]. Also, they can generate all (up to topological equivalency) kinds of structurally stable semiflows with discrete time [28].

We want to restrict ourselves to Pfaffian systems. The functions σ used in practical applications are Pfaffian functions of length 1; moreover, they are solutions of a differential equation $\sigma' = P(\sigma)$, where P is a polynomial for which $P(0) = 0$, $P(1) = 0$, and $P(z) > 0$ for all $z \in (0,1)$. Thus, in this paper, we will consider only such functions σ. Even with this Pfaffian limitation, we can still get both above-mentioned universality properties: with respect to Turing machines and with respect to topological behavior.

2.2 Assumptions on random processes ξ

In this paper, we assume that the fluctuations $\xi_i(t)$ are:

- *Markov processes*, i.e., that the probabilities of different values of $\xi(t)$ depend only on the previous values $\xi_i(t-1)$, and

- are *strong*, in the sense that there is a positive probability to move into a close vicinity of any state.

Formally, this assumption of strong fluctuations can be described as follows.

For $\delta > 0$ let $V(\theta, \delta)$ denote the δ-neighborhood of a point $\theta \in \mathbf{R}^m$.

Assumption 1. *Assume that $\xi_i(t)$ are Markov processes with discrete time, $t = 0, 1, 2, \ldots$. Assume that for each $\theta, \delta > 0$, all positive integers $t > t_0$, and each starting point θ_0 the probability that the process ξ is in the neighborhood $V(\theta, \delta)$ at the moment t is positive:*

$$\mathrm{Prob}(\xi(t) \in V(\theta, \delta) \,|\, \xi(t_0) = \theta_0) > c(\delta) > 0,$$

where a constant $c(\delta)$ is uniform in t, t_0.

Mathematically, this assumption is one of the versions of ergodicity of the Markov process. This assumption holds for many stochastic processes that are used in modeling biological phenomena.

2.3 Evolution

The system (1) is well suited to describe the dynamics of a single individual. Individuals belonging to different species s may have different dynamics. So, a more accurate way to describe the dynamics is to use the equation $u_i(t+1) = f_i(u(t), \xi(t), s)$, where s describes the species.

In biology, different species and subspecies can be characterized by their DNA, i.e., by a sequence of symbols. Without losing generality, we can always encode the 4-values language of DNA codons into a binary code, so we can assume that s is a finite binary sequence.

In mathematical terms, we consider a discrete (finite or countable) set S with $N(S) \leq +\infty$ elements s. We assume that all elements of the set S are binary strings, i.e., that $S \subseteq S_\infty$, where S_∞ denotes the set of all possible finite binary strings.

The fact that the transition functions f_i depend not only on $u(t)$ but also on s can be described by saying that we extend our original phase space H of all the states u to a larger space $H \times S$.

In addition to dynamics within a species, we also have to take into account the possibility of mutation, when s changes. We assume that these transitions follow a Markov chain with transition probabilities $p_{s's}(u)$ to go from s' to s; in line with the biological applications, we take into account that the probability of different transitions (mutations) may depend on the state u.

To take into consideration that only states from a certain set $\Pi \subseteq \mathbb{R}^n$ (called the *viability domain*) are viable, we use the following standard construction: We introduce, formally, an absorbing state a such that $p_{as}(u) = 0$ for each $s \neq a$. If u leaves the viability domain Π, then the system automatically reaches this absorbing state a.

So, our model is defined by:

1. a family of random dynamical systems $u_i(t+1) = f_i(u_i(t), \xi(t), s)$, $i = 1, 2, ..., n$, corresponding to different binary strings $s \in S$;

2. a set $\Pi \subseteq \mathbf{R}^n$;

3. a Markov chain $\mathbf{M}$ with the state space $S \cup \{a\}$ and the transition matrix $\mathbf{W}(u)$ with entries $p_{s's}(u)$ (the transition probability from s' to s depending on u) such that $p_{as}(u) = 0$ (if $s \neq a$).

About dependence of f on $s \in S$ we assume the following. Consider a class $\mathcal{C}$ of dynamical systems (1) with f depending on parameters $r \in \mathcal{P}$, where $\mathcal{P}$ is a set of possible values of the parameters. We assume that the set $\mathcal{P}$ is equipped with a measure ν.

For example, if the functions f_i are defined by a sequence of polynomials (as is the case when f_i are Pfaffian functions), then $\mathcal{P}$ is the set of all tuples of coefficients of all these polynomials, and as ν, we can select the standard Lebesgue measure on this set.

It is reasonable to assume that parameters r are random functions of s. To describe these random functions, we need to introduce, for every natural number l, a probability measure on the set of all possible mappings α from binary strings of lengths $\leq l$ to the set $\mathcal{P}$. It is reasonable to make the following assumption:

Assumption 2. *For every set $A \subseteq \mathcal{P}$ of ν-measure 0, for every integer l, and for every string s of length $\leq l$, the probability that the parameters $\alpha(s)$ are in A is 0:*
$$\mu_l(B_l(s)) = 0, \text{ where } B_l(s) \stackrel{\text{def}}{=} \{\alpha : \alpha(s) \in A\}.$$

For systems from the class $\mathcal{P}$, denote by $P_\Pi(v,r,t)$ the conditional probability that in the next moment of time the system will still be viable ($u(t+1) \in \Pi$) under the condition that its previous state is $u(t) = v \in \Pi$ and that the previous value of the parameters was $r(t) = r$.

Definition 2. *We say that a class of system $\mathcal{P}$ from the general class* **Kh** *is generically unviable in Π, if there exists a function $\kappa(r) > 0$ for which*

$$(3) \quad \sup_{u \in \Pi, t=0,1,2,\ldots} P_\Pi(u,r,t) = 1 - \kappa(r)$$

for ν-almost all values of the parameter r.

This means that at every step there is a non-zero probability $\geq \kappa(r) > 0$ of moving into an unviable state – and since the unviable state is absorbing, we are guaranteed to eventually move into an unviable state.

For every viable state $u_0 \in \Pi$ and for every integer T, by $P_T(\Pi, u_0)$ we denote the conditional probability that $u(t) \in \Pi$ for all $t = 1, 2, \ldots, T$ under the condition that $u(0) = u_0$.

Definition 3. *We say that the evolution is stable if there exists a real number $\delta > 0$ for which $P_T(\Pi, u_0) > \delta$ for all integers $T > 0$ and all states $u_0 \in \Pi$.*

If such a real number δ does not exist, we say that the evolution is *unstable*.

3 Main result

For an arbitrary Turing machine F and for every string s, by $K_F(s)$ we denote the Kolmogorov complexity of the string s relative to F, i.e., the shortest length of the program (= initial configuration) on F for which F generates s [17].

In the following text, by a *Kolmogorov complexity* of a string s, or simply *complexity* (for short), we mean $K_F(s)$ for a fixed Turing machine F.

Theorem 1. *Let F be an arbitrary Turing machine. Consider a class of generically unviable systems. Assume that the Markov chain* **M** *and the system (1) generate strings s with a priori bounded Kolmogorov complexities $K_F(s)$ relative to F. Then, for almost all mappings $\alpha : s \to r(s)$ of strings s to the parameters r, the evolution is unstable and the corresponding system is not viable: $P_T \to 0$ as $T \to \infty$.*

Comment 1. Instead of K_F, one could take any function K' satisfying the following property: for any n there exists finitely many strings s with $K'(s) = n$.

Comment 2. In this analysis, we only consider the Kolmogorov complexity of the *codes* s, and not of the states u themselves. Complexity of the states can also be studied for systems similar to **Kh**; see, e.g., [29].

Comment 3. Theorem 1 says that the evolution is unstable for almost all mappings α, but it does not tell us whether the stable evolution is possible for *some* functions α. The existence of stable evolutions is analyzed in Section 6 for the case of *circuits* – i.e., systems of type (2).

4 Viability and unviability

It is difficult to determine when a given class of systems is generically unviable. Under Assumption 1 on ξ, a natural way for proving unviability is to consider ξ as a control and to use methods from control theory; see, e.g., [19, 24]. This reduction to control leads to complex attainability and controllability problems. We will describe several results that can be thus obtained.

Theorem 2. *Assume that we have a system (1) from the class* **Kp**, *with polynomials* $f_i(u, \xi)$ *of positive degree d, and with* $m \geq 2$. *Assume also that the viability domain* Π *is bounded, i.e.,* $\Pi \subseteq B_R$ *for a ball* B_R *of some radius R. Then for ν-almost all polynomials* f_i, *the system (1) is stochastically not viable, i.e.,* $P_T \to 0$ *as* $T \to \infty$.

Comment. Moreover, for almost all tuples of polynomials f_i, there exists a value $\kappa(f)$ for which

(4) $\quad P_\Pi(u, f, t) \leq 1 - \kappa(f)$

for all $u \in \Pi$ and t.

We present a proof of this theorem at the end of this section. Before that, let us consider other types of systems. Systems from the Class **Kr** can be both viable and not viable. Indeed, let Π be a bounded set.

Example 1. Suppose that, for some $R > 0$, for every u, the range of the map $\xi \to g(u, \xi)$ contains the ball $B_R = \{g : |g| < R\}$. Then one can show that for a sufficiently large $\lambda > 0$, the corresponding system is not viable.

Example 2. Let us consider situations when the functions g_i are uniformly bounded and $m' = 1$. Then, by the definition of m', the corresponding dynamical system $u(t+1) = \tilde{f}(u(t))$ is deterministic (not random). In particular, we can consider the case when this system has an attractor consisting of hyperbolic equilibria points, and that this attractor is contained in the viability domain Π. Then one can show [4] that if the initial state $u(0)$ is sufficiently close to the attractor, then, for sufficiently small values $\lambda > 0$, the corresponding system $u(t+1) = \tilde{f}(u(t)) + \lambda g(u(t), \xi(t))$ is viable.

Let us prove Theorem 2. We start with the following preliminary lemma.

Lemma 1. *Let Π be a compact set. Consider a system of polynomial equations*

(5) $\quad g_i(u) = 0, \quad i = 1, \ldots, N,$

where g_i are polynomials, and the number of equations N is greater than the number of variables n. Then the probability that this system has a solution $u_* \in \Pi$ is equal to 0.

Proof. This lemma easily follows from the resultant theory; see, e.g., [33].

Proof of Theorem 2. Since the set Π is bounded, there exists a real number $R > 0$ such that if $|u| > R$ then $u \notin \Pi$. For systems of class **Kp**, one has $f(u, \xi) = \sum_{l:|l|<d} h_l(u)\xi^l$, where $l = (l_1, \ldots, l_m)$ is a multi-index, $h_l(u)$ are polynomials of u, $|l| \stackrel{\text{def}}{=} l_1 + \ldots + l_m$, and $\xi^l \stackrel{\text{def}}{=} \xi_1^{l_1} \cdot \ldots \cdot \xi_m^{l_m}$. Consider a finite tuple $a = (a_1, a_2, \ldots, a_m)$, where a_j are different positive numbers. Set $\xi_j = a_j \cdot z$, and let $z \to +\infty$.

Suppose $|f(u, \xi)| < C$ for all $\xi(z)$, where $C > 0$. Then one can conclude that $h_l(u) = 0$ for all l for which $|l| < d$. The equations $h_l(u) = 0$ form $\geq n \cdot (d+1)$ polynomial equations with n unknowns u_i.

Now we apply Lemma 1 and conclude that since, in general, such a system has no solutions, in general, the values $|f(u, \xi)|$ are not bounded as $z \to \infty$. Thus, if $u(t) \in \Pi$, for some ξ we have $|u(t+1)| > R$ and consequently $u(t+1) \notin \Pi$. The theorem is proven.

5 Proof of Theorem 1

First, let us show that a stable evolution is possible only when the code length is unbounded in time.

Indeed, suppose that the lengths len(s) of all the codes s are a priori bounded by an integer l. The number of such codes is bounded, and thus, due to Assumption 2, for almost all maps α,

(6) $$\min_{s:\text{len}(s) \leq l} \kappa(r(s)) > \kappa_0 > 0.$$

Indeed, one can observe that the set of all maps α for which $\kappa(r(s)) = 0$ for some string s of length $\leq l$ is contained in the finite union of the sets $B_l(s)$ of measure 0: $\mu_l(B_l(s)) = 0$. Then, since our process is a Markov one, according to Assumption 1, the probability $P_T(\Pi)$ for $\xi(t)$ to be in Π at time moments $0, 1, \ldots, T$ is smaller than $(1 - \kappa_0)^T$, and we conclude that the evolution is unstable. This proves the theorem for the case when all strings have a priori bounded length.

Let us note now that the lengths $l(s)$ of the strings of the relative Kolmogorov complexity $K_F(s)$ not exceeding K are a priori bounded – since there are only finitely many such strings: $l(s) < N_K$ for some N_K. Therefore, all strings of complexity $< K$ are contained in a finite set $\mathcal{B}_K$ of binary strings. The theorem is proven.

Comment. It is worth mentioning that while for every Turing machine F and for every integer K, there exists an upper bound on the length of all the strings s with $K_F(x) \leq K$, this upper bound is not always effectively computable. For example, for a universal Turing machine F, the impossibility of an algorithm computing such upper bounds follows from the well-known theorem of Rabin [22].

6 Stable evolution of circuit population

In this section, we show that for circuits, stable evolution is possible.

Specifically, we consider an evolution of a family ("population") of circuits (2). To simplify our analysis, we consider the boolean case, when the values $u_i(t)$ are always 0s or 1s. In this case, σ is the step function, i.e., $\sigma(z) = 1$ for $z > 0$ and $\sigma(z) = 0$ for $z \leq 0$.

We also assume that for every time t, there is a positive integer $b(t)$ called *connection intensity*. For every i and j, the value $K_{ij}(t)$ is equal either to $b(t)$ or to $-b(t)$ or to 0. In other words, the quantity $u_j(t')$ either "excites" the value $u_i(t)$ at the next moment of time t, or inhibits it, or does not affect this value.

We assume that $h_i = \left(m_i + \dfrac{1}{2}\right) \cdot b$, where m_i are integers, and that the number N, in general, changes with time: $N = N(t)$.

At every moment t, the situation can be described by a directed graph $(V(t), E(t))$ whose vertices correspond to components u_i: $V(t) = \{1, 2, \ldots, N(t)\}$, and where there is an edge $(i \to j) \in E(t)$ if and only if $K_{ij}(t) \neq 0$. Each graph represents a single circuit.

Each step of the evolution of an individual circuit consists of one of the following changes in the graph and in the corresponding values K_{ij}:

1. the graph (V, E) stays the same;

2. one adds a node to V;

3. one adds an edge $i \to j$ to E with a new weight K_{ij};

4. one change a weight K_{ij}; when the new value of K_{ij} is 0, this change deletes an edge $i \to j$.

Steps 2–4 will be called *mutations*. We will assume that mutations occur with a given probability $\mu > 0$.

Based on these individual changes, we can perform the following changes in the population:

- First, at each time step we can simultaneously change many circuits in the population, by performing changes 1–4 on different circuits.

- We can also replicate (make copies of) some circuits and delete ("destroy") some other circuits.

We consider a population consisting of $X(t)$ random circuits (2) of different structure and different depths $d(t)$. We will describe now the set Π. In this description, we will use several ideas from [31]. Suppose that a circuit $Circ_j$, a member of the population, survives at the moment t if and only if it gives a correct output $y(x)$ as an answer to a boolean input x: $y = f(x, t)$, where $f(x, t)$ are given boolean functions depending on t. The output y is the final state of some node: $y(x) = u_1(\tau \cdot d)$, where $u_i(t)$ are computed by the formulas (2) starting with $u(0) = x$. The whole population survives if it contains at least one circuit.

We suppose that f are a priori unknown: to survive, circuits should "learn" correct answers. So, in effect, we are dealing here with the notions from the learning theory [32] – but in a different context.

Suppose that the correct answers are defined by a special piecewise constant sequence of boolean functions

(7) $\quad f(x,t) = f_j(x), \quad N(t) = N_j \quad t \in ((j-1) \cdot T_e, j \cdot T_e],$

where T_e is a positive number (the "length" of the j-th evolution stage) and $x = (x_1, x_2, \ldots, x_{N_j})$. Here we also assume that each function f_j belongs to certain class $\mathcal{C}$ of boolean circuits (2) (naturally, the values N, K, and d can depend on j). Assume that the parameter τ is small enough; thus, we should not take into account the time $\tau \cdot d$ of the circuit reactions.

The problem can be interpreted as a problem of adaptive behavior of a large growing population of evolving circuits under the challenge of a "random environment". Let us now formulate our assumptions about this environment.

Suppose that at the j-th evolution stage, the values x are chosen randomly by a probability distribution $P_j(x)$ on the set Ξ of all possible inputs x. We assume that each circuit obtains the values generated by the same distribution P_j and that the values corresponding to different circuits are independent.

We say that the circuit (2) is *correct* if, whenever the noise is turned off ($\xi(t) = 0$), this circuit returns a correct answer for every input x. For each pair of functions f and f', we can define the probability of error

$$Err(f, f') \stackrel{\text{def}}{=} \text{Prob}\{f(x) \neq f'(x)\},$$

where the probability in the right hand side is defined with respect to P_j. We can then define, for every j, the probability

(8) $\quad Err_j \stackrel{\text{def}}{=} \inf_{f \neq f_j, f \in \mathcal{C}} Err(f_j, f),$

and $\delta_j \stackrel{\text{def}}{=} Err(f_j, 0)$, where 0 denotes a trivial circuit with output 0.

Here, two drastically different situations are possible:

A *Passive environment:* in this case, all the distributions P_j are the same, $P_j = P$. In this case, the environment does not actively interact with the circuits.

B *Active environment*, an environment that tries to create as many difficulties as possible to the circuit population. This may correspond to a predator-prey-type interaction, when a predator tries to learn the prey's behaviour and vice versa. In this case, the probability distributions P_j can be different. (Here, interesting situations appear when for large j, the probabilities corresponding to the distributions P_j are not computable in polynomial time.)

Our objective is to show that a stable evolution is possible. We will show that for the above-described population of circuits, a stable evolution is possible – provided that the circuit growth satisfies some natural conditions. These conditions are listed below.

R1 *We assume that for all time moments $t = 1, 2, \ldots,$*

$$\text{(9)} \quad Res(t) = \sum_C \sum_{i,j,(i,j)\in E_C(t)} |K^C_{ij}(t)| < Poly(t),$$

where the first sum ranges over all circuits involved in the population, and $Poly(t)$ is a polynomial.

This assumption means that, within each time interval $[0,T]$, $t=1,2,\ldots,T$, the evolution process can only use resources whose total amount is bounded by a polynomial of T.

R2 There exists a value $\beta > 0$ such that the noises $\xi_i(t)$ corresponding to different i and t are independent identically distributed (i.i.d.) random quantities for which, for each $a > 0$, we have

$$\text{(10)} \quad 0 < P(|\xi_i(t)| > a) < \exp(-\beta \cdot a).$$

R3 The population size is polynomially bounded: $X(t) < Poly(t)$.

Our main assumption about the functions f_j can be described as follows. Let us assume that a conditional relative complexity of the correct outputs increases slowly in some reasonable sense; for example, we can assume that

$$f_{j+1} = g(f_j, f_{j-1}, \ldots, f_1, x), \quad g \in \mathcal{C},$$

$$\text{(11)} \quad d(g) = depth(g) \leq d_{\max}, \quad Comp(g) < K_{\max},$$

where $d_{\max}$ and $K_{\max}$ are constant (independent on j), and $Comp(g)$ denotes a circuit complexity, i.e., the number of elementary steps necessary to construct g.

Let us first formulate a simple lemma showing that sometimes one can survive without learning.

Lemma 2. (survival without learning) *If the series $\sum_{j=1}^{+\infty} \delta_j$ converges, then for every value $p_0 \in (0,1)$ there exists a circuit population that survives with the probability $\geq p_0$, i.e., for which $P_T > p_0$ for all T.*

Proof. Take X identical circuits with $\xi_i = 0$. For every input, each circuit generates 1. For such individual circuits, the probability P_T to survive within the time interval $[0,T]$ is then equal to $P_T = \prod_{j=1}^{T}(1-\delta_j)^{T_c}$. By taking into account that $\delta_j \to 0$ as $j \to \infty$, we conclude that as $T \to \infty$, the values P_T are bounded from below by some value $\kappa > 0$.

If $\kappa < p_0$ we increase X until we get $\kappa \geq p_0$. The lemma is proven.

Theorem 3. *Assume that for some real number $\rho \in (0,1)$, the functions f_j satisfy the conditions (11) and*

$$\text{(12)} \quad Err_j > \rho$$

for all j. Then there exist values μ and T_e for which there exists an algorithm describing evolution of circuits that satisfies the conditions **R1**, **R2**, and **R3**, and for which $P_T > p_0 > 0$ for all $T > 0$.

In other words, for this algorithm, the system remains stochastically stable for large time intervals.

This theorem can be interpreted as follows: stable evolution is possible even in severe conditions (when a single error leads to destruction) – if the rate of change of the environment complexity is bounded.

Proof. In this proof, we will use two lemmas from [6]. Recall that a *Bernoulli process* is a discrete-time stochastic process consisting of a sequence of independent random variables that take only two values: *success* and *failure*. For each integer M and real number $p \in (0,1)$, we can consider an M-trial Bernoulli process, in which in each of the M trials, the probability of success is equal to p. Let us denote the total number of successes in all M trials by Y.

Lemma 3. *For $k < M \cdot p$, we have*

$$(13) \quad \operatorname{Prob}\{Y < k\} \leq \frac{k \cdot (1-p)}{M \cdot p - k} \cdot C_M^k \cdot p^k \cdot (1-p)^{M-k}.$$

Lemma 4. *For $r > M \cdot p$, we have*

$$(14) \quad \operatorname{Prob}\{Y > M \cdot p + r\} \leq \left(\frac{M \cdot p \cdot e}{r}\right)^r.$$

Since for our choice of K_{ij} and h_i we have $\min |K_{ij} \cdot u_j + h_i| \geq 0.5b$, we can prove the following useful lemma.

Lemma 5. *Let $y(x)$ be a circuit (2) of depth d and complexity K_{Max}, for which $K_{ij} \in \{b, 0, -b\}$ and $\xi = 0$. Let $\tilde{y}(x)$ be the same circuit with the noise ξ (which satisfies the condition **R2**). Then*

$$(15) \quad \sup_x \operatorname{Prob}\{y(x) \neq \tilde{y}(x)\} < \exp(-c \cdot \beta \cdot b), \text{ where } c = c(d, K_{Max}) > 0.$$

Let us now describe a circuit evolution and a population growth that satisfy the conditions **R1**, **R2**, and **R3**. We will proceed in three stages. Our estimates are obtained by induction.

Suppose that at $j = m$ the populations contain correct circuits that give correct answers with probabilities $p'_m = 1 - \exp(-c_1 \cdot m)$, where $c_1 > 0$.

Stage I. Generation of new circuits by random mutations. Consider the time interval $I_m = [m \cdot T_e, m \cdot T_e + K_{max}]$, where $T_e > C \cdot d_{max}$ for some large constant $C > 1$. Denote by $\tilde{x}$ a combined entry (x_0, x), where we use $x_0 = (f_m, f_{m-1}, \ldots, f_1)$ as an additional entry component. Using steps 1–4, we construct all possible circuits of complexity $\leq K_{max}$ and depths $\leq d_{max}$. Among them correct circuits may occur, i.e., circuits coinciding with $g(f_m, x)$. For $t \in I_m$, we set $b(t) = b_*$, where b_* is a large constant independent of m. Such a correct circuit can be obtained with the probability $p_c^+(b_*) \cdot \mu^K$, where $\mu > 0$ is the mutation probability and $p_c^+(b_*)$ is the probability that an incorrect circuit gives a correct answer.

We have already obtained the estimate $p_c^+(b_*) > \exp(-c_2 \cdot b_*)$. Denote $\kappa \stackrel{\text{def}}{=} (\mu \cdot p_c^+(b_*))^K$. Then one can expect that, after $K_{\max}$ steps, we will have at least $X_+ = 0.5\kappa \cdot X_m$ correct circuits, where X_m is the number of circuits at the moment $m \cdot T_e$. Indeed, using Lemma 3, one can prove the following result:

Lemma 6. *Consider the random number $Z_m \stackrel{\text{def}}{=} X_+(m \cdot T_e + K_{max} - 1)$ of correct circuits $X_+(t)$ at the moment $t = m \cdot T_e + K_{max} - 1$. If the parameter β is small enough, then the probability that $Z_m < 0.5\kappa \cdot X_m$ can be bounded by the following expression:*

(16) $\operatorname{Prob}\{X_+(t) < 0.5\kappa \cdot X_m\} < \exp(-c_3 \cdot X_m),$

where $c_3(\mu, p_c^+, K)$ is a positive constant that does not depend on X_m.

Stage II. Removing circuits and increasing b. The following $T_1 = T_e - K_{max} - 1$ time steps we do nothing, no mutations. Many circuits die, as a result of incorrect answers. On this stage, we increase the parameter b in these circuits (Step 4) by setting $b = b_2 = O(m)$. Denote

- by P_1^*, the probability that at the moment $t = T_e(m+1)$, the number $\tilde{X}_+ = X_+(m \cdot T_e + K_{max} - 1)$ of correct circuits is smaller than the number $\tilde{X}_- = X_-(m \cdot T_e + K_{max} - 1)$ of incorrect ones: $\tilde{X}_+ < \tilde{X}_-$, and

- by P_0^*, the probability that there are correct circuits left, i.e., that $\tilde{X}_+ > 0$.

Lemma 7. *There exist values T_1, c_4, and c_5 for which the probabilities P_i^* satisfy the following inequalities for all m:*

(17) $P_0^* < \exp(-c_4 \cdot m), \quad P_1^* < \exp(-c_5 \cdot m).$

Proof. The probability that a correct circuit survives after T_1 trials is larger than q^{T_1}, where $q > 1 - \exp(-c \cdot b_2)$. Thus, the probability that all correct circuits die is $(1 - q^{T_1})^{X_+} < \exp(-c \cdot X_m)$; since $X_m = O(m)$, we get the first estimate (17).

Denote by Z the number of inputs $\tilde{x}$ among T_1 inputs $\tilde{x}$ for which all incorrect circuits give an incorrect answer $y(x) \neq f_{m+1}(x)$. We denote such inputs by x_{inc}. If an incorrect circuit C_{inc} obtains such an input, C_{inc} dies with a probability p_d close to 1: $p_d = 1 - \exp(-c \cdot b_2)$. The probability that a circuit obtains, as an input, some x_{inc} within T_1 independent inputs, is $p_1 = 1 - (1-p)^{T_1}$. Then by Lemma 4, the probability that the number of surviving incorrect circuits is larger than $6(1-p)^{T_1} X_m$, does not exceed $0.8^{O(X_m)}$. The number of the correct circuits will be close to $X_+ = c_9 \cdot X_m$, with a probability $> 1 - \exp(-c \cdot X_m)$, where c_9 depends on μ and b_2 but does not depend on T_1. This observation gives the second estimate (17) for sufficiently large values of T_1 large. This completes the proof of the lemma.

Stage III. Replications. We now come back to the design of the algorithm required in Theorem 3. Notice that it is not a priori known whether a given circuit is correct or not. However, one can investigate structures of circuits and one can find a group of circuits having the same structure. We preserve these circuits and remove all the others. Then, we replicate all the remaining circuits to obtain $X_{m+1} = X(t) = O(m)$ up to the moment $t = (m+1)T_e$. By Lemma 5, it is clear that for new noisy correct circuits, the probability of the incorrect output admits the upper bound $\exp(-c \cdot m))$, where $c > 0$ (we repress the noise by increasing $b(t)$ on Stage II; at the other stages $b(t)$ is a large constant b_* independent of m).

We notice that the probability to survive within I_m is larger than $1 - \exp(-c_1 m)$, $c_1 > 0$. The resources within I_m are $O(m|E(m)|) < O(m^3)$. This completes the proof of Theorem 3.

Theorem 3 has simple intuitive meaning. It describes survival with learning. To survive, the population should learn something about a boolean black box. It is a difficult problem, but the population can recognize a black box step by step, if the box's complexity increases "slowly" (i.e., according to (11)).

Example. An interesting example is given by the sequence of conjunctions

(18) $\quad f_j = D_{\mathbf{i}_1} \wedge D_{\mathbf{i}_2} \ldots \wedge D_{\mathbf{i}_k(j)},$

where each $D_{\mathbf{i}}$ is a disjunction of some literals: $D_{\mathbf{i}} = \tilde{x}_{i_1} \vee \tilde{x}_{i_2} \ldots \vee \tilde{x}_{i_K}$, where $\tilde{x}_i$ is either x_i or $\neg x_i$, $i \in \{1, \ldots, N\}$. The integer K can be interpreted, biologically, as a redundancy parameter. The dependence of the number $k(j)$ on j can be increasing, decreasing, or non-monotonic (it depends on g in (11)). Notice that learning of (18) is hard for large N [31].

In the case of (18) the evolution stability is connected with the K-SAT problem, which for a few decades has been a focus of many important research activities; see, e.g.,, [1, 6, 9, 20]. We can consider our evolution as a "game" of population against an environment which becomes more and more complicated.

1. If the dimension N of inputs x is fixed, one can survive in a simple way (see Lemma 2) if $P_j = P$ and P is uniform. However, the survival probability may be exponentially small in N.

2. Assume now that $N = N(t)$ increases with time. Suppose that each new clause contains a new literal that is not used in the previous clauses. Then for passive environment with uniform P it is possible to survive without learning (Lemma 2). For active environments, Theorem 3 holds if the distributions P_{m+1} have the following property: $\text{Prob}\{1 - \delta_0 > f_m(x) = 1\} > \delta_0$ with δ uniform in m for x chosen randomly according to P_{m+1}.

3. It is natural to assume that conjunctions (18) are constructed randomly, i.e., all indices $i_K \in \{1, 2, \ldots, N(t)\}$ are chosen randomly (*random K-SAT*). For example, at each j we choose a random i, and we add, with probability p, certain L disjunctions to f_m.

In this situation, our problem looks complex, and it is related to the results on phase transitions in hard combinatorial problems [9, 20]. We consider this relation in our forthcoming publications. In this paper, we restrict ourselves to some simple observations.

If $K \leq 2$, one can expect that, inevitably, some new clause will be in a contradiction with previous ones; thus, for the passive case **A**, we can again use Lemma 2.

For $K > 2$ and active case **B**, it is possible that P_m are not computable in polynomial time $Poly(m)$. Indeed, to implement the algorithm from Theorem 3, we should have P_j satisfying the condition (12) (or, at least such that $Err_j > const \cdot j^{-n}$ for some $n > 0$). For large j, it is possible that the number N_j of solutions of the K-SAT problem corresponding to $f_j(x)$ is exponentially small in j; moreover, if $P \neq NP$ and j is a part of the input, there is no polynomial-time algorithm to find x such that $f(x) = 1$ [9, 20]. Nonetheless, even in such a situation, survival is possible if the population always preserves a trivial circuit.

It would be interesting to compare results from this section and Theorem 1 with the real biological situation. A discussion of the problems of species extinction and complexity growth can be found, for example, in [23]. A change of f_j can be interpreted as a variation in ecological conditions. It can be shown that, according to our model, such a change leads to a massive species extinction with an exponential rate (this fact is in good accordance with biological reality, see [23], Ch. 23).

7 Conclusion

There are two fundamental problems of mathematical biology: the morphogenesis problem (emergence of complex structures) and the evolution problem:

a why structures became more and more complicated and

b why Darwin's evolution could generate such structures within "short" time and with "bounded resources" [7, 23, 31].

Mathematical approaches to the first problem started with the seminal work of A. Turing [27]. Now, we can explain the emergence of complicated patterns and describe algorithms to obtain such patterns (for network and circuit models, see [21, 26, 28, 29]). However, the questions about the stability of such emergence are still open.

It seems that the second problem is even more mysterious. In our opinion, the key to this problem can be found in [13, 30]: all biological systems with fixed parameters are unstable, but evolution can stabilize them; in this case, according to our Theorem 1, the Kolmogorov complexity grows (on average). This fact explains why complexity increases in evolution.

Note that complexity here is the complexity of the *genetic code*; the relation between this complexity and the complexity of the organism complexity is not obvious. For some Pffafian models like (2), one can prove that the pattern complexity can be estimated in terms of the Pfaffian chain complexity. Thus, "complex" patterns can be obtained only by using sufficiently "complex" pfaffian models [29]. Probably, both pattern complexity and gene complexity are increasing during the evolution process [23].

Evolution does not necessarily mean "improving". Ch. Darwin avoided the words "higher" and "lower". In fact, following D. Wandschneider ([7], Ch. 10), let us "compare the chance of survival of, say, infusoria with that of humans: risk increases with an increase in capability. A glance at inorganic structures makes this even clearer. The Alps are obviously characterized by considerable stability". Theorem 1 explains this paradox: the evolution of unstable structures has no goals and stability is not necessarily increasing: simply, if evolution stops, destruction is inevitable.

We think (following [31]) that the problem **b** can be correctly posed mathematically only by NP-hard ideas: short time means polynomially bounded, and energetic resources also should be polynomially bounded. Here we show that, at least in certain cases, such fast evolution is possible (even under severe restrictions, but these restrictions should evolve sufficiently slowly in time). One can hope that recent ideas on phase transitions in hard combinatorial problems (for example, [9, 20]) can help us understand efficiency of the Darwin evolution and the Red Queen law (extinction of species when the number of ecological restrictions become too large) [30].

Acknowledgments

The authors are thankful to Prof. O. Radulescu for discussions and to the referees for many useful remarks. One of the authors was supported by the grants RFBR 10-01-00627-a and CDRF NIH Grant RR07801.

BIBLIOGRAPHY

[1] D. Achlioptas, Lower bounds for random 3-SAT via differential equations, Theor. Comp. Sci. 265 (2001) 159185.

[2] J. P. Aubin, A. Bayen A., N. Bonneuil and P. Saint-Pierre (2005) Viability, Control and Games: Regulation of complex evolutionary systems under uncertainty and viability constraints, Springer-Verlag.

[3] R. Albert and A. L. Barabási, (2002) Statistical mechanics of complex networks, Rev. Modern Physics, **74**, pp. 47-97

[4] R.Bhattacharya, M.Majumdar, Random Dynamical Systems. Theory and Applications., Cambridge, 2007

[5] W. B. Cannon: The Wisdom of the Body, W. W. Norton Co., New York, 1932.

[6] T. H. Cormen, C. E. Leiserson, R. L. Rivest, and C. Stein, Introduction to Algorithms (Second Edition), MIT Press, 2001.

[7] V. Hisle and Ch. Illies (eds), Darwinism and Philosophy University of Notre Dame Notre Dame, Indiana, 2005.

[8] P. Erdos, A. Rényi, (1960) On the evolution of random graphs, Publ. Math. Inst. Hungarian Academy of Sciences, **5**, pp. 17-61

[9] E.Friedgut, Sharp thresholds of graph properties, and the k-SAT problem. J. Amer. Math. Soc.,12, 1999, no. 4, p.1017-1054.

[10] A. Gabrielov, N. Vorobjov, Complexity of computations with Pfaffian and Noetherian functions, in: Normal Forms, Bifurcations and Finiteness Problems in Differential Equations, 211-250, Kluwer, 2004.

[11] L. Glass and S. Kauffman, The logical analysis of continuous, nonlinear biochemical control networks, (1973) J. Theor. Biology, **34**, pp. 103-129

[12] D. Grigoriev, Deviation theorems for solutions of linear ordinary differential equations and applications to parallel complexity of sigmoids, (1995) St.Petersburg Math. J., **6**, pp. 89-106

[13] M. Gromov and A. Carbone, Mathematical slices of molecular biology, Preprint IHES/M/01/03, 2001.

[14] J. E. Hutchinson (1981). Fractals and self similarity" Indiana Univ. Math. J. **30**, pp. 713747.

[15] A. Khovanskii, Fewnomials, Translations of Mathem. Monographs, Amer. Math. Soc., **88**, 1991.
[16] P. Koiran and C. Moore, Closed-form analytic maps in one and two dimensions can simulate Turing machines. (1999) Theoretical Computer Science, **210(1)** 217-223
[17] Ming Li and P. Vitanyi, An Introduction to Kolmogorov Complexity and Its Applications, Second Edition Springer Verlag, 1997.
[18] A.Lesne. Complex networks: from graph theory to biology. (2006) Letters in Math. Phys., **78**, pp. 235-262
[19] C. Lobry, Une proprieté generique des couples de champs de vecteurs, (1972) Chechoslovak Mathematical Journal, **22 (97)** 230-237.
[20] S. Mertens, M. Mézard, R. Zecchina, Threshold values of Random K-SAT from the cavity method, Random Structures and Algorithms 28 (2006) 340-373
[21] E. Mjolness, D. H. Sharp and J. Reinitz, A connectionist Model of Development, (1991) J. Theor. Biol. **152**, pp. 429-453.
[22] M. Blum, A machine-independent theory of the complexity of recursive functions, (1967) J. Assoc. Comput. Machin., 14, , pp. 322-336
[23] M. Ridley, Evolution, 2nd ed. (Blokwell Scientific Publications Ltd, Oxford, 1996)
[24] H. J. Sussmann, A general theorem on local contollability SIAM J. Control and Optimization Vol. 25 N1, pp. 158 -194
[25] P. Smolen, D. Baxter, J. H. Byrne, Mathematical modelling of gene networks, Review in Neuron, **25**, (2000) 247-292.
[26] D. Thieffry and R. Thomas, Dynamical behaviour of biological regulatory networks, II.Immunity control in bacteriophage lambda, Bull. Math. Biology, **57**, (1995) 277-295.
[27] A. M. Turing, The chemical basis of morphogenesis, (1952) Phil. Trans. Roy. Soc. B, **237** pp. 37-72.
[28] S. Vakulenko, Complexité dynamique de reseaux de Hopfield, (2002) C. R. Acad. Sci. Paris Sér. I Math., t.335.
[29] S. Vakulenko, D. Grigoriev, (2006) Algorithms and complexity in biological pattern formation problems, Annales of Pure and Applied Logic **141**, pp. 421-428
[30] L. Van Valen, (1973) A new evolutionary law, Evolutionary Theory, **1**, 1-30
[31] L. G. Valiant, Evolvability. Lect. Notes Comput. Sci., v. 4708, 2007, pp. 22-43.
[32] L. G. Valiant, A theory of learnable. (1984) Comm. ACM 27, pp. 1134-1142
[33] B.L. van der Waerden, Algebra, Volumes I, II. New York, NY: Springer (2003).

Frame Correspondences in Modal Predicate Logic

JOHAN VAN BENTHEM

ABSTRACT. Understanding modal predicate logic is a continuing challenge, both philosophical and mathematical. In this paper, I study this system in terms of frame correspondences, finding a number of definability results using substitution methods, including new analyses of axioms in intermediate intuitionistic predicate logics. The semantic arguments often have a different flavour from those in propositional modal logic. But eventually, I hit boundaries to first-order definability of frame conditions. I then relate these findings to the known incompleteness theorems for modal predicate logic, and point out some new directions for further research, including the use of strengthened higher-order proof systems for the basic modal language.[1]

1 Introduction

In Grisha Mints' work on modal logic, objects and predication are never far around the corner. That reminds me of my student days, when Hughes and Cresswell 1968 [18] was the reigning textbook, with propositional modal logic only a stepping stone toward modal predicate logic, the vehicle for the real philosophical applications. But gradually, modal propositional logic has stolen the show: in the *Handbook of Modal Logic* (Blackburn, van Benthem & Wolter, eds., 2006 [12]) modal predicate logic gets only one chapter out of twenty-one.[2] In my own research, van Benthem 1983 [4] does only slightly better, according it one chapter out of nineteen. There are reasons for this change of fortunes. Propositional systems are convenient and application-rich, and their mathematical theory has turned out elegant and challenging. By contrast, since the 1960s, deep difficulties have come to light regarding the very design of modal predicate logic: enough to fill a Black Book.

But my topic here is not dark, but light! I will explore how my original interests in Correspondence Theory cross over from modal propositional logic to modal predicate logic, yielding new theorems and observations.[3] The presentation will be fast-paced, geared toward an intended audience of experts reading this book. My feeling is that there is much more to the subject that I broach here, but I must leave that to the reader.

---

[1] This paper is in honor of the many great qualities of my colleague Grisha Mints.
[2] Garson 2001 [16] and Brauner & Ghilardi 2006 [14] are key references for modal predicate logic. See also the book Gabbay, Shehtman, & Skvortsov 2005 [15], forthcoming soon.
[3] For Correspondence Theory, van Benthem 1984 [5] is probably still the best source.

2 Modal Predicate Logic: The Basics

We start by recapitulating some well-known notions and results.

DEFINITION 1. The *language of modal predicate logic* arises from the standard formation rules for first-order predicate logic plus a construction clause for the modality, yielding a format

$$Px \mid \neg \mid \vee \mid \wedge \mid \exists x \mid \forall y \mid \Diamond \mid \Box$$

Formulas with free variables $\varphi = \varphi(x, y, \ldots)$ then express modal predication.

One might just combine models here for the two components, producing a family of first-order models ordered by a modal accessibility relation. But a better generalization has turned out to be this:

DEFINITION 2. *Models* for the language of modal predicate logic are structures $\mathbf{M} = (W, R, D, V)$, where W is a set of possible worlds, R an accessibility relation, and D a domain map assigning sets of objects to each possible world. Finally, V is a valuation function interpreting each predicate letter P at each world w as a predicate $V(P, w)$ of the right arity.

Now we must combine the semantics of predicate logic, using assignments taking variables to objects, with the earlier one for modal propositional logic. The following stipulation explains when a formula φ is true at world w under assignment a, where we assume that the values $a(x)$ for the free variables x in φ belong to the domain D_w. Here and in what follows, bold face letters $\mathbf{x}$ (and later on also $\mathbf{d}$, $\mathbf{e}$) stand for finite sequences:

$\mathbf{M}, w, a \models P\mathbf{x}$ iff the tuple of objects $a(\mathbf{x})$ from D_w belongs to the predicate $V(P, w)$,

$\mathbf{M}, w, a \models \neg \varphi$ iff not $\mathbf{M}, w, a \models \varphi$

$\mathbf{M}, w, a \models \varphi \vee \psi$ iff $\mathbf{M}, w, a \models \varphi$ or $\mathbf{M}, w, a \models \psi$

$\mathbf{M}, w, a \models \exists x\, \varphi$ iff for some $d \in D_w$, $\mathbf{M}, w, a[x := d] \models \varphi$

$\mathbf{M}, w, a \models \Diamond \varphi$ iff for some v with Rwv where $a(x) \in D_v$ for all free variables x in φ, D, $\mathbf{M}, v, a \models \varphi$

Here individual quantifiers range over the local domain of the objects that exist at the current world. The clause for the modality makes sure that all objects used by a to evaluate $\Diamond \varphi$ in w are also available for evaluating φ in the world v. On the basis of this truth definition, Boolean conjunction $\wedge$, modal box $\Box$, and universal quantifiers $\forall$ are then defined as usual.

Often, the modal clause of this semantics is simplified by making a further structural assumption that object domains grow along accessibility:

For all w, v, $Rwv \rightarrow D_w \subseteq D_v$ Domain Cumulation

These models validate a minimal modal predicate logic fusing standard predicate logic with the minimal propositional modal logic K, where Domain Cumulation ensures validity of the modal distribution axiom. In what follows, we keep this structural property as an optional extra, as it is quite strong. Our preference is to analyze what axioms mean in terms of frame correspondence, and the weaker the base used then, the better.

3 Translation and Invariance for World-Object Bisimulations

The expressive power of this system can be analyzed with the same techniques that have become standard for modal propositional logic. The first-order correspondence language L_{corr} has two sorts of 'worlds' and 'objects', basic binary relations Rwv for world accessibility and Ewx for object x being in the domain of world w, and $(k+1)$-ary predicates $Pw\mathbf{x}$ for each k-ary predicate $P\mathbf{x}$ in the first-order language of the system.

DEFINITION 3. The *standard translation* $trans(\varphi)$ takes formulas φ in the language of modal predicate logic to L_{corr}-formulas that have the same free object variables as φ plus one free world variable w:

$$\begin{aligned}
trans(Px) &= Pwx \\
trans(\neg\varphi) &= \neg trans(\varphi) \\
trans(\varphi \vee \psi) &= trans(\varphi) \vee trans(\psi) \\
trans(\exists x\, \varphi) &= \exists x\, (Ewx \wedge trans(\varphi)) \\
trans(\Diamond\varphi) &= \exists v\, (Rwv \wedge \&_i Evx_i (x_i \text{ free in } \varphi) \\
&\quad \wedge [v/w] trans(\varphi))?
\end{aligned}$$

Any model $\mathbf{M}$ for modal predicate logic is at the same time a model for the correspondence language L_{corr}, and indeed the following equivalence tightly connects modal semantics with standard first-order evaluation:

THEOREM 4. (Translation Theorem) *For each model $\mathbf{M}$ and each formula φ of modal predicate logic, $\mathbf{M}, w, a \models \varphi$ iff $\mathbf{M}, \alpha \models trans(\varphi)$, where the assignment α sends object variables to their a-values, while the single free world variable of $trans(\varphi)$ goes to the world w.*

Thus, syntactically, modal predicate logic may be seen as a fragment of the full two-sorted first-order language L_{corr}. In this setting, its characteristic semantic invariance can then be defined as a mixture of two well-known structural relations between models: *modal bisimulation*, plus the notion matching it for a full first-order language, namely, *potential isomorphism*:[4]

DEFINITION 5. A *world-object bisimulation* between models $\mathbf{M}, \mathbf{N}$ for modal predicate logic is a relation Z between tuples $w\mathbf{d}$ in $\mathbf{M}$ and $v\mathbf{e}$ in $\mathbf{N}$ of the same length, where all objects in tuples belong to the domain of the initial world.[5]

Here, the relation Z satisfies the following three properties:

---

[4] Van Benthem 1996 [7] and van Benthem & Bonnay 2008 [10] explore the analogy between potential isomorphism and bisimulation in more general settings.

[5] This assumption will hold throughout this section: when we write a tuple $w\mathbf{d}$, we always assume that the objects mentioned occur in its leading world.

(a) matching corresponding objects $(\mathbf{d})_i$, $(\mathbf{e})_i$ of the tuples matched by Z induces a partial isomorphism[6] between $\mathbf{M}$ and $\mathbf{N}$,

(b) if wRw' in $\mathbf{M}$ and $w'\mathbf{d}$ exists, then there is also a world v' in $\mathbf{N}$ with vRv' and $w'\mathbf{d}Zv'\mathbf{e}$; and the same clause holds in the direction from $\mathbf{N}$ to $\mathbf{M}$, and

(c) if d in $\mathbf{M}$, then there is also an object e in $\mathbf{N}$ with the pair $w\mathbf{d}d$, $v\mathbf{e}e$ in Z – and again, also vice versa.[7]

In terms of this invariance relation, here is an analogue for modal predicate logic of a well-known characterization for the propositional modal fragment of a pure first-order language over worlds:

THEOREM 6. (Invariance Theorem) *The following statements are equivalent for formulas φ in L_{corr}:*

(a) φ is invariant for world-object bisimulations,

(b) φ is definable by a formula of modal predicate logic.

Proof. The crucial step is essentially as that in the Invariance Theorem for propositional modal logic (van Benthem 1996 [7], Theorem 3.12, p. 57; Blackburn, de Rijke & Venema 2000 [13], Theorem 2.68, p. 103). One starts from two models with assignments verifying the same modal predicate-logical formulas (i.e., these models are indistinguishable in the language), and then extends these models to ω-saturated ones. And between models of the later kind, the tuple-to-tuple relation of verifying the same modal predicate-logical formulas turns out to be a world-object bisimulation. ∎

4 Frame Correspondences for Special Axioms

Beyond the minimal core, modal axioms impose constraints on models that can be determined by correspondence arguments, as in propositional modal logic. These involve the following notion (van Benthem 1983 [4], Chapter 12):

DEFINITION 7. A formula φ of modal predicate logic *holds in a frame* $\mathbf{F} = (W, R, D)$ (here a 'frame' is a model stripped of its valuation) iff φ is true at each world in $\mathbf{F}$ under all valuation functions V.

Examples. Here are some examples, proved by straightforward arguments:

PROPOSITION 8. *The following modal axioms have the listed correspondents:*

$\exists x \Diamond Px \to \Diamond \exists x\, Px$	$True$	Tautology
$\exists x \Box Px \to \Box \exists x\, Px \forall w$	$\forall v\,(Rwv \to \forall x\,(Exw \to Exv))$	Cumulation
$\Diamond \exists x\, Px \to \exists x \Diamond Px$	$\forall w \forall v\,(Rwv \to \forall x\,(Exv \to Exw))$	Anti – Cumulation

---

[6] A *partial isomorphism* between two models is any isomorphism (for the relevant first-order vocabulary) between sub-models of these models – usually *finite* ones.

[7] There are other ways of merging modal bisimulation and potential isomorphism.

Proof. As an illustration, we prove the third correspondence.[8] Reverse domain inclusion guarantees truth of $\Diamond \exists x\, Px \to \exists x\, \Diamond Px$ in a frame, no matter for which objects in which worlds the predicate P holds. Conversely, let d be any object in any successor world v of w. Make P true only for d in v, and nowhere else. Under this interpretation V, the antecedent $\Diamond \exists x\, Px$ is true at w, and so by frame truth of the implication for any choice of valuation, under this same V, the world w also has $\exists x\, \Diamond Px$ true. But then by our truth definition, that can only happen if the object d already exists at w. ∎

A general method. Behind this result lies a generalization of the Sahlqvist Theorem for propositional modal logic. Given the right syntactic form of modal axioms, first-order definable equivalents exist on frames, and these can even be computed (van Benthem 1983 [4], Theorem 9.10, p. 105):

THEOREM 9. *There is an effective translation into first-order frame properties for all modal predicate-logical axioms of the syntactic form* $\alpha \to \beta$, *where* α *has the inductive syntax rule* $\exists \mid \wedge \mid \vee \mid \Diamond \mid \gamma$, *with γ having the simpler syntax* $P\mathbf{x} \mid \forall \mid$ *Box, while β is a wholly positive formula with a syntactic construction rule* $P\mathbf{x} \mid \forall \mid \exists \mid \wedge \mid \vee \mid \Diamond \mid \Box$.

Proof. The argument goes just like that for the propositional Sahlqvist theorem via the 'method of substitutions' (cf. Blackburn, de Rijke & Venema 2000 [13], Theorem 3.54, p. 165). The crucial point, as explained in these references, is that the special syntactic shape of the modal-quantificational antecedent α allows for first pulling out all existential quantifiers and modalities into a universal prefix, after which the special universal syntactic form of the remaining antecedent formula allows for verification by a first-order definable *minimal* valuation for its atomic predicates.[9] To get the right first-order frame property for $\alpha \to \beta$, it suffices to substitute the minimal valuation extracted effectively from α into the standard first-order L_{corr}-translation of the positive consequent formula β. Correctness uses the semantic monotonicity of the latter. ∎

Illustrations are earlier quantifier/modality interchanges for $\forall \mid \exists \quad \Box \mid \Diamond$.

Limits. Here is a principle beyond the method of minimal substitutions:

THEOREM 10. $\Box \exists x\, Px \to \exists x\, \Box Px$ *has no first-order frame correspondent.*

Proof.[10] Consider the following family of finite frames $\mathbf{F}_n$, each consisting of an irreflexive root world w pointing at n successor worlds $v_1, \ldots, v_n$:

---

[8]This is a Barcan Axiom that, assuming Cumulation, imposes 'constant domains'.

[9]Papers [8, 9] state precisely what syntax for first-order antecedents support minimal valuations defining the intersection of all verifying predicates. They also contain extensions to LFP(FO): first-order logic with *fixed-points*.

[10]What follows is a sketch: the book van Benthem 1983 [4], p. 139, has details.

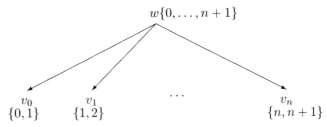

The domain of w is the finite set of natural numbers $\{0,\ldots,n+1\}$, and each successor world v_i only has two objects in its domain, as follows: v_0 has $\{0,1\}$, v_1 has $\{1,2\}$, ..., v_n has $\{n,n+1\}$. On each such frame, $\Box \exists x\, Px \to \exists x\, \Box Px$ is true, whatever the valuation for the predicate P. It is clearly true in all the v_i, since these do not have successors. Next, suppose the antecedent $\Box \exists x\, Px$ holds at w, under an arbitrary valuation. Either object 1 has property P in v_1, and we get $\Box P1$, since v_1 is the only world where 1 occurs, or it does not. Then 2 has the property P in v_1, and either it also has P in v_2, and w has $\Box P2$, or we go on. If we never satisfy $\exists x\, \Box Px$ in this way, we reach the final point $n+1$ in v_n as a witness for $\Box P(n+1)$ at w.

Next, assume that our modal predicate-logical axiom $\Box \exists x\, Px \to \exists x\, \Box Px$ has a first-order frame equivalent α in L_{corr}: we will derive a contradiction. First, we write a set Σ of first-order sentences true in all the above models, describing their main world order and object features. Choose a new relation symbol S imposing an order on the successor worlds. Σ says that S makes the successor worlds v lie in a discrete linear order with a unique beginning and endpoint, and no 'limit points'. Next, each successor world v contains two objects, while each of these objects occurs in exactly two adjacent worlds – except for the endpoints of S, at each of which one additional isolated object occurs. Also, the objects in the root world are precisely those that occur in some successor world. Finally, we let Σ say that, for each natural number n, there are at least n successor worlds v.[11]

The above family of frames $\mathbf{F}_n$ clearly shows that the infinite set of formulas $\{\alpha\} \cup \Sigma$ is finitely satisfiable. But then, by Compactness for first-order logic, there is a model $\mathbf{M}$ for all of $\{\alpha\} \cup \Sigma$, with an infinite set of successor worlds v of the following form:

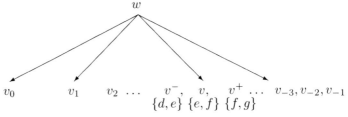

That is, the ordering S of the successor worlds v is like the natural numbers followed by a number of copies of the integers, and ending in a copy of the negative integers. But on such a model, we can refute our modal predicate-logical principle $\Box \exists x\, Px \to \exists x\, \Box Px$ as follows:

---

[11] This mouthful is a routine exercise in writing down a lot of first-order formulas.

Let P be false of the isolated objects at the start and the end of the order S, and, using the special domain structure of the worlds v as described, alternating across S-adjacent v-worlds, make P true for just one object in each world in a way that avoids ever giving the same object the property P across two different adjacent worlds.

As a result of this stipulation for predicate P, the antecedent $\Box \exists x\, Px$ holds in the model $\mathbf{M}$ at the initial world w, but the consequent $\exists x\, \Box Px$ does not.

But the first-order sentence was true in the model $\mathbf{M}$ by construction, and it was to be equivalent to our modal axiom: a contradiction. ∎

Special outcomes in special settings. Here is an interesting counterpoint to known results from the literature. In modal propositional logic, apparently quite mild structural conditions on relational frames can change modal correspondences drastically. In particular, the 'McKinsey Axiom'

$$\Box \Diamond p \to \Diamond \Box p$$

is not first-order,[12] but it becomes first-order on transitive frames – where it expresses the property of 'atomicity' – though van Benthem 1983 shows that this equivalence cannot be proved by the substitution method. We find a similar effect of seemingly mild structural conditions here:

FACT 11. *On frames satisfying Domain Cumulation, the modal predicate-logical axiom $\Box \exists x\, Px \to \exists x\, \Box Px$ is first-order definable.*[13]

Proof. The equivalent is the conjunction of two first-order properties:

(a) Domain Anti-Cumulation,

(b) each world whose domain has more than one object has at most one world successor.

First, if both of these first-order properties hold in a frame, then so does $\Box \exists x\, Px \to \exists x\, \Box Px$. If a world has just one object d, and we have both Domain Cumulation and Anti-Cumulation, all successors have just that object d, and the antecedent implies that this d has property P throughout. And if a world has at most one accessible successor, then truth of $\Box \exists x\, Px$ implies that of $\exists x\, \Box Px$, either trivially since there are no successor worlds, or because some object d in the unique successor world satisfies P, and that same object d will then satisfy $\Box Px$ in w.

Next, we show that frame truth of $\Box \exists x\, Px \to \exists x\, \Box Px$ implies the two stated first-order conditions. First consider Domain Anti-Cumulation (a). Suppose that wRv where v has an object d not occurring in w. We can use any such situation to refute our modal axiom:

In world v, make the predicate P true for d only, and in all other successor worlds of w, make P true for all the objects.

---

[12] The McKinsey Axiom is not even definable in LFP(FO): cf. [8]. Despite the analogy with $\Box \exists x\, Px \to \exists x\, \Box Px$, the proofs work really differently.

[13] Domain Inclusion crucially did not hold in our preceding counter-example.

By domain inclusion, the stipulation about world v alone refutes $\exists x \,\Box Px$ at w, while the two stipulations together make $\Box \exists x \, Px$ true at w.

Next, take condition (b) of 'partial function'. Let world w have at least two objects 1, 2 and more than one successor, say v_1, v_2 and perhaps others. Now define a valuation for the predicate P as follows:

> P holds of 1 and of no other object in v_1, P holds of 2 and no other object in v_2, and P holds of all objects in all other successor worlds.

This makes the formula $\Box \exists x \, Px$ true at the world w while there is no object at w that has the property P in all successor worlds. Contradiction, and hence w has at most one successor world. ∎

The preceding correspondence argument, though quite elementary, cannot work in the earlier Sahlqvist substitution style. To see this, consider the following model where $\Box \exists x \, Px \to \exists x \,\Box Px$ fails at w:

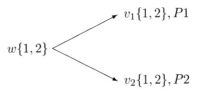

Given the symmetry between the two successor worlds, there is no uniform definition for the predicate P purely within the language L_{corr} that witnesses this failure. The above correspondence proof in fact used an interpretation for the predicate P that is not definable in this uniform manner.

5 Excursion: Joys of Intuitionistic Predicate Logic

Correspondence analysis is especially vivid in intuitionistic predicate logic. This special area has stronger constraints on frames. Modal accessibility is a *pre-order* satisfying Domain Inclusion, while atomic predicates Pd are *hereditary*: once true at a world, they stay true in all its successor worlds.

Even the propositional version has many surprising features here, witness the thorough study in Rodenburg 1986 [24]. Here is a result on the predicate logic, essentially from Meijer Viol 1995 [20], but in a streamlined presentation.

Intuitionistic semantics gives 'individuality' to quantifier principles that are just lumped together as 'valid' in classical logic. A nice example is the frame correspondence for what Beth 1959 [11] called 'Plato's Law':[14]

$$\exists x \,(\exists y \, Py \to Px) \quad \mathbf{P}$$

Even in the realm of intuitionistic frames, this law defines an interesting second-order mix of purely relational conditions with object occurrence:

THEOREM 12. *On frames, Plato's Law defines the conjunction of the three conditions: (a) Constant Domains, (b) Accessibility is a linear order if the domain D has at least two objects, (c) If the object domain D is infinite, then the accessibility relation is also well-founded.*

---

[14] Such principles were studied earlier by Casari and Minari: cf. [21].

Proof. First, Plato's Law holds under the stated conditions. For a start, any constant domain with just one object clearly validates the principle. Next, consider a finite domain with objects $d_1, \ldots, d_n$ ($n \geq 2$) with a linear world pre-order. Suppose that none of these objects d witnesses Plato's Law: that is, there exists a successor world where property P fails for d while some object other than d satisfies P. But then consider any finite sequence of successor worlds where each object in the initial world finds such a refutation. The minimal world v of this linear order leads to a contradiction. Each object d lacks P in v, by Heredity. But then $\exists y\, Py$ cannot be true in v to refute Plato's Law for 'its' object d_v. The same argument holds in infinite object domains, by an appeal to the well-foundedness of the linear order.

Conversely, we show that frame truth of Plato's Law implies the stated principles. Consider any world w in a frame. (a) First, if Constant Domain fails, we have a successor world with some new object d not occurring in w. Making P true only for d in w and all its successor worlds in the frame (so as to satisfy Heredity) will refute Plato's Law at w. (b) Next, assume Constant Domain with a set of at least two objects 1, 2, and suppose that Linearity fails: i.e., the world w has two incomparable successors v_1, v_2. Now make P true of object 1 only in v_1 and all its successor worlds, and likewise for object 2 in v_2. Again, this satisfies the condition of Heredity while refuting Plato's Law at w. (c) Finally, assume Constant Domain with infinitely many objects in w. Divide up these objects into countably many disjoint sets $D_1, D_2, \ldots$ Now, suppose that the order of the successor worlds violates the condition of well-foundedness. That is, the frame contains a countable descending sequence of successor worlds $\ldots v_n R \ldots v_2 R v_1$. But then, in the frame, define a predicate P as follows: for each n,

make P true for the objects in D_n at v_n and all its successor worlds.

This stipulation guarantees Heredity. Moreover, it is easy to verify that with this predicate P, Plato's Law gets refuted at the world w: for each object d in w, the descending sequence has a stage where it still lacks the property P while some other objects already have P. ∎

REMARK 1. This correspondence argument has interesting analogies with known ones from modal propositional logic. For instance, the linearity of the accessibility order is also expressed by this consequence:

$$\exists xy \neg x = y \to ((A \to B) \vee (B \to A)),$$

where the consequent can be turned into propositional Sahlqvist form. But *well-foundedness* itself is not definable in propositional modal logic.

6 From Correspondence Arguments to Formal Derivations

Correspondence arguments are semantic and do not necessarily imply the existence of matching formal derivations purely inside some given system of modal (predicate) logic. Indeed, there are some tricky features in setting up the right proof systems in modal predicate logic. Notoriously, putting together predicate logic with propositional modal logic in an 'obvious union' validates a general schema of modal predicate-logical distribution

$$\Box(\varphi \to \psi) \to (\Box\varphi \to \Box\psi)$$

that implies the modal law $\exists x \Box Px \to \Box \exists x\, Px$ for Domain Cumulation. This surprising observation from Hughes & Cresswell 1968 [18] shows that innocent-looking combinations of axiom schemata can really be much stronger than merely taking the union of modal logics as 'theories'.[15]

These are not minor issues, and they relate to the discovery of mathematical deficiencies of deduction in modal predicate logic, as early as Ono 1973 [22]. Contributions by many authors, like Shehtman & Skvortsov and Ghilardi, are surveyed in Brauner & Ghilardi 2006 [14]. We will side-step these issues, and just make a few points about formal proofs. First, correspondence arguments like the ones in this paper are proofs of a kind, stated in an informal mathematical meta-language of models – and they can be natural and perspicuous. Thus it makes sense to look for matching deductions in formal systems: in particular, proofs inside modal predicate logic. And indeed, the latter can be really elegant when available.

Here is an illustration, relating Plato's Law **P** of the preceding section to Grisha Mints' recent work on intermediate logics relevant to logic programming, presented at the Stanford logic seminar in the spring of 2008.

Excursion: A surprising alternative form of Plato's Law. Grisha Mints has recently proved that the existential quantifier, not definable in terms of the universal one in intuitionistic logic, does become definable in the modal predicate logic of the two-world Kripke model with constant domains, by the following nice equivalence:

$$\exists x\, Px \leftrightarrow \forall y\, ((\forall x\, (Px \to Py)) \to Py) \qquad \mathbf{M}$$

Here we add a further observation tying this up with known principles:

FACT 13. *As a schema over intuitionistic predicate logic, **M** is equivalent to **P**.*

Proof. We only need one half of Mints' principle **M**, viz.

$$\forall y\, ((\forall x\, (Px \to Py)) \to Py) \to \exists x\, Px,$$

because the other direction is provable in intuitionistic predicate logic anyway. This is easiest to see when replacing **M** by its intuitionistic equivalent

$$\exists x\, Px \leftrightarrow \forall y\, ((\exists x\, Px \to Py) \to Py).$$

---

[15]To see the difference, consider the following 'proof' that addition is explicitly definable in first-order arithmetic with 0 and successor S, by a prima facie valid appeal to Beth's Theorem. Show that addition is implicitly definable, as follows. Consider Peano Arithmetic PA for 0, S plus the usual recursion equations for +. Take a copy PA′ of this theory with 0, S and an operation +′. The combined theory PA + PA′ derives that + and +′ are the same, by an easy induction. E.g., $x + Sy = S(x + y) = S(x +' y) = x +' Sy$. But of course, addition is not explicitly definable using just 0 and S! Explanation: we illicitly used the induction axiom of Peano Arithmetic for the *combined language*, which is not in the union of the theories. This shows how powerful schemas can be when used in a richer language.

Staying with the latter version, the relevant half of the equivalence **M** that we need in what follows is

$$\forall y \, ((\exists x \, Px \to Py) \to Py) \to \exists x \, Px.$$

Now here is our equivalence argument:

(a) From **P** to **M**. Assume that $\forall y \, ((\exists x \, Px \to Py) \to Py)$ and take a witness for **P**: $\exists y \, Py \to Pd$. We get $(\exists x \, Px \to Pd) \to Pd$, and so Pd. But this implies the consequent $\exists x \, Px$.

(b) From **M** to **P**. First define $Qu := \exists x \, Px \to Pu$. Now plug this into the relevant half of **M**:

$$\forall y \, ((\exists x \, Qx \to Qy) \to Qy) \to \exists x \, Qx = \forall y \, ((\exists x \, (\exists x \, Px \to Px)$$
$$\to (\exists x \, Px \to Py)) \to (\exists x \, Px \to Py)) \to \exists x \, (\exists x \, Px \to Px).$$

Finally, it suffices to prove intuitionistically (a routine exercise) that

$$\forall y \, ((\exists x \, (\exists x \, Px \to Px) \to (\exists x \, Px \to Py)) \to (\exists x \, Px \to Py)).$$

∎

But many further results in earlier sections suggest formal proofs.

EXAMPLE 1. Turning correspondence arguments into formal proofs. Assuming Domain Cumulation, the earlier $\Box \exists x \, Px \to \exists x \, \Box Px$ implied Domain Anti-Cumulation, the characteristic property for the Barcan Axiom

$$\forall x \, \Box Px \to \Box \forall x \, Px, \text{ or in an equivalent form: } \Diamond \exists x \, Px \to \exists x \, \Diamond Px.$$

Can we formally derive the latter principle from the former using only the minimal modal predicate logic? This is not clear, as our semantic argument in Section 4 involves first-order substitutions with parameters that are not expressible inside the modal language. Roughly speaking, to closely match that argument, we would need a formal proof proceeding as follows:

> "Given that $\Diamond \exists x \, Px$, pick a successor world v with an object d satisfying the predicate P, and then definably change the denotation of P to P^+ on all other successor worlds to make $\exists x \, P^+ x$ true there. This definition validates $\Box \exists x \, P^+ x$, and hence $\exists x \, \Box P^+ x$, which gives us an object d at w that has the property P^+ in v. But given how we just defined P^+, this can only mean that d has the old property P at v."

It is not clear that this proof, no matter how simple, can be made to work with substitutions purely inside the language of modal predicate logic.

Similar challenges arise with deriving the intuitionistic constant domain axiom $\exists xy \, \neg x = y \to (\forall x \, (A \lor \forall x \, Bx) \to (A \lor \forall x \, Bx))$ from Plato's Law, or with the propositional linearity axiom $\exists xy \, \neg x = y \to ((A \to B) \lor (B \to A))$.

The general situation is discussed in van Benthem 1979 [3] on incompleteness in modal propositional logic. Correspondence arguments are naturally formalizable in logical systems, but often these proof systems are not purely modal, they need additional second-order comprehension axioms involving first-order definable predicates with object parameters.

FACT 14. Except for the case of Well-foundedness, all correspondence proofs in this paper live in a weak second-order logic with only first-order substitution instances for second-order quantifiers.[16]

But van Benthem 1979 [3] also shows that even such simple correspondence arguments relating purely modal axioms via first-order arguments may still lack a counterpart in pure bottom-level modal deduction – by presenting a particularly simple 'frame-incomplete' propositional modal logic. In line with this observation, the gap between correspondence and completeness analysis seems even larger in the area of modal predicate logic.

7 Conclusion and Open Problems

We have explored some correspondence theory for modal predicate logic, showing how some promising results can be found. Even so, this foray raises more questions than it answers. Here are a few further directions.

Van Benthem 1983 [4] relates purely relational frame properties definable by axioms in modal predicate logic to such frame properties already defined by purely *propositional modal axioms*, and proves a partial 'conservativity property'. What is the general link between correspondences in both areas?

Next, the issue of the existence of *formal modal proofs* matching semantic correspondence arguments has merely been raised here, but not analyzed in depth. In the propositional case, this gap has been narrowed using richer 'hybrid' modal languages that address the expressive deficit of the modal base language (Venema 1991 [25]). This is also relevant here, and Areces & ten Cate 2006 [1] have the state of the art. Another approach would use yet richer modal languages that arise when analyzing higher-order correspondence arguments. Van Benthem 2006 [9] shows how Löb's Axiom is frame-definable in first-order logic with *fixed-points* LFP(FO), which can define properties such as the earlier Well-foundedness. It might be interesting to take a similar look at the model predicate-logical case, which can merge fixed-point features of accessibility with recursively defined properties of objects.

Finally, our analysis has worked with just the simplest traditional semantics of modal predicate logic. But as we said at several places, there are open problems in defining this system as a mathematically perspicuous 'merge' of modal propositional logic and first-order predicate logic. Indeed, the design of this system has long been under discussion, with many suggested re-modelings making the logic weaker, and hence the gap between validity and modal deduction smaller. Ghilardi proposed category-theoretic 'functional models' (van Benthem 1993 [6]; Brauner & Ghilardi 2006 [14]), while Ono 1999 [23] shows

---

[16]'First-order' here refers to the *first-order language of frames* with accessibility and elementhood, not to the built-in formal first-order language of objects.

how incompleteness phenomena will sometimes go away when one moves to algebraic models for modal predicate logic. There are also the neighbourhood models of Arlo Costa & Pacuit 2006 [2], or the general frames of Goldblatt & Mares 2006 [17]. In such settings, our correspondence arguments would have to be redone. In this connection, it also seems relevant that predicate logic *itself* may be defined and analyzed as a modal logic (van Benthem 1996 [7], Chapter 9). Thus, modal predicate logic may also be viewed as a product of two modal logics (Kurucz 2006 [19]). Again, our correspondence analysis makes sense here – but it still has to be done.

So few results, so many questions. I will have to join forces with Grisha!

Acknowledgment

I thank Grisha Mints for a Stanford logic seminar that sparked off my renewed interest. I also thank the anonymous referee for this volume, as well as Hiroakira Ono during a visit to JAIST for comments on this paper.

BIBLIOGRAPHY

[1] C. Areces & B. ten Cate, "Hybrid Logics", in P. Blackburn, J. van Benthem, & F. Wolter, eds., *Handbook of Modal Logic*, Elsevier Science Publishers, Amsterdam, 2006, pp. 821–868.
[2] H. Arlo Costa & E. Pacuit, "First-Order Classical Modal Logic", *Studia Logica*, 2006, Vol. 84, No. 2, pp. 171–210.
[3] J. van Benthem, "Syntactic Aspects of Modal Incompleteness Theorems", *Theoria*, 1979, Vol. 45, No. 2, pp. 63–77.
[4] J. van Benthem, *Modal Logic and Classical Logic*, Bibliopolis, Napoli, 1983.
[5] J. van Benthem, "Correspondence Theory", in: D. Gabbay and F. Guenthner, eds., *Handbook of Philosophical Logic*, Vol. II, Reidel, Dordrecht, 1984, pp. 167–247. Expanded version in *Handbook of Philosophical Logic*, 2nd edition, vol. III, Kluwer, Dordrecht, 2001, pp. 325–408.
[6] J. van Benthem, "Beyond Accessibility: Functional Models for Modal Logic", in: M. de Rijke, ed., *Diamonds and Defaults*, Kluwer, Dordrecht, 1993, pp. 1–18.
[7] J. van Benthem, *Exploring Logical Dynamics*, CSLI Publications, Stanford, California, 1996.
[8] J. van Benthem, "Minimal Predicates, Fixed-Points, and Definability", *Journal of Symbolic Logic*, 2005, Vol. 70, No. 3, pp. 696–712.
[9] J. van Benthem, "Modal Frame Correspondences and Fixed-Points", *Studia Logica*, 2006, Vol. 83, No. 1, pp. 133–155.
[10] J. van Benthem & D. Bonnay, 2008, "Modal Logic and Invariance", *Journal of Applied Non-Classical Logics*, 2006, Vol. 18, No. 2/3, pp. 153–173.
[11] E. W. Beth, *The Foundations of Mathematics*, North-Holland, Amsterdam, 1959.
[12] P. Blackburn, J. van Benthem, & F. Wolter, eds., *Handbook of Modal Logic*, Elsevier Science Publishers, Amsterdam, 2006.
[13] P. Blackburn, M. de Rijke, & Y. Venema, *Modal Logic*, Cambridge University Press, Cambridge, 2000.
[14] T. Brauner & S. Ghilardi, "First-Order Modal Logic", in P. Blackburn, J. van Benthem, & F. Wolter, eds., *Handbook of Modal Logic*, Elsevier Science Publishers, Amsterdam, 2006, pp. 549–620.
[15] D. Gabbay, V. Shehtman, & D. Skvortsov, *Quantification in Non- Classical Logic*, book manuscript, King's College, London, 2005.
[16] J. Garson, "Quantification in Modal Logic", in: D. Gabbay & F. Guenthner, eds., *Handbook of Philosophical Logic*, 2nd edition, Volume III, Kluwer, Dordrecht, 2001, pp. 267–323.
[17] R. Goldblatt & E. Mares, "A General Semantics for Quantified Modal Logic", *Proc. Advances in Modal Logic 2006*, Melbourne, 2006, pp. 227–246.
[18] G. Hughes & M. Cresswell, *An Introduction to Modal Logic*, Methuen, London, 1968.

[19] A. Kurucz, "Combinations of Modal Logics", in P. Blackburn, J. van Benthem, & F. Wolter, eds., *Handbook of Modal Logic,* Elsevier Science Publishers, Amsterdam, 2006, pp. 869–924.
[20] W. Meijer Viol, *Instantial Logic,* Dissertation, ILLC Amsterdam, 1995.
[21] P. Minari, "'Completeness Theorems for some Intermediate Predicate Calculi", *Studia Logica,* 1983, Vol. 42, No. 4, pp. 431–441.
[22] H. Ono, "Incompleteness of Semantics for Intermediate Predicate Logics", *Proceedings Japan Academy,* 1973, Vol. 49, No. 9, pp. 711–713.
[23] H. Ono, "Algebraic Semantics for Predicate Logics an their Completeness", in: E. Orlowska, ed., *Logic at Work, To the Memory of Elena Rasiowa,* Physica Verlag, Heidelberg, 1999, pp. 637–650.
[24] P. Rodenburg, *Intuitionistic Correspondence Theory,* Dissertation, Mathematical Institute, University of Amsterdam, 1986.
[25] Y. Venema, *Many-Dimensional Modal Logics,* Dissertation, Institute for Logic, Language and Computation, University of Amsterdam, 1991.

CPSIA information can be obtained at www.ICGtesting.com
Printed in the USA
BVOW04s0039080414

349969BV00005B/22/P